Silage production and utilisation

XIV
ISC
2005

Silage production and utilisation

Proceedings of the XIVth International Silage Conference,
a satellite workshop of the XXth International Grassland
Congress, July 2005, Belfast, Northern Ireland

edited by:
R.S. Park
M.D. Stronge

Subject headings:
Grass ensiling
Silage microbiology
Ruminant production

ISBN 9076998752

First published, 2005

Wageningen Academic Publishers
The Netherlands, 2005

Organising Committee

Dr C S Mayne, Chairperson, ARINI/DARDNI
Dr R E Agnew, ARINI/DARDNI
Dr A F Carson, ARINI/DARDNI
Dr R C Binnie, ARINI
Dr P J Frost, ARINI/DARDNI

Dr T W J Keady, ARINI/DARDNI
Dr P O'Kiely, Teagasc
Dr A Wylie, DARDNI
R S Park, DARDNI
M G Porter, ARINI

Reviewers

Dr C S Mayne
Dr R E Agnew
Dr R C Binnie
Dr A F Carson
Dr J P Frost

Dr T W J Keady
Dr R J Merry
Dr C Ohlsson
Dr P O'Kiely
Dr D C Patterson

Dr C Thomas
Dr A R G Wylie
Dr T Yan
R S Park
M G Porter

Main Sponsors of the XIV International Silage Conference

Department of Agriculture and Rural Development for Northern Ireland (www.dardni.co.uk)

ALLTECH (www.alltech.com)

LALLEMAND (www.lallemand.com)

Foreword

This book presents the proceedings of the XIV International Silage Conference, a Satellite Workshop of the XX International Grassland Congress, held in Belfast, Northern Ireland from 3-6 July 2005. This Workshop was one of five satellite meetings held immediately after the main Congress, designed to enable in depth presentation and discussion on current practices and the science underpinning silage production and utilisation.

A total of 122 papers are included in these proceedings, presented in five main sections. The first section examined the effects of conserved feeds on milk and meat production. Papers addressed topics from effects of harvesting system, chop length and additive use on intake and animal performance through to effects of silage quality and supplementation on meat quality, milk conjugated linoleic content and microbiological quality of milk.

The increasing interest in ensiling "alternative forage" crops was highlighted in Section 2 with a number of papers examining effects of conserving wheat, forage maize, legumes, sorghum and *Lotus* species on the resulting forage composition and animal performance. New developments in ensiling techniques were considered in Section 3, including reviews of recent advances in big bale silage production and techniques to improve silage fermentation, aerobic stability and ultimately nutritive value of silages.

Section 4 examined challenges and new developments in the ensilage of tropical forages, ranging from tropical grasses to whole crop rice and sugar cane. The final section included a large number of papers on the chemical and biological characteristics of silages and in particular the use of near infrared spectroscopy to predict intake potential and nutritive value of conserved forage.

This is an important reference book, which provides an excellent overview of current developments in forage conservation and utilisation of conserved forage in animal production systems.

Table of contents

Foreword   7

Keynote presentations   17

An overview of silage production and utilisation in Ireland (1950-2005)   19
C.S. Mayne and P. O'Kiely

Grass silage: factors affecting efficiency of N utilisation in milk production   35
P. Huhtanen and K.J. Shingfield

Recent developments in feeding beef cattle on grass silage-based diets   51
M. McGee

Ensiled maize and whole crop wheat forages for beef and dairy cattle: effects on animal performance   65
T.W.J. Keady

Update on technologies for producing and feeding silage   83
P.D. Forristal and P. O'Kiely

Silage production from tropical forages   97
L.G. Nussio

Recent developments in methods to characterise the chemical and biological parameters of grass silage   109
R.S. Park, R.E. Agnew and M.G. Porter

Advances in silage quality in the 21st Century   121
D.R. Davies, M.K. Theodorou, A.H. Kingston-Smith and R.J. Merry

Section 1: A. Effect of conserved feeds on milk production   135

The effect of grass silage chop length on dairy cow performance   137
Å.T. Randby

Whole crop silage from barley fed in combination with red clover silage to dairy cows   138
J. Bertilsson and M. Knicky

Responses to grass or red clover silages cut at two stages of growth in dairy cows   139
A. Vanhatalo, K. Kuoppala, S. Ahvenjärvi and M. Rinne

The effect of chop length and additive on silage intake and milk production in cows   140
V. Toivonen and T. Heikkilä

Effect of supplementing grass silage with incremental levels of water soluble carbohydrate on *in vitro*
rumen microbial growth and N use efficiency   141
D.R. Davies, D.K. Leemans and R.J. Merry

Effects of access time to feed and sodium bicarbonate in cows given different silages   142
T. Heikkilä and V. Toivonen

Dairy cow performance associated with two contrasting silage feeding systems   143
C.P. Ferris, D.C. Patterson, R.C. Binnie and J.P. Frost

Pea-barley bi-crop silage in milk production   144
M. Tuori, P. Pursiainen, A.-R. Leinonen and V. Karp

Conjugated linoleic acid content of milk from cows fed different diets   145
E. Staszak and J. Mikołajczak

Feeding with badly preserved silages and occurrence of subclinical ketosis in dairy cows   146
F. Vicente, B. de la Roza, A. Argamentería, M.L. Rodríguez and M. Peláez

Modelling contamination of raw milk with butyric acid bacteria spores   147
M.M.M. Vissers, F. Driehuis, P. de Jong, M.C. te Giffel and J.M.G. Lankveld

Use of a dairy whole farm nutrient balance education tool to teach the importance of forages in the
context of nutrient management concepts at the whole-farm level   148
J.H. Harrison and T.D. Nennich

Feeding mixed grass-clover silages with elevated sugar contents to dairy cows   149
J. Bertilsson

Section 1: B. Effect of conserved feeds on meat production   **151**

An evaluation of grain processing and storage method, and feed level on the performance and meat quality of beef cattle offered two contrasting grass silages   153
T.W.J. Keady, F.O. Lively and D.J. Kilpatrick

Nutritive value for finishing beef steers of wheat grain conserved by different techniques   154
P. Stacey, P. O'Kiely, A.P. Moloney and F.P. O'Mara

Effect of feeding red clover, lucerne and kale silage on the voluntary intake and liveweight gain of growing lambs   155
R. Fychan, C.L. Marley, M.D. Fraser and R. Jones

The effects of alfalfa silage harvesting systems on dry matter intake of Friesland dairy ewes in late pregnancy   156
H.F. Elizalde

Replacement of maize/soybean meal concentrate by high moisture maize grain plus wholeseed soybean silage for cattle   157
C.C. Jobim, A.F. Branco, V.F. Gai and U. Cecato

Effect of additive treatment on meat quality   158
V. Vrotniakiene and J. Jatkauskas

Blood meal as a source of histidine for cattle fed grass silage and barley   159
R. Berthiaume and C. Lafrenière

An evaluation of the inclusion of alternative forages with grass silage-based diets on carcass composition and meat quality of beef cattle offered two contrasting grass silages   160
T.W.J. Keady, F.O. Lively, D.J. Kilpatrick and B.W. Moss

Section 2: Alternative forages   **161**

Effects of feeding legume silage with differing tannin levels on lactating dairy cattle   163
U.C. Hymes Fecht, G.A. Broderick and R.E. Muck

NDF digestion in dairy cows fed grass or red clover silages cut at two stages of growth   164
K. Kuoppala, S. Ahvenjärvi, M. Rinne and A. Vanhatalo

The effects of maize and whole crop wheat silages and quality of grass silage on the performance of lactating dairy cows   165
D.C. Patterson and D.J. Kilpatrick

The feeding value of conserved whole-crop wheat and forage maize relative to grass silage and *ad-libitum* concentrates for beef cattle   166
K. Walsh, P. O'Kiely and F. O'Mara

Sustained aerobic stability of by-products silage stored as a total mixed ration   167
N. Nishino, H. Hattori and H. Wada

Evaluation of narrow-row forage maize in field-scale studies   168
W.J. Cox, J.H. Cherney and D.J.R. Cherney

Ensiling safflower (*Carthamus tinctorius*) as an alternative winter forage crop in Israel   169
Z.G. Weinberg, S.Y. Landau, A. Bar-Tal, Y. Chen, M. Gamburg, S. Brener and L. Devash

Effect of variety and species on the chemical composition of *Lotus* when ensiled   170
C.L. Marley, R. Fychan and R. Jones

Effect of additives at harvest on the digestibility in lambs of whole crop barley or wheat silage   171
S. Muhonen, I. Olsson and P. Lingvall

Effects of varying dietary ratios of lucerne to maize silage on production and microbial protein synthesis in lactating dairy cows   172
G.A. Broderick and A.F. Brito

Effects of two different chopping lengths of maize silage on silage quality and dairy performance   173
K. Mahlkow and J. Thaysen

Use of silage additives in ensiling of whole-crop barley and wheat - A comparison of round big bales and precision chopped silages   174
M. Knický and P. Lingvall

Cob development in forage maize: influence of harvest date, cultivar and plastic mulch   175
E.M. Little, P. O'Kiely, J.C. Crowley and G.P. Keane

Yield and composition of forage maize: interaction of harvest date, cultivar and plastic mulch   176
E.M. Little, P. O'Kiely, J.C. Crowley and G.P. Keane

Parameters of ensiled maize with biological and chemical additives 177
J. Grajewsk, A. Potkański, K. Raczkowska-Werwińska, M. Twarużek and B. Miklaszewska

Ensiling of tannin-containing sorghum grain 178
E.M. Ott, Y. Acosta Aragón and M. Gabel

Fermentation characteristics of maize/sesbania bi-crop silage 179
M. Kondo, J. Yanagisawa, K. Kita and H. Yokota

The influence of crop maturity and type of baler on whole crop barley silage production 180
P. Lingvall, M. Knicky, B. Frank, B. Rustas and J. Wallsten

Effect of stage of maturity on the nutrient content of alfalfa 181
Y. Tyrolova and A. Vyborna

Field beans and spring wheat as whole crop silage: yield, chemical composition and fermentation
characteristics 182
L. Ericson, K. Arvidsson and K. Martinsson

Utilisation of whole-crop pea silages differing in condensed tannin content as a replacement for soya
bean meal in the diet of dairy cows 183
K.J. Hart, R.G. Wilkinson, L.A. Sinclair and J.A. Huntington

Ensiled high moisture barley or dry barley in the grass silage-based diet of dairy cows 184
S. Jaakkola, E. Saarisalo and R. Kangasniemi

Effects of species, maturity and additive on the feed quality of whole crop cereal silage 185
E. Nadeau

Comparison of different maize hybrids cultivated and fermented with or without sorghum 186
Sz. Orosz, Z. Bellus, Zs. Kelemen, E. Zerényi and J. Helembai

Utilisation of coffee grounds for total mixed ration silage 187
C. Xu, Y. Cai, N. Hino, N. Yoshida and M. Ogawa

Forage preferences of horses 188
C.E. Müller

Section 3: Developments in ensiling techniques **189**

The effect of silage harvester type on harvesting efficiency 191
J.P. Frost and R.C. Binnie

Harvesting silage with two types of silage trailer (feed rotor with knives and precision chop) 192
H. Arvidsson and P. Lingvall

The effects of a new plastic film on the microbial and fermentation quality of Italian ryegrass bale silages 193
G. Borreani and E. Tabacco

Section 3: Developments in ensiling techniques
A. Silage fermentation **195**

Influence of different alfalfa-grass mixtures and the use of additives on nutritive value and
fermentation of silage 197
P. Lättemäe and U. Tamm

The effect of neutralising formic acid on fermentation of fresh and wilted grass silage 198
E. Saarisalo and S. Jaakkola

Effects of inoculation of LAB on fermentation pattern and clostridia spores in easily ensilable grass silages 199
J. Thaysen, G. Pahlow and E. Mathies

Effect of biological additives in red clover – timothy conservation 200
A. Olt, H. Kaldmäe, E. Songisepp and O. Kärt

Application of a new inoculant "Chikuso-1" for silage preparation of forage paddy rice 201
Y. Cai, C. Xu, S. Ennahar, N. Hino, N. Yoshida and M. Ogawa

Synergism of chemical and microbial additives on sugarcane (*Saccharum officinarum* L.) silage fermentation 202
T.F. Bernardes, G.R. Siqueira, R.P. Schocken-Iturrino, A.P.T.P. Roth and R.A. Reis

The influence of the application of a biological additive on the fermentation process of red clover
silage 203
Ľ. Rajčáková, R. Mlynár and M. Gallo

Inoculant effects on ensiling and *in vitro* gas production in lucerne silage 204
R.E. Muck, I. Filya and F.E. Contreras-Govea

Effects of stage of growth and inoculation on fermentation quality of field pea silage 205
G. Borreani, L. Cavallarin, S. Antoniazzi and E. Tabacco

A novel bacterial silage additive effective against clostridial fermentation 206
E. Mayrhuber, M. Holzer, W. Kramer and E. Mathies

In vitro gas production and bacterial biomass estimation for lucerne silage inoculated with one of three
lactic acid bacterial inoculants 207
F.E. Contreras-Govea, R.E. Muck, I. Filya, D.R. Mertens and P.J. Weimer

Correlation between epiphytic microflora and microbial pollution and fermentation quality of silage
made from grasses 208
B. Osmane and J. Blūzmanis

Hygienic value and mycotoxins level of grass silage in bales for horses 209
A. Potkański, J. Grajewski, K. Raczkowska-Werwińska, B. Miklaszewska, A. Gubała, M. Selwet and
M. Szumacher-Strabel

Polyphenol oxidase activity and in vitro proteolytic inhibition in grasses 210
J.M. Marita, R.D. Hatfield and G.E. Brink

The effect of dry matter content and inoculation with lactic acid bacteria on the residual water soluble
carbohydrate content of silages prepared from a high sugar grass cultivar 211
D.R. Davies, D.K. Leemans, E.L. Bakewell and R.J. Merry

Using the red clover polyphenol oxidase gene to inhibit proteolytic activity in lucerne 212
R.D. Hatfield, M.L. Sullivan and R.E. Muck

New results on inhibition of clostridia development in silages 213
E. Kaiser, K. Weiß and I. Polip

Ensilability and silage quality of different cocksfoot varieties 214
U. Wyss

Ensiling characteristics of sudangrass silage treated with green tea leaf waste or green tea polyphenols 215
M. Kondo, K. Kita and H. Yokota

Effects of silage preparation and microbial silage additives on biogas production from whole crop
maize silage 216
M. Neureiter, C. Perez Lopez, H. Pichler, R. Kirchmayr and R. Braun

A 16S rDNA-based quantitative assay for monitoring Lactobacillus plantarum in silage 217
M. Klocke, K. Mundt, C. Idler, P. O`Kiely, S. Barth

A comparison of the efficacy of an ultra-low volume applicator for liquid-applied silage inoculants
with that of a conventional applicator 218
G. Marley, G. Pahlow, H.-H. Herrmann and T.R. Owen

Section 3: Developments in ensiling techniques
B. Aerobic stability 219

Improving the aerobic stability of whole-crop cereal silages 221
I. Filya, E. Sucu and A. Karabulut

Aerobic stability and nutritive value of low dry matter maize silage treated with a formic acid-based
preservative 222
I. Filya, E. Sucu and A. Karabulut

Microbial changes and aerobic stability in high moisture maize silages inoculated with Lactobacillus buchneri 223
R.A. Reis, E.O. Almeida, G.R. Siqueira, T.F. Bernardes, E.R. Janusckiewicz and M.T.P. Roth

Effect of residual sugar in high sugar grass silages on aerobic stability 224
G. Pahlow, R.J. Merry, P. O'Kiely, T. Pauly and J.M. Greef

An in vitro study on the influence of residual sugars on aerobic changes in grass silages 225
S.D. Martens, G. Pahlow and J.M. Greef

The effects of the growth stage and inoculant on fermentation and aerobic stability of whole-plant
grain sorghum silage 226
E. Tabacco and G. Borreani

Perennial ryegrasses bred for contrasting sugar contents: manipulating fermentation and aerobic
stability using wilting and additives (1) (EU FP V -Project 'SweetGrass') 227
P. O'Kiely, H. Howard, G. Pahlow, R. Merry, T. Pauly and F.P. O'Mara

Perennial ryegrasses bred for contrasting sugar contents: manipulating fermentation and aerobic
stability of unwilted silage using additives (2) (EU-Project 'SweetGrass')   228
H. Howard, P. O'Kiely, G. Pahlow and F.P. O'Mara

Perennial ryegrasses bred for contrasting sugar contents: manipulating fermentation and aerobic
stability of wilted silage using additives (3) (EU-Project 'SweetGrass')   229
H. Howard, P. O'Kiely, G. Pahlow and F.P. O'Mara

The effect of additive containing formic acid on quality and aerobic stability of silages made of
endophyte-infected green forage   230
L. Podkówka, J. Mikołajczak, E. Staszak and P. Dorszewski

The effect of acetic acid on the aerobic stability of silages and on intake   231
B. Ruser and J. Kleinmans

Effectiveness of *Lactobacillus buchneri* to improve aerobic stability and reducing mycotoxin levels in
maize silages under field conditions   232
A. Bach, C. Iglesias, C. Adelantado and M.A. Calvo

The effect of Lalsil Dry inoculant on the aerobic stability of lucerne silage   233
J.P. Szucs and Z. Avasi

Section 3: Developments in ensiling techniques
C. Nutritive value   **235**

Ensiling characteristics and ruminal degradation of Italian ryegrass with or without wilting and added
cell wall degrading enzymes   237
Y. Zhu, H. Jianguo, Z. He, X. Qingfang, B. Chunsheng and N. Nishino

Quality and nutritive value of grass-legume ensiled with inoculant Lactisil 300   238
J. Jatkauskas and V. Vrotniakiene

Effect of additives in grass silage on rumen parameters in Rusitec   239
A. Potkański, A. Cieślak, K. Raczkowska-Werwińska, M. Szumacher-Strabel and A. Gubała

The quality and nutritive value of big bale silage harvested from bog meadows   240
H. Żurek, B. Wróbel and J. Zastawny

The aerobic stability and nutritive value of grass silage ensilaged with bacterial additives   241
B. Wróbel and J. Zastawny

Section 3: Developments in ensiling techniques
D. Big bale silage production   **243**

Factors affecting bag silo densities and losses   245
R.E. Muck and B.J. Holmes

Transport of wrapped silage bales   246
Å.T. Randby and T. Fyhri

Wrapping rectangular bales with plastic to preserve wet hay or make haylage   247
D. Undersander, T. Wood and W. Foster

Bacteria and yeast in round bale silage on a sample of farms in County Meath, Ireland   248
J. McEniry, P. O'Kiely, N.J.W. Clipson, P.D. Forristal and E.M. Doyle

Schizophyllum on baled grass silage in Ireland: national farm survey 2004   249
M. O'Brien, P. O'Kiely, P.D. Forristal and H. Fuller

Bagged silage: Mechanical treatment applied by packing rotor improves fermentation   250
M. Sundberg and T. Pauly

Carbon dioxide permeation properties of polyethylene films used to wrap baled silage   251
C. Laffin, G.M. McNally, P.D. Forristal, P. O'Kiely and C.M. Small

National survey to establish the extent of visible mould on baled grass silage in Ireland and the identity
of the predominant fungal species   252
M. O'Brien, P. O'Kiely, P.D. Forristal and H. Fuller

Section 4: Ensilage of tropical forages   **253**

Effect of ensiling temperature, delayed sealing, and simulated rainfall on the fermentation and aerobic
stability of maize silage grown in a sub-tropical climate   255
A.T. Adesogan and S.C. Kim

Effect of different densities on tropical grass silages 256
T.F. Bernardes, R.C. Amaral, G.R. Siqueira and R.A. Reis

Sugarcane silage compared with traditional roughage sources on performance of dairy cows 257
O.C.M. Queiroz, L.G. Nussio, M.C. Santos, J.L. Ribeiro, P. Schmidt, M. Zopollatto, M.C. Junqueira, M.S. Camargo, S.G.T. Filho, L.G. Vieira, M.O. Trivelin, L.J. Mari and D.P. Souza

Moisture control, inoculant and particle size in tropical grass silages 258
S.F. Paziani, L.G. Nussio, D.R.S. Loures, L.J. Mari, J.L. Ribeiro, P. Schmidt, M. Zopollatto, M.C. Junqueira and A.F. Pedroso

Stability of silage wrapped round bales in Réunion Island 259
P. Grimaud, V. Barbet-Massin, P. Thomas and D. Verrier

Ensilage of tropical grasses and legumes using a small-scale technique 260
M. Delacollette, S. Adjolohoun, R. Agneesens and A. Buldgen

The use of *Lactobacillus buchneri* inoculation to decrease ethanol and 2,3-butanediol production in
whole crop rice silage 261
N. Nishino and H. Hattori

Microorganism occurrence in Tanzânia (*Panicum maximum* Jacq. cv. Tanzânia) grass silage exposed
to the environment 262
R.M. Coan, R.A. Reis, G.R. Garcia, R.P. Schocken-Iturrino and E.D. Contato

Forage variety and maturity on fermentative losses of sugarcane silages added with urea 263
P. Schmidt, L.G. Nussio, C.M.B. Nussio, A.A. Rodrigues, P.M. Santos, J.L. Ribeiro, L.J. Mari, M. Zopollatto, M.C. Santos, O.C.M. Queiroz and D.P. Souza

Effect of moisture on the fermentation and the utilisation by cattle of silages made from tropical
grasses 264
M. Niimi, O. Kawamura, K. Fukuyama and S. Sei

Section 5: Chemical and biological characterisation of silages 265

Estimation of legume silage digestibility with various laboratory methods 267
A. Olt, M. Rinne, J. Nousiainen, M. Tuori, C. Paul, M.D. Fraser and P. Huhtanen

Evaluation of prediction equations for metabolisable energy concentration in grass silage used in
different energy feeding systems 268
T. Yan and R.E. Agnew

The effect of fermentation quality on voluntary intake of grass silage by growing steers 269
S.J. Krizsan and Å.T. Randby

Determination of toxic activity of mould-damaged silage with an *in vitro* method 270
A. Solyakov and T. Pauly

Butyric acid bacteria spores in whole crop maize silage 271
F. Driehuis and M.C. te Giffel

Effects of the stage of growth and inoculation on proteolysis in field pea silage 272
L. Cavallarin, G. Borreani, S. Antoniazzi and E. Tabacco

Ruminal proteolysis in forages with distinct endopeptidases activities 273
G. Pichard, C. Tapia and R. Larraín

Effects of particle size in forage samples for protein breakdown studies 274
G. Pichard and C. Tapia

A new system for the evaluation of the fermentation quality of silages 275
E. Kaiser and K. Weiß

Prediction of indigestible NDF content of grass and legume silages by NIRS 276
L. Nyholm, M. Rinne, M. Hellämäki, P. Huhtanen and J. Nousiainen

Analysis of silage fermentation characteristics using transflectance measurements by
near infrared spectroscopy 277
A. Martínez, A. Soldado, R. García, D. Sánchez and B. de la Roza-Delgado

A simple method for the correction of fermentation losses measured in laboratory silos 278
F. Weissbach

Development of a method for the fast and complete assessment of quality characteristics in undried
grass silages by means of an NIR-diode array spectrometer 279
H. Gibaud, C. Paul, J.M. Greef and B. Ruser

Prediction of red clover content in mixed swards by near-infrared reflectance spectroscopy 280
B. Deprez, D. Stilmant, C. Clément, C. Decamps and A. Peeters

Keyword index 281

Author index 285

Keynote presentations

An overview of silage production and utilisation in Ireland (1950-2005)

C.S. Mayne[1] and P. O'Kiely[2]
[1]Agricultural Research Institute of Northern Ireland, Hillsborough, Co Down BT26 6DR
Email: sinclair.mayne@dardni.gov.uk
[2]Teagasc, Grange Research Centre, Co Meath, Ireland.

Key Points

1. The seasonal nature of grass growth in Ireland necessitates effective integration of grazing and grass conservation to fully manage and utilise grass within meat and milk production systems.
2. Silage now accounts for 87% of the total grass conserved in Ireland.
3. The rapid expansion of silage making in Ireland between 1950 and 2000 was facilitated by significant advances in mechanisation (forage harvesters, mower conditioners and stretch film wrapping of big bales) and by improved understanding of the preservation process. The expansion was required to support the major increase in livestock numbers.
4. Excellent silage-making practices can result in grass silages with similar nutritive values to those of the grasses from which they were made and these silages can sustain high levels of performance in cattle and sheep.
5. Key challenges for the future include: the development of lower cost, reduced labour harvesting systems; the improved prediction of silage feeding value based on analysis of the standing crop; the development of feeding strategies to improve the efficiency of nutrient capture in silage-based systems; and the production of meat and milk of enhanced nutritional value.

Keywords: grass silage, forage conservation systems, feeding value, silage feeding systems

Introduction

Conservation of grass as hay or silage has been a feature of grassland management in Ireland for centuries. This reflects the fact that the grass growing season varies from less than 280 days to approximately 320 days (depending on geographical location) and in addition, the utilisation of grass by grazing may not be feasible for up to 5 months in some wetter areas. Furthermore, the highly seasonal nature of grass growth in Ireland (Figure 1) means that it is often difficult to manage grass growth early in the season using grazing animals.

Effective integration of grazing and grass conservation enables conservation of grass surpluses as hay or silage, facilitating improved management and utilisation of grass during the grazing season, whilst also providing high quality forage for winter feeding of livestock.

The aim of this paper is to review developments in silage production and utilisation in Ireland, particularly in relation to advances in science and technology from 1950 to the present day, and to illustrate how these advances have assisted the development of efficient silage production systems.

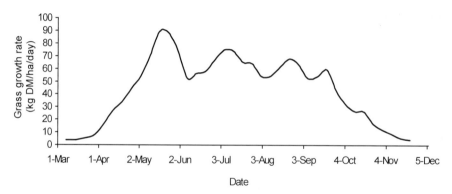

Figure 1. Average growth curve for perennial ryegrass (1999-2003) - Northern Ireland (AgriSearch Grass Check)

Trends in forage conservation in Ireland (1950-2005)

Data presented in Figure 2 illustrate the major change in emphasis in conserved forage production in Ireland since 1950. In addition to a 64% increase in the total area of land used for conserved forage production between 1950 and 2000, there has also been a continuous increase in the area conserved as silage and a concurrent decline in the area conserved as hay. The area conserved as grass silage, and the total area conserved, reached their respective peaks in 2000. The small reduction since 2000 reflects current concerns regarding the costs of silage production, particularly in the context of lower prices for animal products, and an increased emphasis on grazing and opportunities to extend the grazing season.

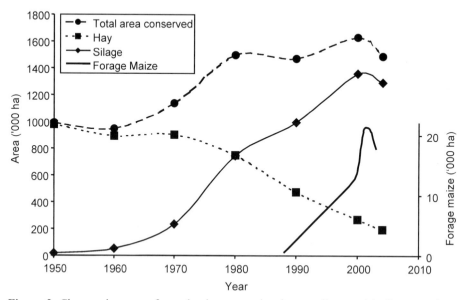

Figure 2 Changes in areas of grassland conserved as hay or silage and in forage maize area, in Ireland 1950-2004
Source: Central Statistics Office (Eirestat) and Economics and Statistics Division, DARD

Alongside the change in conservation practice over the last 50 years, there has been a very significant increase in livestock numbers (Figure 3). Between 1950 and 2000, cattle numbers increased by 65% while sheep numbers trebled over the same period. The increase in animal numbers has necessitated a drive towards more efficient grass conservation practices.

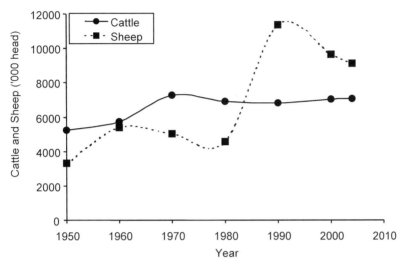

Figure 3 Change in livestock numbers (cattle and sheep) in Ireland 1950-2004
Source: Central Statistics Office (Eirestat) and Economics and Statistics Division, DARD

The data presented in Figure 2 also highlights the rapid expansion of forage maize use throughout Ireland, with over 2000 ha of maize now being grown in Northern Ireland. Recent developments in plant breeding, coupled with improvements in agronomic practices - particularly the development of the complete-cover, plastic mulch system - have considerably increased the yield potential and feeding value of forage maize. For example, Easson (personal communication) has reported forage maize yields of up to 18 t DM/ha in Northern Ireland, with starch contents commonly in the range of 250-300 g/kg DM. Key drivers of the increased interest in forage maize and other alternative forage crops such as whole crop wheat, have been the reduced labour and expertise required in growing and harvesting the crop and the improved predictability of feeding value relative to that achievable with grass silage.

Keady *et al.* (2002), determined the cost of producing and feeding a range of conserved forages in Northern Ireland and concluded that, relative to a 3-cut grass silage system costing £85/t utilised DM and a 4-cut grass silage system costing £95/t utilised DM, fermented whole crop wheat cost £88/t and forage maize under plastic, £91/t utilised DM. Given the competitive costs of these alternative forage crops, a major challenge for grass silage in the future is the need to develop harvesting systems that require less labour input, reduced overall costs and improved predictability of final product feeding value.

Harvesting systems

The increase in the area of grassland conserved as silage since 1950 was facilitated by major changes in harvesting systems used in silage making. In the early 1950's, virtually all silage made in Ireland was produced using tractor-mounted buckrakes or stationary green-crop loaders. Today, precision-chop harvesters dominate the scene. Survey data from the Republic of Ireland over the last 15 years (Table 1) highlights the increased reliance on precision chop harvesters and big bale systems and the demise of single and double chop harvesting systems. There are no comparable data for Northern Ireland but similar trends in harvesting system have been observed here. However, it is estimated that big bale silage accounts for less than 20% of the total silage produced in Northern Ireland.

Table 1 Trends in choice of harvesting system for silage in the Republic of Ireland (% of silage conserved for each system)

| Year | Harvesting system | | |
	Big bale	Single/Double chop	Precision chop
1991	23	40	37
1996	32	17	51
1999	35	9	56
2002	32	8	59

Another significant change in silage-making in Ireland has been the move from on-farm harvesting using the farmer's own equipment to a greater reliance on specialised silage contractors. It is estimated that more than 80% of silage produced on Irish farms is now undertaken by specialist contractors, who have made a very significant investment in specialised machinery – mainly high-output, self-propelled, precision-chop harvesters. Whilst this innovation undoubtedly speeds up the rate of harvesting and can assist in controlling costs on small livestock farms, there are a number of disadvantages. Firstly, the high output of these machines can cause operational difficulties in effective filling and compaction of silos, particularly where silos are relatively small or have difficult access. Secondly, a reliance on contractors reduces the farmer's flexibility to harvest the crop at the optimum stage of plant growth or in appropriate weather conditions. As a result, some farmers are now considering investing in their own equipment, with recent interest in self-loading forage wagons or, in some cases, increased reliance on big bale silage production systems.

Developments in mechanisation of silage production

Silage-making in Ireland in the 1950's largely relied on tractor-powered reciprocating mowers to cut the grass crop and tractor-mounted buckrakes or stationary green-crop loaders to collect the cut grass. Most grass was ensiled in small bunker silos, with tower silos being used on only a few larger farms. Generally, the grass was relatively mature and of low to moderate digestibility at the time of harvest, but this assisted the fermentation process in the absence of chopping and/or additive use. Following the introduction of flail or single-chop harvesters in the USA in the early 1950's, these were rapidly introduced in Ireland during the 1960's and this innovation drove the rapid increase in silage-making in Ireland through the 1960's and 1970's. The main advantages of the flail harvester were the saving in labour and the increased liberation of plant cell contents which accelerated the fermentation process in the silo, as demonstrated by Murdoch et al. (1955) in a series of classical experiments (Table 2). Chopping reduced silage pH and increased lactic acid content and amino acid N content.

These research findings were followed by further research studies throughout Europe which highlighted the important factors associated with the production of high quality silage.

One of the major advantages of conserving grass as silage rather than as hay is the ability to cut grass for silage at an earlier stage of growth and it was soon recognised that the most important factor affecting the nutritive value of silage was the stage of maturity of herbage at harvest (McIlmoyle & Steen, 1979).

Table 2 Effect of chopping grass prior to ensiling on fermentation quality (Murdoch *et al.* 1955)

	Unchopped	Chopped
Herbage dry matter at harvest (g/kg)	234	230
Silage		
pH	5.4	4.7
Lactic acid (g/kg DM)	2.0	26.0
Butyric acid (g/kg DM)	46	24
Amino acid N (g/kg N)	188	296

Further developments in mechanisation resulted in the design of improved flail harvesters and these became commonplace on livestock farms throughout Ireland, and remain in use on some smaller farms to the present day. Research data demonstrating the beneficial effects of chopping grass prior to ensiling on silage preservation and feeding value led to the introduction of precision-chop forage harvesters in the early 1970's. These machines had a higher tractor power requirement but could also achieve high work rates whilst chopping grass to as little as 5-7 mm particle size. However, in an extensive review of the literature, Marsh (1978) concluded that there was little evidence of benefits from fine chopping on fermentation in farm scale silos, except where fine chopping improved the consolidation of heavily wilted silages. More recent research suggests little benefit of fine chopping on animal performance, except in sheep. Nevertheless, the higher harvesting rate of precision chop machines has resulted in these machines largely replacing the flail type harvester.

Conservation efficiency

Irish research has always emphasised the importance of efficient forage conservation systems that minimise quantitative and qualitative losses during harvesting, storage and feeding out. The variable, but often damp, Irish weather has a major impact on the practice and efficiency of silage-making systems. Ultimately, weather, directly or indirectly, influences factors such as the choice of crop, the physical, chemical and microbiological characteristics of the crop at harvest, the harvesting and ensiling practices used, and the cumulative losses due to fermentation, effluent and aerobic deterioration. Weather and management practices interact in their effects on a range of factors. For example, yields of up to 40 tonnes of wet grass per hectare are not uncommon and are, in part, facilitated by relatively high inputs of inorganic and organic N. Grass dry matter (DM) and water soluble carbohydrate (WSC) concentrations (O'Kiely & Muck, 1998) and buffering capacities (Muck *et al.*, 1991) vary widely in Ireland, while counts of lactic acid bacteria are relatively high, with mean values in excess of 100,000 colony forming units/g herbage (Moran *et al.,* 1991).

Recent developments in silage-making equipment have included the introduction of mower conditioners which, when used optimally, can improve wilting efficiency by between 55 and 109% (Merry *et al.*, 2000) and this has led to renewed interest in the use of rapid wilting systems. However, wilting is not always compatible with contractor systems or suited to prevailing weather conditions.

Another major development in silage-making in Ireland was the arrival of big bale silage production systems in the late 1970's. This technique enabled silage production on small livestock farms, without a need for significant investment in silos or silage-making machinery. Whilst the initial approach to producing big bale silage involved placing individual bales in large plastic bags, this was soon superseded by the development of stretch-wrapping of bales in the mid 1980's. Irish machinery manufacturers were quick to adopt the new technology and continue to lead the market in terms of the manufacture of stretch-wrapped baled silage systems. An important feature of big bale silage is the potential to reduce DM losses compared to those from precision-chop silage ensiled in bunker silos (Kennedy, 1989). Further developments in the design of big balers led to the introduction of chopping mechanisms prior to the bale chamber. This has enabled the production of wrapped silage of higher density that is more appropriate for sheep production, given the importance of chop length on intake and performance in sheep.

Crop growth and management

The rate of reseeding of grassland in Ireland is quite low at less than 3% per annum. Perennial ryegrass is the main constituent of seed mixtures because it has the potential for superior yield, nutritive value and ensilability within the prevailing systems. Wilson & Collins (1980) compared a range of grasses found in permanent swards and concluded that perennial and Italian ryegrasses were considerably easier to preserve successfully compared to other grasses, mainly due to their higher concentration of available water-soluble carbohydrates (WSC). Furthermore, when Keating & O'Kiely (2000) compared silages made from a perennial ryegrass sward and from a previously well-managed and agronomically productive old pasture of diverse botanical composition, the perennial ryegrass silages produced more beef carcass per hectare, mainly due to their inherently higher digestibility and their better preservation.

Farmers strive to maintain satisfactory soil P, K and pH statuses on land used for silage production, and then supply sufficient N to promote economically justifiable yields. Whereas P fertilisers (O'Kiely & Tunney, 1997) and K fertilisers (Keady & O'Kiely, 1998) do not negatively impact on grass ensilability, applied N can reduce grass DM and WSC contents and increase buffering capacity. These effects increase both with increasing rates of N addition and as the interval after N application decreases (O'Kiely & Muck, 1998). Advice to farmers therefore seeks to recommend total rates of N application that will promote superior yields without unduly compromising the ensilability of the crop. Manures recycled from housed livestock are an important source of crop nutrients and, on integrated grassland farms, they are recycled mainly onto fields managed for silage production. Clearly, it is essential that they are applied in a manner that prevents contamination of the herbage. Research has demonstrated that where slurry is judiciously applied in an even and timely manner, at an appropriate rate and with the inorganic fertiliser input modified to take account of the estimated N contribution from slurry, then silage fermentation and environmental criteria need not be compromised (O'Kiely *et al.*, 1994; Frost & Stevens, 2000).

Controlling fermentation and reducing effluent loss

Early studies confirmed that warm fermentation, which was favoured in the early 1950's (Brown & Kerr, 1965a), incurred greater losses during ensilage than cold fermentation (Brown & Kerr, 1965b). The importance of adequately sealing ensiled forage against contact with air was demonstrated by Brown & Kerr (1965c), while Jackson (1969) showed that it was not necessary to physically evacuate the air trapped in a sealed silo. The ready availability of suitable plastic sheeting for horizontal silos (Brown & Kerr, 1965c) and of stretch-film for baled silage (Forristal *et al.*, 1999) made achieving anaerobic conditions economically feasible. Similarly, the move to most grass being harvested by well-equipped contractors operating high-output machinery greatly facilitated the rapid achievement of anaerobic storage conditions.

Flynn (1981), Cushnahan & Mayne (1995) and Keady & Murphy (1995) each showed that excellent silage-making practices could result in silages with nutritive values quite similar to those of the grasses from which they were made. This highlighted the importance of limiting all sources of conservation losses.

In order to remain relatively independent of frequently wet weather conditions, silage-making in Ireland has tended to rely on direct-harvesting or minimal wilting of grass. In experiments where wilted silage was compared with a poorly-preserved unwilted control silage, successful wilting reduced losses (Kormos & Chestnutt, 1966) and improved preservation (Wilson & Flynn, 1979). However, where the unwilted silage was well preserved - sometimes due to the application of formic acid - then traditional wilting techniques did little to improve the conservation (Jackson & Anderson, 1968). Under conditions where wet weather prevented a mown crop from drying effectively, the wilting process caused a significant increase in losses (Kormos & Chestnutt, 1968) and produced a poorer silage preservation (Wilson & Flynn, 1979). Overall, Mayne & Gordon (1986a, b) concluded that both unwilted and wilted silage systems were capable of conserving forage effectively provided that satisfactory meteorological and management conditions prevailed.

A key issue when ensiling low dry matter herbage is the production of silage effluent, and Irish farmers have always had to be very careful to ensure they managed silage effluent appropriately. Stewart (1980) and Binnie & Frost (1995) developed protocols for safely spreading silage effluent on grassland to capitalise on its fertiliser value for the growing crop but without scorching the new grass regrowth. Patterson & Walker (1982) and Steen (1986) quantified the nutritive value that could be obtained if effluent were cleanly collected and fed to farm livestock, while Ferris & Mayne (1994) and O'Kiely (1992) demonstrated that the co-ensilage of dry concentrate feedstuffs with wet grass could significantly reduce effluent output and enhance the value of the resultant silage. Finally, O'Donnell *et al.* (1995a, b) established guidelines to limit the corrosion of concrete silos by acidic silage effluent.

Silage additives and their use

Silage additives have elicited much interest through the years. Molasses was the focus of early attention due to its perceived palatability attributes and the relative ease with which it could be applied to wet grass of low WSC content, at the silo, in low-output harvesting systems. Under conditions where unwilted silage, made without additive, usually preserves badly, the addition of molasses, at 10 to 20 litres per tonne of herbage, can improve the fermentation and reduce in-silo losses. In contrast, the response to molasses was minimal when the control silage was well preserved (McCarrick, 1963). With the advent of direct-cut harvesting systems and the introduction of on-harvester application of additives, the use of formic acid became possible. Research by McCarrick (1963) and Flynn (1981) showed the potential of formic acid to aid preservation under a range of difficult ensiling conditions.

A wide variety of additives were evaluated for their efficacy as preservatives but none surpassed those mentioned above. During the 1980's, sulphuric acid was used as a low cost alternative to formic acid but, from the early 1990's onwards, the market rapidly became resistant to acid additives which were considered corrosive to machinery and concrete and dangerous to farm operatives who had to use them. Bacterial inoculants were topical during the 1990's, and their potential to improve animal productivity was widely demonstrated (Gordon, 1989a; Mayne, 1990, O'Kiely, 1996). However, since most dairy and beef cows calve in spring, with cows grazing grass from early lactation, the scope to obtain an economic return is often limited. In more recent years, the use of additives has decreased considerably as farmers seek to reduce costs and contractors seek to operate high throughput systems unimpeded by the delays associated with additive application. A reduction in additive use was also made possible by the significant improvement in overall silage-making standards that has occurred over the years.

Great improvements have been made also in the management of the exposed silage face in opened silos. Considerable differences exist among silages in their susceptibility to aerobic deterioration (O'Kiely, 1989) and the most important consideration in reducing aerobic losses at feeding out is minimising the duration of exposure of the silage to air. Some problems with mould growth on baled silage continue on many farms (O'Brien et al., 2005), indicating that improvements in the application of technology are still required.

Factors influencing the potential feeding value of silage

Feeding value for dairy cattle

The key factors influencing the feeding value of grass silage for dairy cattle include the stage of development of the crop at ensiling, the mechanical treatment of the crop and the extent and type of fermentation achieved within the silo. In a comprehensive review, Gordon (1989b) concluded that, on average, a 10 g/kg reduction in D-value (digestible organic matter in the dry matter) resulted in a decline in milk yield of 0.37 kg/cow/day. Other studies (Givens et al., 1989) have shown that herbage D-value in primary growth perennial ryegrass-based swards declines linearly from 1 May, with a mean decrease of 2.5 g/kg per day. Accordingly, each one-week delay in harvesting grass for silage after 1 May results in a depression in milk yield of approximately 0.65 kg/cow/day.

Developments in silage mechanisation over the last 50 years have been accompanied by major changes in the degree of laceration and/or chopping prior to ensiling. However, the effects of chop length on the performance of dairy cows are quite variable. Murphy (1983) showed no difference between forage wagon and precision-chop silages when both were easy fed, even though differences in chop length were very large (230 vs 52 mm). In contrast, Castle et al. (1979) compared three chop lengths and observed increases in both silage intake and milk yield with the shorter chop material. Whilst part of this response was due to improved silage fermentation with short chopped grass, particle length per se also had some effect. It is worth noting that the work of Castle et al. (1979) was undertaken with cows offered very low levels of supplement (2.0 kg DM/cow/day). Overall, it appears that chop length has a relatively limited effect on silage intake and animal performance providing the long chop material is well fermented and cows are offered moderate levels of concentrates.

The importance, for animal performance, of achieving good fermentation during ensilage has been well documented in many studies. For example, Baker *et al.* (1991) observed a 56% reduction in intake of a poorly-preserved silage compared to a well-preserved silage of similar digestibility. Given the relatively low DM content of grass in Ireland during the growing season, there is a high risk of poor preservation during ensilage, particularly in grasses with a WSC content of less than 30 g/kg fresh weight (O'Kiely *et al.*, 1986). A number of approaches have been adopted to improve the likelihood of achieving a good fermentation during ensilage. These include wilting, adding sugar to the crop, reducing pH by the addition of organic or inorganic acids, applying homofermentative lactic acid bacteria and/or enzymes to the crop and the addition of absorbents to increase crop DM content.

Formic acid was widely used as the main silage additive on dairy farms in Ireland from the early 1970's to the early 1990's and was applied directly into the delivery chute or chopping chamber of the forage harvester at rates of between 2.5 and 4.0 l per tonne of fresh crop. Steen (1991), in a review of 17 comparisons of untreated and formic acid-treated silages, observed significant improvements in silage fermentation, silage intake, milk yield and milk composition with formic acid treatment. The early 1990's saw renewed interest in the concept of restricting silage fermentation through the application of high levels (6 to 9.5 l/t grass) of either formic or mixed organic acids. Large responses in both silage intake and milk production were recorded, particularly with low dry matter crops (Chamberlain *et al.*, 1990), with milk production responses of up to 1.9 kg/day. However difficulties in applying high rates of additive and the increased cost of organic acids limited the commercial uptake of this approach in Ireland.

The increased cost of formic acid in the mid 1980's led to interest in the use of sulphuric acid as an alternative silage additive. Initial studies indicated that sulphuric acid compared well with formic acid in terms of effects on silage fermentation and animal performance (Murphy, 1986). However Steen (1991) concluded that the use of sulphuric acid resulted in poorer animal performance compared to untreated silage, possibly through detrimental effects on liver copper status.

The use of bacterial inoculants as silage additives came to the fore in the 1980's primarily because they were safer to handle and were less corrosive to machinery than the acids. Whilst some early inoculant additives produced disappointing results (Done, 1986), improved understanding of silage microbiology led to the development of more effective inoculants, most of which were based on *Lactobacillus plantarum*. Gordon (1989a) reported large increases in silage intake (+ 1.2 kg DM/day) and milk yield (2.0 kg/cow/day) with inoculant treatment, even though the inoculant had little effect on conventional measures of silage fermentation. This stimulated new research into the development of new laboratory procedures to evaluate the feeding value of silage.

The silage additive market in Ireland is now dominated by inoculants, with a wide range of products available. However, formic acid is still used on some farms, particularly where the grass being harvested has low DM and WSC contents.

Given the prevailing climatic conditions in Ireland, pre-wilting of grass prior to ensiling is a high risk venture. Results of the "Eurowilt" programme (Zimmer & Wilkins, 1984), which involved a co-ordinated programme of experiments throughout Western Europe, indicated little difference in animal performance between unwilted and wilted silage. Other work in Northern Ireland (Small & Gordon, 1988) reported lower milk output per hectare with wilted than with direct-cut material. Consequently, there was little Irish interest in field wilting until the early 1990's. At this stage, the development of improved high-output mower conditioners, coupled with the advent of self-propelled, precision-chop, forage harvesters, and increasing concerns over the environmental impact of silage effluent led to renewed interest in field wilting. Further research in the late 1990's demonstrated improved intake and performance of dairy cows offered wilted silage compared with those offered direct-cut material (Patterson *et al.*, 1996), with the extent of the increased intake positively correlated with the extent and rate of water loss in the field (Wright *et al.*, 2000).

Silage production systems on intensive dairy farms in Ireland are, today, largely characterised as follows:
- Grass is pre-cut at a leafy stage using a mower or mower conditioner and pre-wilted for a few hours (up to a maximum of 36-48 hours) depending on weather conditions.
- Herbage is collected by self-propelled precision chop harvester (average DM content of 230 g/kg).
- Additive is applied according to conditions:
  o Very wet, low sugar grass – formic acid.
  o Moderate DM grass with water soluble carbohydrate content greater than 20 g/kg fresh grass – either no additive (81% of farms), or inoculant.

Supplementation of silage for milk production

Significant changes in the approach to supplementation of grass silage diets over the last 50 years primarily reflect changes in the economics of milk production and advances in mechanisation. Milk production systems in the 1970's were largely based on high digestibility direct-cut silage fed with concentrates in flat rate feeding systems (5-10 kg/day) twice daily during milking. In the mid to late 1980's, there was some interest in silage-only systems (Reeve, 1989), largely reflecting the growing constraints on milk production imposed by European Union milk quotas.

Recent trends have seen a significant increase in the feeding of total mixed rations in which grass silage is incorporated with other forages and relatively high levels of concentrate feeds. However, many dairy farms in Ireland continue to operate very successfully with grass silage easy-fed as the sole forage and with concentrate supplements fed through in-parlour or out-of-parlour feeders. The key issue in today's feeding systems is to formulate supplements incorporating ingredients which complement the nutrients supplied from silage and which minimise adverse effects on the environment, particularly with respect to nitrogen and phosphorus.

Feeding value for beef cattle

Some of the earlier experiments on the feeding value of silage for beef cattle compared it to hay, which had previously been the standard conserved winter forage. McCarrick (1966) showed that despite higher intakes and liveweight gains with hay, carcase gains from comparable, well-preserved, grass silages were superior, and these effects were more evident for leafy, highly digestible crops (Table 3).

Table 3 Intake and growth performance by cattle fed hay or silage

Cutting date	27 May		10 June	
Forage	Silage	Hay	Silage	Hay
Forage DM intake (kg/day)	6.32	7.60	5.58	7.24
Liveweight gain (kg/day)	0.61	0.73	0.32	0.50
Carcase gain (kg/day)	0.45	0.38	0.19	0.20
Carcase gain (kg)/tonne DM intake	71.2	50.0	24.1	27.6

Source: McCarrick (1966)

Flynn (1981) reported a series of regression equations that quantified the magnitude of the increase in intake and growth rate, and thus the improvement in feed conversion efficiency, that can occur when well-preserved, unwilted, silages of increasing digestibility are offered without supplementation to finishing beef cattle. Subsequently, Drennan & Keane (1987) and Steen et al. (2002) clearly showed that the benefits to animal growth from feeding cattle with silages of superior digestibility are most evident at lower levels of concentrate supplementation and that the scale of the benefits decrease as the level of concentrates offered increases and the proportion of silage in the diet therefore decreases. Regression equations have been generated to quantify total feed intake, carcass gain and carcass fat content as the proportion of concentrates in the diet increases with a silage of either high or low digestibility (Drennan & Keane, 1987; Steen, 1998). This information allied to previous data from Steen (1989) that identified the optimal crude protein concentrations for supplementary concentrates with silage of medium to high digestibility to finishing steers, heifers and bulls, provides solid guidelines to beef farmers seeking to optimise rations for finishing cattle.

Each silage-making system has its own advantages and disadvantages and thus each system differs in the circumstances to which it is more or less suited. Steen (1985) found that direct-cut, flail-harvested silage and pre-cut, unwilted, precision-chop silage and wilted, precision-chop silage were each capable of supporting similar levels of growth in beef cattle and similar carcass output per hectare provided each system was operated correctly. Similarly, O'Kiely et al. (1999) found that baled silages and precision-chop silages could support broadly similar levels of animal performance. In contrast, poorly preserved, unwilted silage (Flynn, 1981) or silage saved from grass that has laid for an extended period on the ground during field wilting (Steen, 1985) could depress animal performance.

In general, finishing cattle fed well-preserved silage or hay have fatter carcasses compared to those fed silages of lower dry matter content (McCarrick, 1966). Increasing levels of supplementation with energy-rich concentrates increases carcass fat content (Drennan & Keane, 1987) and increasing the protein content of the supplement beyond the level at which a growth response is obtained can also increase carcass fat content (Steen, 1996). Silage fermentation (O'Sullivan et al., 2004) and wilting (Moloney et al., 2004; Noci et al., 2004) can also impact on meat quality, as can supplementation with concentrates (Steen et al., 2002).

Feeding value for sheep

In common with the dairy and beef sectors in Ireland, grass silage has generally replaced hay as the principal forage for sheep during the winter feeding period. Key factors influencing the feeding value of silage for sheep include digestibility and fermentation characteristics. However, chop length effects are more important in sheep than in cattle (Apolant & Chestnutt, 1982). Well preserved, high digestibility silage of short chop-length can sustain high levels of sheep and lamb performance without the need for supplementation. A particular issue of concern in relation to silage quality for sheep is the carry over of soil and/or other contamination during ensilage, as this can result in listeriosis in sheep (Low & Donachie, 2000). In this context it is worth highlighting that current research on silage additives is examining the use of species and/or strains of lactic acid bacteria that produce anti-microbial agents such as bacteriocins, to inhibit pathogenic bacteria and spoilage (Merry et al., 2000).

Feeding systems for grass silage

The increased reliance on silage for winter feeding of livestock in Ireland, coupled with increases in the numbers of livestock per farm, has necessitated the development of improved lower labour systems of feeding. In the 1950's, most dairy cows were tethered individually in traditional cow byres with silage manually brought to the cows. As herd size increased, "self feeding" systems were introduced in which cows physically removed silage from the silo face, usually from behind a mechanical barrier or electric wire. These systems generally performed well and continue to be used on a number of dairy farms throughout Ireland. Many of the key management recommendations for self-feed silage were based on results of studies undertaken at Experimental Husbandry Farms in England (Phipps, 1986). Key recommendations included a maximum silo face height of 1.8 m, with a silage feed-face width of 150 mm/cow.

The development of silage shear grabs and block cutters in the early 1980's facilitated the introduction of easy-feed and total mixed ration (TMR) feeding systems. Use of shear grabs enabled removal of silage from the silo, with minimal disturbance to the silo face, and facilitate transport of silage to the feed manger. The development of easy-feed systems, in which silage is presented to animals behind a feed rail or barrier, provides considerable flexibility in developing management strategies to maximise forage intake. However, given the importance of this topic, surprisingly little detailed research has been undertaken to examine the effects of factors such as frequency of feeding, trough space per animal, feed barrier design etc on food intake and animal performance.

The results of a recent study by Ferris et al. (2002) indicates that forage intake and the performance of dairy cows offered forages in relatively simple easy-feed systems, with new blocks of forage offered twice weekly, was similar to that of cows offered a TMR once daily. Further research is needed to develop low cost feeding systems which enable livestock to maximise forage intake, whilst maintaining high levels of animal welfare and performance.

Finally, accurate prediction of silage feeding value is an important prerequisite if the full potential of silage is to be achieved in livestock feeding systems. Early silage analysis in Ireland involved the determination of fermentation parameters (pH, ammonia N and lactic acid) and fibre and protein fractions. Whilst these techniques provided a general indication of potential intake and nutritive value, they were laborious to undertake, prone to inter-laboratory variation and lacked precision. A series of studies at the Agricultural Research Institute of Northern Ireland in the early 1990's, involving intake potential and digestibility measurements on over 130 grass silages, resulted in the development of prediction equations for nutritive value and intake potential through near infrared spectroscopic (NIRS) analysis of fresh silage samples (Park et al., 1998). This resulted in a rapid, cheap and reliable method for predicting a wide range of chemical and biological parameters. A commercial service has been developed at the Institute based on NIRS and this is now the main centre for silage analysis in Ireland, with over 14,000 farm silages analysed each year.

Present and future challenges in silage production in Ireland

There have been major advances in the science and practice of silage making and feeding to livestock in Ireland over the last 55 years. However, a number of major challenges remain, particularly in relation to the increased cost of conserved forage relative to grazing and grain and by-product feeds, and the relative unpredictability of silage-making in relation to the feeding value of the final product. The key challenges for research and development are to develop lower cost harvesting systems with reduced labour and fuel requirement. The lack of effect of chop length on animal performance in cattle feeding systems indicates that longer chop length harvesting methods, including self-loading forage wagons, may have a role.

The unpredictability of silage feeding value also needs to be addressed. There is an urgent need to develop systems which will enable rapid prediction of silage feeding value, and the impact of factors such as delayed harvesting, effect of wilting and impact of additive on feeding value of the final product, based on the analysis of the cut herbage in the field.

Finally, a major challenge facing farmers throughout Europe at present is the need to develop strategies which will enable more effective utilisation of nutrients - particularly nitrogen and phosphorus - both within the animal and from slurry applied to grassland. Slurry application to grazed swards poses particular problems in terms of intake and animal performance, but there are opportunities to make more effective use of slurry nutrients in silage-making systems, through reductions in inorganic N applications, without compromising silage fermentation.

References

Apolant, S.M. & D.M.B. Chestnutt (1982). An evaluation of silage and pregnant and lactating ewes. *55th Annual Report, Agricultural Research Institute of Northern Ireland*, pp. 30-37.

Baker, R.D., K. Aston, C. Thomas & S.R. Daley (1991). The effect of silage characteristics and level of supplement on intake, substitution rate and milk constituent output. *Animal Production*, 52, 586-587.

Binnie, R.C. & J.P. Frost (1995). Some effects of applying undiluted silage effluent to grassland. *Grass and Forage Science*, 50, 272-285.

Brown, W.O. & J.A.M. Kerr (1965a). Losses in the conservation of grassland herbage in lined trench silos. 1. Comparison of long and lacerated silages made by the warm fermentation process. *Journal of Agricultural Science*, 64, 135-141.

Brown, W.O. & J.A.M. Kerr (1965b). Losses in the conservation of grassland herbage in lined trench silos. 2. Comparison of lacerated silages of low and high dry-matter content made by the cold fermentation process. *Journal of Agricultural Science*, 64, 143-149.

Brown, W.O. & J.A.M. Kerr (1965c). Losses in the conservation of heavily-wilted herbage sealed in polythene film in lined trench silos. *Journal of the British Grassland Society*, 20, 227-232.

Castle, M.E., W.C. Retter, & J.N. Watson (1979). Silage and milk production : comparisons between grass silage of three different chop lengths. *Grass and Forage Science*, 34, 293-301.

Chamberlain, D.G., S. Robertson, P.A. Martin & D.A. Jackson (1990). Effects of the addition of ammonium salts of methanoic and propanoic acids and octanoic acid at ensiling on the nutritional value of silage for milk production. *Proceedings of the Ninth Silage Conference, University of Newcastle Upon Tyne*, 120-122.

Cushnahan, A. & C.S. Mayne (1995). Effects of ensilage of grass on performance and nutrient utilisation by dairy cattle. 1. Food intake and milk production. *Animal Science*, 60, 337-345.

Done, D. (1986). Silage inoculants – A review of experimental work. *Research and Development in Agriculture*, 3, 2, 83-87.

Drennan, M.J. & M.G. Keane (1987). Responses to supplementary concentrates for finishing steers fed silage. *Irish Journal of Agricultural Research*, 26, 115-127.

Ferris, C.P., R.C. Binnie, J.P Frost & D.C. Patterson (2002). A comparison of two silage feeding systems involving different labour inputs for dairy cows. In: *Proceedings XIII International Silage Conference*, Auchincruive, Scotland, 382-383

Ferris, C.P. & C.S. Mayne (1994). Effects on milk production of feeding silage and three levels of sugar-beet pulp either as a mixed ration or as an ensiled blend. *Grass and Forage Science*, 49, 241-251.

Flynn, A.V. (1981). Factors affecting the feeding value of silage. *Recent Advances in Animal Nutrition*, (Ed. W. Haresign), Butterworths, p. 81-89.

Forristal, P.D., P. O'Kiely & J.J Lenehan (1999). The influence of the number of layers of film cover and film colour on silage preservation, gas composition and mould growth on big bale silage. *Proceedings XII International Silage Conference*, Uppsala, Sweden, 251-252.

Frost, J.P. & R.J. Stevens (2000) Productive use of slurry on farms. *73rd Annual Report, Agricultural Research Institute of Northern Ireland*, pp. 50-60.

Givens, D.I., J.M. Everington & A.H. Adamson (1989). The nutritive value of spring grown herbage produced on farms throughout England and Wales over four years. 1. The effect of stage of maturity and other factors on chemical composition, apparent digestibility and energy values measured *in vivo*. *Animal Feed Science and Technology*, 27, 157-172.

Gordon, F.J., (1989a). An evaluation through lactating cattle of a bacterial inoculant as an additive for grass silage. *Grass and Forage Science*, 44, 169-179.

Gordon, F.J. (1989b). The principles of making and storing high quality, high intake silage. In: C.S. Mayne (ed.) *Silage for Milk Production, Occasional Symposium No 23, British Grassland Society*, 3-19.

Jackson, N. & B.K. Anderson (1968). Conservation of fresh and wilted grass in air-tight metal containers. *Journal of the Science of Food and Agriculture*, 19, 1-4.

Jackson, N. (1969). Losses in the nutritive value of heavily-wilted herbage ensiled in evacuated and non-evacuated polythene containers. *Journal of the British Grassland Society*, 24, 17-22.

Keady, T.W.J., C.M. Kilpatrick, A. Cushnahan & J.A. Murphy (2002). The cost of producing and feeding forages. *Proceedings of the XIIIth International Silage Conference, Auchincruive, Scotland*, 322-323.

Keady, T.W.J. & J.J. Murphy (1995). An evaluation of ensiling *per se* and addition of sugars and fishmeal on the rate of forage intake and performance of lactating dairy cattle. *Irish Journal of Agricultural and Food Research*, 34, 96-97.

Keady, T.W.J. & P. O'Kiely (1998). An evaluation of potassium and nitrogen fertilisation of grassland and date of harvest on fermentation, effluent production, dry matter recovery and predicted feeding value of silage. *Grass and Forage Science*, 53, 326-337.

Keating, T. & P O'Kiely (2000). Comparison of old permanent grassland, *Lolium perenne* and *Lolium multiflorum* swards grown for silage. 1. Effects on beef production per hectare. *Irish Journal of Agricultural and Food Research* 39, 1-24.

Kennedy, S.J. (1989). Methods of making and feeding silage. *Annual Report on Research and Technical Work of the Department of Agriculture for Northern Ireland, 1989*, p. 285.

Kormos, J. & D.M.B. Chestnutt (1966) A study of ensiling wilted and unwilted grass at two stages of maturity. 1. Nutrient losses. *Record of Agricultural Research by the Ministry of Agriculture in N. Ireland*, 15, 11-22.

Kormos, J. & D.M.B. Chestnutt (1968). Measurement of dry matter losses in grass during the wilting period. 2. The effects of rain, mechanical treatment, maturity of grass and some other factors. *Record of Agricultural Research by the Ministry of Agriculture in N. Ireland*, 17, 59-65.

Low, J.C. & W. Donachie (2000). Listeriosis. In: W.B. Martin and I.D. Aitken (eds.). *Diseases of Sheep.* 3rd Edition, Blackwell Science Limited, Oxford, UK.

McCarrick, R.B. (1963). Silage-making investigations by An Foras Taluntais. *Proceedings of a Grass Conservation Conference, An Foras Taluntais*, p. 195-216.

McCarrick, R.B. (1966). Effect of method of grass conservation and herbage maturity on performance and body composition of beef cattle. *Proceedings of 10th International Grassland Congress*, Helsinki, p. 575-580.

McIlmoyle, W.A. & R.W.J. Steen (1979). The potential of conserved forage for beef production. In: C Thomas (ed.) *Forage Conservation in the 80's. Proceedings Occasional Symposium No 11*, British Grassland Society.

Marsh, R. (1978). A review of the effects of mechanical treatment of forages on fermentation in the silo and on the feeding value of the silages. *New Zealand Journal of Experimental Agriculture*, 6, 271-278.

Mayne, C.S. (1990). An evaluation of an inoculant of *Lactobacillus plantarum* as an additive for grass silage for dairy cattle. *Animal Production*, 51, 1-13.

Mayne, C.S. & F.J. Gordon (1986a). The effect of harvesting system on nutrient losses during silage making. 1. Field losses. *Grass and Forage Science*, 41, 17-26.

Mayne, C.S. & F.J. Gordon (1986b). The effect of harvesting system on nutrient losses during silage making. 2. In-silo losses. *Grass and Forage Science*, 41, 341-351.

Merry, R.J., R. Jones & M.K. Theodorou, (2000). The conservation of grass. In: A. Hopkins (ed.) *Grass : its production and utilisation,* (Third Edition), Blackwell Science, Oxford, UK, 196-228

Moloney, A.P., B. Murray, D.J. Troy, G.E. Nute & R.I. Richardson (2004). The effects of fish oil inclusion in the concentrate and method of silage preservation on the colour and sensory characteristics of beef. *Proceedings of the Agricultural Research Forum,* Tullamore, p. 13.

Moran, J.P., P. O'Kiely, R.K. Wilson & M.B. Crombie-Quilty (1991). Lactic acid bacteria levels on grass grown for silage in Ireland. In: *Proceedings of a conference on Forage Conservation towards 2000, Landbauforschung Volkenrode, Sonderheft* 123, 283-286.

Muck, R.E., P. O'Kiely, & R.K. Wilson (1991). Buffering capacities in permanent pasture grasses. *Irish Journal of Agricultural Research,* 30, 129-142.

Murdoch, J.C., D.A. Balch, M.C. Holdsworth & M. Wood (1955). The effect of chopping, lacerating and wilting of herbage on the chemical composition of silage. *Journal of the British Grassland Society,* 10, 181-188.

Murphy, J.J. (1983). Silage for dairy cows – conservation on method of feeding. *Irish Grassland and Animal Production Association Journal,* 18, 50-58.

Murphy, J.J. (1986). A comparative evaluation of the feeding value for dairy cows of silages treated with formic and sulphuric acids. *Irish Journal of Agricultural Research,* 25, 1-9.

Noci, F., A.P. Moloney & F.J. Monaghan (2004). The effects of fish oil inclusion in the concentrate and method of silage preservation on fatty acid composition of muscle from steers. *Proceedings of the British Society of Animal Science,* p. 86.

O'Brien, M., P O'Kiely, P.D. Forristal & H. Fuller (2005). National survey to establish the extent of visible mould on baled grass silage in Ireland and the identity of the predominant fungal species. *Proceedings of the International Silage Conference, Belfast.*

O'Donnell, C., V.A. Dodd, P. O'Kiely, & M. Richardson (1995a). Corrosion of concrete by silage effluent. 1. Concrete characteristics. *Journal of Agricultural Engineering Research,* 60, 83-92.

O'Donnell, C., P. O'Kiely, V.A. Dodd & M. Richardson (1995b). Corrosion of concrete by silage effluent. 2. Environmental aspects. *Journal of Agricultural Engineering Research,* 60, 93-97.

O'Kiely, P. (1989). Aerobic stability of farm silages. *Irish Journal of Agricultural Research,* 28, 102-103.

O'Kiely, P. (1992). The effect of ensiling sugarbeet pulp with grass on silage composition, effluent production and animal performance. *Irish Journal of Agricultural and Food Research,* 31, 115-128.

O'Kiely, P. (1996). Performance of beef cattle offered grass silages made using bacterial inoculants. *Irish Journal of Agricultural and Food Research,* 35, 1-15.

O'Kiely, P., O.T. Carton & J.J. Lenehan (1994). Effect of time, rate and method of slurry application to grassland grown for silage. F.A.O. Network on Animal Waste Utilisation – 7th consultation, Germany.

O'Kiely, P., A.V. Flynn, & R.I. Wilson, (1986). Predicting the requirement for silage preservative. *Farm and Food Research,* 17, 2, 42-44.

O'Kiely, P., J.J. Lenehan, & P.D. Forristal. (1999). Big bale and precision-chop silage systems: conservation characteristics and nutritive value for beef cattle. *Irish Journal of Agricultural and Food Research,* 37, 107.

O'Kiely, P. & R.W. Muck (1998). Grass silage. In: *Grass for dairy cows* (eds. J.H. and D.J.R. Cherney), CABI Publications, p. 223-251.

O'Kiely, P. & H. Tunney (1997). Silage conservation characteristics of grass that received a range of rates of phosphorus fertiliser. *Irish Journal of Agricultural and Food Research,* 36, 104.

O'Sullivan, A., K. O'Sullivan, K. Galvin, A.P. Moloney, D.J. Troy & J.P. Kerry (2004). Influence of concentrate composition and forage type on retail packaged beef quality. *Journal of Animal Science,* 82, 2384-2391.

Park, R.S., R.E. Agnew, F.J. Gordon & R.J. Barnes (1998). The development and transfer of undried grass silage calibrations between near infrared spectroscopy instruments. *Animal Feed Science and Technology,* 78, 325-340.

Patterson, D.C. & N. Walker (1982). Some factors affecting the voluntary intake by pigs of diets containing effluent from the ensilage of grass. *Journal of Agricultural Science, Cambridge,* 98, 123-129.

Patterson, D.C., T. Yan, & F.J. Gordon. (1996). The effect of wilting grass prior to ensiling on the response to bacterial inoculants. 2. Intake and performance by dairy cows over three harvests. *Animal Science,* 62, 419-429.

Phipps, R.H. (1986). The role of conserved forage in milk production systems with particular reference to self and easy-feed silage systems. In: W.H. Broster, R.H. Phipps & C.L. Johnson (eds.) *Principles and Practice of Feeding Dairy Cows, NIRD Technical Bulletin No 8,* 133-162.

Reeve, A. (1989). What can silage produce? – an R and D view. In: C S Mayne (ed.) *Silage for milk production. Occasional Symposium No 23, British Grassland Society,* 31-41.

Small, J.C. & F.J. Gordon. (1988). The effect of systems of harvesting grass for silage on the output of silage and milk per hectare. *Journal of Agricultural Science, Cambridge,* 111, 369-383.

Steen, R.W.J. (1985). The effect of field wilting and mechanical treatment on the feeding value of grass silage for beef cattle and on beef output per hectare. *Animal Production,* 41, 281-291.

Steen, R.J.W. (1986). An evaluation of effluent from grass silage as a feed for beef cattle offered silage based diets. *Grass and Forage Science*, 41, 39-45.

Steen, R.W.J. (1989). Recent research on protein supplementation of silage-based diets for growing and finishing beef cattle. *62nd Annual report, Agricultural Research Institute of Northern Ireland*, pp. 12-20.

Steen, R.W.J. (1991). Recent advances in the use of silage additives for dairy cattle. In: C.S. Mayne (ed.) *Management Issues for the Grassland Farmer in the 1990's. Occasional Symposium No. 25, British Grassland Society*, 87-101.

Steen, R.W.J. (1996). Effects of protein supplementation of grass silage on the performance and carcass quality of beef cattle. *Journal of Agricultural Science, Cambridge,* 127, 403-412.

Steen, R.W.J. (1998). Effect of the proportion of grass silage and concentrates in the diet on the performance of beef cattle. *Proceedings of a Research Seminar on Ruminant Production, Agricultural Research Institute of Northern Ireland*, April, pp. 49-63.

Steen, R.W.J., D.J. Kilpatrick & M.G. Porter (2002). Effects of the proportions of high or medium digestibility grass silage and concentrates in the diet of beef cattle on liveweight gain, carcass composition and fatty acid composition of muscle. *Grass and Forage Science*, 57, 279-291.

Stewart, T.A. (1980). Prevention of damage to grass swards from applications of silage effluent by neutralisation or dilution of effluent acids. *Grass and Forage Science*, 35, 47-54.

Wilson, R.K. & C.P Collins (1980). Chemical composition of silages made from different grass genera. *Irish Journal of Agricultural Research*, 19, 75-84.

Wilson, R.K. & A.V. Flynn (1979). Effects of fertiliser N, wilting and delayed sealing on the chemical composition of grass silages made in laboratory silos. *Irish Journal of Agricultural Research*, 18, 13-23.

Wright, D.A., J.P. Frost, D.C. Patterson & D. Kilpatrick (2000). The influence of weight of ryegrass per unit area and treatment at or after mowing on rate of drying. *Grass and Forage Science,* 82, 86-98.

Zimmer, E. & R.J. Wilkins, (1984). Efficiency of silage systems : A comparison between unwilted and wilted silages. *Landbauforschung Volkenrode, Sanderheft,* 69, 5-12.

Grass silage: factors affecting efficiency of N utilisation in milk production

P. Huhtanen and K.J. Shingfield

MTT Agrifood Research Finland, Animal Production Research, FIN-31600 Jokioinen, Finland, Email: pekka.huhtanen@mtt.fi

Key points

1. Low efficiency of N utilisation for milk production in cows fed grass silage-based diets is mainly due to excessive N losses in the rumen.
2. The type and extent of *in silo* fermentation can alter the balance of absorbed nutrients.
3. There is very little experimental evidence that the capture of N in the rumen can be improved by a better synchrony between energy and N release in the rumen. Nitrogen losses in the rumen can be reduced by decreasing the ratio between rumen degradable N and fermentable energy.
4. Rapeseed meal has increased milk protein output more than isonitrogenous soybean meal supplementation, probably due to higher concentration of histidine in rapeseed protein.
5. Efficiency of N utilisation for milk production is not necessarily lower for the grass silage-based diets compared to other diets.

Keywords: grass silage, dairy cow, microbial protein synthesis, protein degradation, nutrient balance

Introduction

Grass and maize silage comprise a major proportion of conserved forages used in dairy cow rations in Central and Northern Europe. Development of early maturing varieties, lower labour requirements, improved nutrient management on dairy farms and some nutritional benefits has increased the amount of maize silage being used at the expense of grass silage. It can be expected that the use of forage maize will continue to expand into northern climates if the predicted increases in global climate warming occur. In regions that are not suitable for maize production whole-crop cereal silages represent a viable alternative or a complimentary forage for grass and legume silages.

Dairy farming is known to contribute to both atmospheric and hydrospheric pollution (Tamminga, 1992). Excessive nitrogen (N) excretion is a major environmental concern especially for intensive grass and grass silage-based systems, since the efficiency of nitrogen (N) utilisation for milk production on grass silage-based diets is lower compared to maize silage-based diets (Givens & Rulquin, 2004). Grass silage provides a substantial proportion of crude protein the (CP) in the diet of high producing dairy cows in many regions of the EU. Despite a relatively high concentration of CP in dairy cow rations containing grass silage and cereal grains, protein supplementation has consistently elicited positive production responses. Responses to supplementary protein are indicative of inefficient utilisation of N in the basal diet. Protein supplementation can alleviate certain nutritional limitations of CP in grass silage, but does result in reduced overall efficiency of N utilisation and increased N emissions. Increases in N emissions into the environment are generally a result of greater excretion of N in urine N, which is more of a concern than an increase in faecal N output due to greater exposure to evaporative losses and leaching into water supplies. Therefore other strategies to improve N utilisation of grass silage-based diets are urgently required, not only to reduce environmental N emissions, but also to reduce feed costs. The overall objective of this paper is to review research examining N metabolism of dairy cows fed grass silage-based diets, with particular emphasis on nutritional concepts that may facilitate improved N utilisation for milk production from grass silage.

Utilisation of silage nitrogen

Microbial protein synthesis

Givens & Rulquin (2004) recently concluded that the utilisation of grass and legume CP for milk production is low compared with other forages. The poor utilisation of N from grass silage N was attributed primarily to the low efficiency of capture of rumen degradable N (RDN). However, the generally accepted conclusion that the efficiency of microbial protein synthesis (MPS) is lower for grass silage-based diets compared with other diets is largely based on indirect evidence with few direct comparisons (ARC, 1984; Givens & Rulquin, 2004). The values for the energetic efficiency of MPS (EMPS) reported in the literature are highly variable, which may, in part, reflect differences in experimental techniques and difficulties in the determination of MPS (Shingfield, 2000). It should also be noted that errors in estimating digesta flow potentially have a two-fold influence on estimates of EMPS values. Overestimating DM flow from the rumen overestimates microbial N supply, but underestimates the amount of organic matter apparently digested in the rumen. Consequently, the differences in EMPS are greater than those in total microbial N. A direct comparison of silage and barn-dried hay harvested from the grass same sward at the same time did not suggest a lower ruminal N utilisation in the rumen for silage compared to hay (Jaakkola & Huhtanen, 1993) (Table 1). Microbial N flow at the duodenum and EMPS were significantly (P<0.05) higher for the silage diets compared with the hay diets, but undegraded feed protein flow was higher for the hay diets. Rumen ammonia N concentration was marginally higher in cattle fed silage-based diets (12.9 *vs.* 11.6 mmol/l), but the difference can be attributed to the higher CP content of the silage based diets (164 *vs.* 152 g/kg DM for silage and hay, respectively) rather than to an effect of forage conservation method on EMPS. Similarly, the flow of N to the duodenum has been shown to be unaffected when lucerne is conserved by ensiling compared with drying (Hristov & Broderick, 1996). Furthermore, studies have shown no significant differences in the yields of milk and milk protein between grass hay and silage harvested at the same stage of maturity (Bertilsson, 1983).

Table 1 Effects of forage conservation method and proportion of concentrate on ruminal N metabolism in growing cattle (Jaakkola & Huhtanen, 1993)

Forage	Silage			Hay		
Concentrate (g/kg DM)	250	500	750	250	500	750
CP (g/kg DM)	168	165	161	148	152	155
Rumen ammonia N (mmol/l)	13.9	12.8	12.0	11.3	11.7	12.0
N intake (g/d)	178	181	174	161	165	173
Duodenal flow (g/d)						
Non-ammonia N	142	152	150	132	146	150
Microbial N	77	89	85	64	76	79
Feed N	53	52	54	56	59	59
N degradability	0.71	0.72	0.68	0.65	0.65	0.65

Most of the evidence from measurements of duodenal N flow, rumen ammonia N concentration and milk production responses are not consistent with the widely held view that ensiling of grass necessarily results in reduced efficiency of N utilisation for milk production.

A low efficiency of silage N utilisation often reported in dairy cows fed grass silage-based diets can be attributed both to poor ensiling techniques and/or extensive *in silo* fermentation rather than being a result of conservation method *per se*. On both a theoretical basis (Chamberlain, 1987) and from consideration of experimental evidence (Jaakkola & Huhtanen, 1992), it is clear that the end-products of silage fermentation are not a significant source of energy for rumen microbial growth. Although the review of Dewhurst *et al.* (2000) suggested that there are no consistent effects of silage additives on EMPS, most of the evidence from within study comparisons suggests increased ruminal MPS in response to restricting *in silo* fermentation by using high levels of acid-based additives (van Vuuren *et al.*, 1995). Earlier data indicating a strong positive association between residual water soluble carbohydrate content and MPS (Jaakkola *et al.*, 1993) support the concept that MPS is related to the supply of fermentable energy for rumen microbes (Table 2). Analysis of a large data set (230 treatment means) from production trials examining the effects of silage fermentation characteristics on milk production, indicated significant inverse relationships between both silage total acid and silage ammonia-N concentrations, and the concentration and yield of milk protein (Huhtanen *et al.*, 2003).

Table 2 The effects of the rate of formic acid application on grass silage fermentation characteristics and duodenal N flow (Jaakkola *et al.*, 1993)

	Formic acid (l/t)			
	0.0	2.0	4.0	6.0
In silage (g/kg DM)				
WSC	3	19	37	92
Lactic acid	62	78	47	17
VFA	63	18	13	10
Ammonia N (g/kg total N)	103	46	28	21
Duodenal flow (g/d)				
Non-ammonia N	114.5	126.1	128.4	136.9
Microbial N	49.0	57.3	58.4	65.4
Feed N	53.4	56.7	57.9	59.4

Ruminal production of microbial N *in vitro* was lower from the soluble N in lucerne silage than from the corresponding fraction in lucerne hay (Peltekova & Broderick, 1996) suggesting that utilisation of soluble N from silage is reduced due to a shortage of available protein and peptides. Evaluation of *in vivo* data from studies in sheep and growing cattle has consistently shown an increase in MPS in response to protein supplementation (Weiss *et al.*, 2003). In contrast, measurements of nutrient flow into the omasum in cows fed a range of protein supplements (soybean meal, fishmeal, maize gluten meal, rapeseed feeds, and urea) do not support the hypothesis that MPS in dairy cows fed grass silage-based diets is limited by the supply of protein or peptides available to rumen microbes (Figure 1). Efficiency of MPS tended to decrease in response to protein supplementation, which may reflect ruminally degraded protein supplying less energy for rumen microbes than fermentable carbohydrates.

Two reasons for the lack of response in EMPS to protein supplementation in dairy cow studies could be suggested: (1) the feeding level was substantially higher and the supply of peptides and amino acids was probably more constant due to a more consistent eating pattern and (2) the silages in dairy cow studies were restrictively fermented, whereas the silages in the studies as reviewed by Weiss *et al.* (2003) were more extensively fermented. Restricting the extent of *in silo* fermentation is known to increase the proportion of peptide N in silage soluble N (Nsereko & Rooke, 1999) which can result in a more optimal supply of N fractions to rumen microbes.

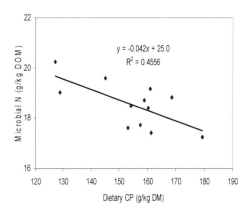

Figure 1 Relationship between dietary CP concentration and efficiency of MPS (data from Ahvenjärvi *et al.*, 1999; 2002; Korhonen *et al.*, 2002b). The values are adjusted for a random study effect.

Ruminal degradability of silage crude protein

Due to extensive proteolysis *in silo* and the high relative amount of N in the form of NPN, a large proportion of silage N is degraded to ammonia in the rumen. The values reported for the effective ruminal degradability (EPD) of silage N typically varies between 0.70 and 0.90. Factors such as forage species, maturity, wilting and the use of ensiling additives have been associated with changes in ruminal degradability of silage N. However, there is very little evidence demonstrating that variations in EPD of silage N are of any nutritional importance. Wilting has consistently reduced the EPD values of silage (Givens & Rulquin, 2004), while production experiments have shown rather small positive production responses to wilting, which were linked to increases in silage DM intake which are mainly derived from the higher DM of the wilted silages. Verbic *et al.* (1999) reported proportionally 0.19 and 0.27 higher metabolisable protein (MP) values for extensively wilted silage and hay compared with formic acid-treated silage. Such differences as determined *in situ* are in contrast with measurements of duodenal N flow and data from production studies examining the effects of forage conservation method.

The lack of progress in this area can be attributed to the inadequacies of current methods to provide reliable estimates of ruminal degradability of silage CP. Some differences in EPD estimated by the *in situ* technique may simply reflect differences in microbial contamination of undegraded residues and particle losses from the bags. Recent evidence strongly suggests that the assumptions of the Ørskov & McDonald (1979) model used to estimate effective protein EPD are not valid. Implicit in the *in situ* method is that the rapidly degradable fraction (*a*-fraction) is instantaneously degraded or that degradation rate is infinite with no escape. Recent studies (Choi *et al.*, 2002; Volden *et al.*, 2002) have clearly demonstrated that variable proportions of dietary N can escape from the rumen in the liquid phase as non-ammonia non-microbial N. To overcome this problem, Hvelplund and Weisbjerg (2000) proposed a model in which the escape of the *a*-fraction is incorporated into the model. The second problem with the model proposed by Ørskov & McDonald (1979) is the assumption of random first-order passage kinetics of feed particles from the rumen. However passage kinetic studies based on duodenal sampling have shown mechanisms of selective retention of feed particles in the rumen (Huhtanen *et al.*, 2005).

It is clear that methodological limitations in the *in situ* technique result in an overestimation in the range of EPD values for forages. For example, accounting for the passage of amino N in the liquid phase (escape of *a*-fraction) and retention of feed particles in the rumen non-escapable pool would reduce the effects of conservation method, e.g. wilting and restricting the extent of *in silo* fermentation on EDP. Similarly, the reduction in the EPD value with advance in the maturity of ensiled grass is more likely to be related to greater microbial contamination of undegraded residues than true differences in N degradability. Possible decreases in the EPD of silage with advancing maturity are at least partly compensated for by an increase in the proportion of ADF-bound N (Rinne *et al.*, 1997) and reduced intestinal digestibility of undegraded feed protein. Before clear progress can be made in this area, methods for estimating quantitative – not just qualitative - differences in the supply of undegraded protein from forages are urgently required. The lack of a simple and reliable reference method has delayed progress in developing reliable tools for estimating ruminal CP degradability. At present, using a constant EPD value probably results in at least as good a prediction of silage MP value as utilising values obtained by the *in situ* technique.

Amino acid supply from grass silage-based diets

Theoretically N utilisation for milk production could be improved by increasing the supply of regarded as the first limiting amino acid (AA). Lysine (Lys) and methionine (Met) have long been limiting or co-limiting AA in cows fed diets based on maize silage and maize (NRC, 2001). However, Lys and Met infusions have not increased milk protein yield on grass silage-based diets. The studies conducted at the Hannah Research Institute and at MTT have demonstrated that histidine (His) is the first limiting AA for milk protein synthesis when grass silage is the basal forage in the diet. Abomasal infusion of 6.5 g His per day increased milk protein yield in cows fed grass silage and a cereal grain-based supplement, but Met and Lys AA did not elicit any further response in addition to His when infused either individually or as a mixture (Vanhatalo *et al.*, 1999). Use of feather meal to further increase the imbalance in AA supply from grass silage-based diets demonstrated a substantial milk protein yield response to abomasal His infusion (Kim *et al.* 1999, 2001a). Subsequent studies also demonstrated a positive response to His infusion in cows fed a grass-silage based diet without feather meal (Kim *et al.*, 2001b). Increases in milk protein yield to His infusions have been shown to be linear up to the levels of 6 g/d (Korhonen *et al.*, 2000). A transfer efficiency from the abomasum to milk of 0.28 was clearly below default values of MP utilisation used in current feed protein evaluation systems (e.g. 0.67; NRC, 2001).

That His is the first limiting AA for milk protein synthesis in cows fed grass silage-based diets can be attributed to the high contribution of microbial protein to total MP supply. Since the concentration of His in milk protein is markedly higher than microbial protein, whereas

Lys and Met content in milk protein is very similar to that in microbial protein. Low plasma His concentrations in cows fed grass silage and cereal grain based diets (Vanhatalo *et al.*, 1999; Korhonen *et al.*, 2000) and numerically smaller amounts of His than Met in digesta flow to the small intestine compared with Met (Korhonen *et al.*, 2000; 2002b) also support the view that His is the first limiting AA with diets based on grass silage and cereal grains. Application of the NRC (2001) AA sub-model to data from production experiments conducted at MTT (72 diets) indicated that only the His content of MP was positively associated with milk protein yield (Huhtanen, 2005).

Attempts to identify the second limiting AA for grass silage-based diets have not been particularly successful, and in some cases, even the ranking of the first limiting AA has been variable (Kim *et al.*, 2000). The AA profile of MP in cows fed grass silage-based diets is relatively well balanced, except for His, and therefore even small changes in the AA profile of CP available for absorption can change the ranking order of the next limiting AA. Even if the second limiting AA could be identified, its potential to increase milk protein output is probably rather limited. However, if protein supplements of low ruminal degradability are limited in the supply of certain AA, then substantial milk protein yield responses could be obtained by balancing AA supply as evidenced by the responses to His when feather meal is included in grass silage-based diets (Kim *et al.*, 1999; 2001a).

Protein supplementation

The CP content of concentrate supplements can be increased by replacing cereal grains or other energy rich ingredients with protein feeds. Statistical analysis of data from studies conducted mainly in Finland demonstrated positive effects of protein supplements on total DM intake (Figure 2). Increased DM intake in response to protein has been attributed to improved cell wall digestibility (Oldham, 1984). Whilst OM digestibility was improved with protein supplementation in these studies (Figure 2), the effect is too small to explain increases in DM intake. It is likely that the effects of protein supplementation are mediated partially via metabolic mechanisms, probably related to the protein or AA energy ratio. It is evident that constraints on rumen fill are not the sole factor regulating the intake of highly digestible grass silage based diets as indicated by the decrease in rumen NDF fill with increases in silage OM digestibility (Rinne *et al.*, 2002). Protein supplementation improves nutrient balance (amino acid: ME ratio), which increases milk yield, and thereby increases energy requirements which would stimulate an increase in DM intake.

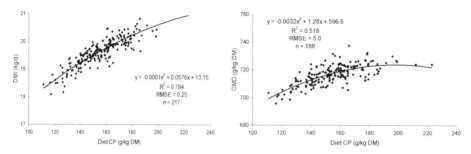

Figure 2 Effects of increasing dietary crude protein (CP) concentration on DM intake and organic matter digestibility (OMD). The values are adjusted for a random study effect.

A metabolic mechanism of intake response regulation, potentially mediated through improvements in the ratio of amino acids:energy at the tissue level, has been known for some time, and is clearly demonstrated in studies involving post-ruminal casein infusions which showed that increases in hay intake were accompanied by changes in rumen fill (Egan, 1970). Similar responses in dairy cows fed grass-red clover silages were observed by Khalili & Huhtanen (2002). Duodenal casein infusions resulted in higher DM intake, increased rumen NDF fill and total chewing time compared with infusions of the same amount of casein in to the rumen.

The effect of protein supplementation on OM digestibility is curvilinear with greater incremental responses being observed at low dietary CP concentrations. This suggests that the supply of rumen degradable protein may limit cell wall digestion of grass silage-based diets at low CP concentrations. Digestibility of digestible NDF determined by 12 d *in situ* incubations was increased when the CP content of diets fed to dairy cows was increased by replacing cereal grains with soybean meal or rapeseed expeller (Shingfield *et al.*, 2003). Whether this response was related to the stimulation of cellulolytic bacteria by an increased supply of preformed amino acids and peptides, or to differences in the intrinsic digestion properties of grain and protein supplement or to reduced starch content of the diet remains unclear.

Milk and milk protein yields are both enhanced in a quadratic manner with increased protein supplementation (Figure 3). Due to an increase in both DM intake and OM digestibility, milk yield responses can be attributed to increased energy intake as well as to enhanced AA supply. The effects of protein supplementation in an individual experiment have seldom been significant, but analysis of a large number of studies have shown that protein supplementation had a significant (P<0.001) linear effect on milk protein content (0.17 g/kg per 10 g CP/kg diet DM). However, when milk protein content was corrected for the contribution from urea N, the response decreased to 0.06 indicating that proportionally 0.65 of the increase in milk protein content was due to elevated urea levels. Even though protein supplementation has produced substantial milk yield responses, these responses have consistently been associated with a reduction in the efficiency of dietary N utilisation. In this MTT data set an increase of 10 g/kg DM in dietary CP concentration reduced N efficiency by 0.014 units. In addition to less efficient N utilisation, protein supplements will often increase phosphorus emissions, since protein sources are generally rich in this mineral.

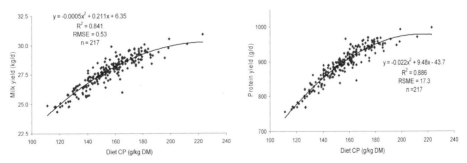

Figure 3 Effects of increasing dietary crude protein (CP) concentration on yields of milk and protein. The values are adjusted for a random study effect.

In addition to level, the source of protein supplementation is an important determinant of milk production responses, since both ruminal protein degradability and the balance between essential amino acids influence the supply of the most limiting amino acids to the mammary gland. Chamberlain *et al.* (1989) reported higher milk yield responses to fishmeal compared with soybean meal (0.15 vs. 0.12 g per 1 g increase in CP intake). However, because protein supplements of animal origin are not permitted for use in ruminant diets in the EU due to the incidence of bovine spongiform encelphalopathy, the importance of high quality sources of vegetable protein for intensive milk production has further been highlighted. Indirect evidence has suggested that utilisation of rapeseed protein is at least equal to that of soybean protein (Korhonen, 2003; Lock & Shingfield, 2004). Marginal milk protein yield responses to additional protein (g milk protein per g additional CP intake) were lower (0.12) for soybean meal compared to fishmeal and rapeseed feeds (0.15 and 0.16, respectively) based on 13, 10 and 13 studies, and 33, 21 and 38 treatment means, respectively (Korhonen, 2003).

Shingfield *et al.* (2003) fed incremental amounts of either solvent extracted soybean meal or heat-moisture treated rapeseed expeller cake with a fixed amount of a cereal based concentrate (10 kg/d on a fresh weight basis) containing 120, 150, 180 or 210 g CP/kg DM and grass silage *ad libitum*. Inclusion of rapeseed elicited slightly higher milk protein output responses than soybean meal (Table 3). Feeding rapeseed expeller was associated with higher plasma concentrations of His and branched-chain amino acids (BCAA). The changes in plasma amino acids were indicative of an increased supply of both total amino acids and His which, when coupled with a lower concentration of milk urea, are indicative of more efficient utilisation of absorbed AA for diets containing rapeseed than soybean protein supplements. Other studies have shown that rapeseed containing feeds result in similar milk protein yields compared with a mixture of fishmeal and soybean meal (Dewhurst *et al.*, 1999) and higher milk protein output than a mixture of soybean meal and maize gluten meal (Khalili *et al.*, 2001; Vanhatalo *et al.*, 2003).

Table 3 Effects of inclusion of soybean meal and rapeseed expeller on feed intake and milk production of cows fed grass silage *ad libitum* and 10 kg concentrate per day (Shingfield *et al.*, 2003)

	Soybean meal					Rapeseed cake		
Concentrate CP (g/kg DM)	120	150	180	210		150	180	210
Intake (kg DM/d)	20.0	20.3	20.4	20.5		20.6	20.9	20.8
Yield								
Milk (kg/d)	26.2	26.9	27.0	28.5		28.2	29.3	30.1
ECM (kg/d)	29.8	30.3	30.4	32.1		31.2	31.6	32.8
Fat (g/d)	1325	1335	1342	1404		1354	1343	1410
Protein (g/d)	859	902	889	954		930	967	993
Concentration								
Fat (g/kg)	51.2	50.8	50.7	49.9		48.9	46.9	47.3
Protein (g/kg)	33.5	34.3	33.6	33.8		33.9	33.7	33.4
Urea (mmol/l)	3.30	4.37	5.19	5.87		3.90	4.71	5.24

Improving N utilisation of grass silage based diets

There is now increasing concern about excessive emissions of N into the environment from milk production and feeding grass silage-based diets to dairy cows has been implicated as one of the main sources for the low efficiency of N utilisation. However, it should be noted that attempts to alter N efficiency in the animal is a much less effective strategy for improving on-farm N utilisation than restricting fertiliser and concentrate N use, and increasing the efficiency of N uptake from the soil (Van Bruchem et al., 1999). The greatest potential to improve N utilisation is to reduce rumen N losses, but improving the balance between absorbed nutrients (e.g. amino acid balance and glucose supply) can also result in substantial improvements in the efficiency of N utilisation. Rumen N losses in cows fed grass silage-based diets are probably higher than current estimates based on duodenal sampling. Analysis of nutrient flows at the omasum indicated that zero ruminal N balance occurs at a diet CP concentration of 130 g/kg DM (Schwab et al., 2005), a level that is markedly lower than is typically fed on-farm.

Forage management and feeding

Intensive dairy farming based on grass silage, typically uses high fertiliser N inputs and harvests grass at an early stage of maturity. While these strategies both increase silage N concentration, the effects on N efficiency are markedly different. Shingfield et al. (2001) investigated the opportunities to improve the efficiency of N utilisation by comparing the effects of additional N derived from silage due to increased N fertilisation or from urea, wheat gluten meal and rapeseed expeller supplements. Interestingly, urea and increased N fertilisation decreased the efficiency of N utilisation more than protein supplements, particularly when compared with rapeseed expeller.

In contrast to N derived from increased N fertilisation, incremental protein from earlier harvesting is utilised at a higher efficiency for milk protein secretion. Rinne et al. (1999a) harvested timothy-meadow fescue swards at one week intervals, which resulted in silages with a wide range in CP content (113 to 172 g/kg DM). Even though the efficiency of N utilisation decreased from 0.333 to 0.283, as harvesting date progressed, the marginal milk protein response of 0.16 was comparable to that attained with fishmeal or rapeseed feeds. Higher utilisation of additional N from earlier harvesting compared with that derived from N fertilisation is related to increased supply of fermentable energy and higher capture of N in the rumen. Rumen N losses could also be reduced by replacing early-harvested grass silage with maize silage and whole-crop cereal silages. Because grass silage provides excess RDP, whereas maize and whole-crop silages are often deficient in degradable N, and therefore mixtures of these forages could also be used to improve the balance between RDP and fermentable energy supply.

Carbohydrate supplementation

The effects of synchronisation of energy and N release in the rumen on MPS have been intensively investigated during recent decades. However, few benefits in terms of improved efficiency of MPS from a more synchronous release of N and energy release in the rumen have been reported. Often, the effects of synchronisation and diet composition have been confounded. Sugar supplements have sometimes been more efficient than starch in promoting MPS, but these effects may be related to higher numbers of rumen protozoa when starch is fed (Chamberlain et al., 1985) and increased intra-ruminal N recycling. Part of the difference between sugar and starch supplements may be attributed to a partial escape of starch from ruminal fermentation, thereby providing less ATP for microbial growth. When the synchrony between energy and N release in the rumen of cattle fed a grass silage-based diet was manipulated by feeding sucrose either twice daily or as continuous infusion, the latter promoted a higher MPS and lower rumen ammonia N concentration than feeding sucrose twice daily (Khalili & Huhtanen, 1991). Twice-daily feeding would have been expected to provide more energy during the post feeding period of extensive proteolysis and to stimulate microbial growth. Reviewing the available literature, Chamberlain & Choung (1995) concluded that there is no convincing evidence of a close synchronisation of energy and N release in the rumen having beneficial effects on MPS. It can be concluded that the ratio between RDP and fermentable energy is much more important determinant of ruminal N losses than synchrony in the rates of N and energy release.

Protein supplementation

High producing dairy cows require an abundant supply of absorbed AA. Without accurate predictions of the absorption and utilisation of AA, an over supply of protein often occurs to ensure that the genetic potential for milk production is expressed. However, this approach reduces the efficiency of N utilisation. In cows fed grass silage-based diets the supply of RDP, which is often in excess of microbial N requirements, is further increased by protein supplementation and a major proportion of the additional N will be excreted as urea-N in the urine.

Theoretically, decreasing ruminal protein degradability by physical or chemical treatment should be an effective strategy to improve the efficiency of N utilisation for milk production. Treatments used to reduce ruminal protein degradability have often resulted in lower EPD values as determined by *in situ* incubation, but production responses to reductions in EPD are inconclusive (Santos *et al.*, 1998). The absence of a response to reduced degradability could be due to the untreated protein supplement already satisfying AA requirements. To test this hypothesis, Rinne *et al.* (1999b) fed incremental levels of solvent extracted rapeseed meal and heat-moisture treated rapeseed expeller, which differed in EPD by 0.18 units (0.82 *vs.* 0.64). While both protein supplements linearly increased milk protein output, the rapeseed expeller with lower EPD did not produce a greater increase in yield of protein.

Preventing a potential deficiency of the most limiting AA in the mammary gland is another reason for feeding protein in excess. It is probable that His is the first limiting AA when the supply of absorbed AA is largely a reflection of the profile in microbial protein, as often is the case for grass silage-based diets. With the limitations in the use of animal proteins for ruminants, the scope to balance the limited His supply by high quality plant proteins is rather limited. Rapeseed meal, which is one of the best sources of His amongst vegetable proteins, has the same amount of His per g protein as milk. Rumen protected lysine and methionine supplements have none or only small positive effects on N utilisation in cows fed diets based mainly on maize-silage or maize (Satter *et al.*, 1999), and it is unlikely that protected sources of His would markedly improve the efficiency of N utilisation on grass silage diets due to the relatively balanced profile of absorbed AA under these circumstances. Increasing His concentration in rapeseed protein by gene technology (Wahlroos, 2004) provides a new and interesting opportunity to supply a more balanced profile in absorbed AA in the future.

Glucose supply

Modifications to the chemical composition of herbage during ensilage can have a major impact on the balance of nutrients absorbed from the gastro-intestinal tract of ruminant animals. Diets based on restrictively-fermented grass silage containing relatively high concentrations of water soluble carbohydrates and minimal amounts of lactate are characterised by a rumen fermentation pattern rich in lipogenic volatile fatty acids (Chamberlain & Choung, 1993), associated with a relatively high EMPS (van Vuuren *et al.*, 1995). Diets based on high lactate silages are associated with a rumen fermentation pattern rich in propionate (Figure 4) and a lower EMPS compared with restrictively fermented silages. An increased supply of microbial protein at the duodenum with restrictively fermented silages is seldom associated with a corresponding increase in milk protein secretion (Huhtanen *et al.*, 2003). These general considerations of the effects of silage fermentation characteristics suggest that diets based on restrictively-fermented silages are more limited in the supply of glucose, while the available AA are more limiting in diets based on high lactate silages. For diets based on restrictively-fermented silages, the benefits of increased microbial protein may be negated by a concomitant reduction in glucose supply from ruminal propionate production. Higher AA and lower glucose concentrations in the plasma of cows fed diets based on restrictively-fermented silage compared with high lactate silage (Miettinen & Huhtanen, 1997) indicate that influencing silage fermentation characteristics has the potential to alter the balance of absorbed nutrients.

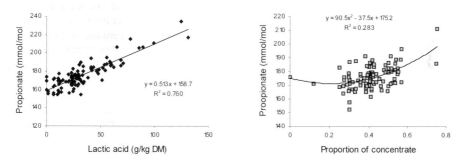

Figure 4 Effects of dietary concentration of lactic acid and proportion of concentrate on molar proportion of propionate in rumen VFA (data from Finnish studies). The values are adjusted for a random study effect.

Increasing the supply of glucose by infusing propionate into the rumen or infusing glucose post-ruminally has on average resulted in small positive production responses (Huhtanen 1998; Rigout *et al.*, 2003). Higher milk protein yield responses to infusions of His plus glucose (Huhtanen *et al.*, 2002) or casein plus glucose (Vanhatalo *et al.*, 2003a) as combined infusions rather than single infusions are consistent with the utilisation of AA for milk protein synthesis being compromised by limited glucose supply from the basal diet.

Because increasing the proportion of concentrate in the diet has only small effects on the proportion of glucogenic VFA in cattle fed grass-silage-based diets (Lock & Shingfield, 2004, Figure 4), increased starch digestion in the small intestine is an alternative strategy to enhance milk protein production by sparing glucogenic AA from catabolism in the liver. However,

Nocek & Tamminga (1991) concluded that production studies show no clear evidence that post-ruminal starch digestion increases milk production. Keady *et al.* (1996; 1999) obtained increases in the concentration of milk protein, but with no effects on total DM intake or milk production, by increasing starch in the concentrate with grass silage-based diets. In a recent study, milk protein yield tended to increase linearly to post-ruminal starch infusions (Reynolds *et al.*, 2001), but at best, only proportionally 0.175 of the gross energy in starch was recovered as milk energy. Small increases in milk energy output suggest that a large proportion of energy in starch was used for body tissue synthesis or oxidation. Replacing barley with maize in dairy concentrates would be a practical approach to increase glucose supply from increased digestion of starch in the small intestine, but milk protein yield responses (+20 g/d) have been minimal in high-yielding (35 kg/d) cows fed grass silage-based diets (Khalili *et al.*, 2001). Increased post-ruminal starch supplies tend to result in greater increases in lactose yield than in protein yield (Reynolds *et al.*, 2001; Khalili *et al.*, 2001). A small positive increase in milk protein output (30 g/d) with abomasal glucose infusions (300 g/d) compared with the control diet and corresponding infusions of starch (Vanhatalo *et al.*, 2003b) indicate that the use of post-ruminal glucose infusions may not necessarily mimic situations of increased supply of starch in the small intestine.

In addition to limited glucose supply, an imbalanced AA profile may limit milk protein yield responses from improved microbial protein synthesis with restrictively fermented silages. The concentration of His in microbial protein is markedly lower than in milk protein (21 *vs.* 27 g/kg AA). Plasma AA profile measurements (Miettinen & Huhtanen, 1997) suggest that improved microbial supply associated with silages of restricted fermentation may exacerbate the imbalance in AA supply to the mammary gland of cows fed grass silage-based diets. Restricting *in silo* fermentation and post-ruminal infusions of casein similarly increased plasma concentrations of branched-chain AA and lysine compared with extensively fermented high-lactate silage, but only casein infusion increased plasma His concentration and significantly enhanced milk protein output.

Conclusions and future perspectives

This review challenges the perception that the efficiency of N utilisation for milk production on grass silage-based diets is particularly poor. Part of this perception may have originated from observations based on indirect comparisons of EMPS, an overestimation in the range of ruminal protein degradability determined by current methods and kinetic models (*in situ*) and responses found with poorly-preserved silages being accepted as typical of all silages. Well-preserved restrictively-fermented silages and dried forages harvested from the same sward have generally resulted in comparable animal performance, despite estimates of MP supply sometimes being 0.20-0.25 lower for silage.

The AA profile of absorbed protein is relatively well-balanced in animals fed grass silage-based diets due to the high contribution of microbial protein and therefore reducing rumen ammonia-N losses represents the most promising means for improving overall efficiency of N use (milk N/dietary N). The easiest way to decrease N excess in the rumen is to alleviate an oversupply of RDP, but this is very difficult to achieve without compromising milk production. Partial replacement of early harvested grass silages with maize or other whole crop silages, forage management factors including N fertilisation and maturity at harvest all have a major impact on the N efficiency of milk production, but it should be noted that manipulating silage CP concentration by these two strategies results in very distinct effects on N utilisation.

Because modifications of the chemical composition of parent herbage during ensiling can have a major impact on the balance of nutrients available to rumen microbes and the host animal, dynamic and mechanistic substrate-based models could provide a more accurate prediction of nutrient supply than currently available models. The new Nordic dairy cow model Karoline (Danfær et al., 2004) suggests that cell wall characteristics (e.g. indigestible NDF, rate of NDF digestion) and fermentation characteristics (e.g lactic acid, VFA) of silage have a greater impact on the total supply of AA available for absorption than silage protein parameters. To monitor the efficiency of N utilisation at farm level, analysis of dietary CP and assessment of milk urea concentration are useful diagnostic tools. Low milk urea concentrations (<160 mg/l) are indicative of an RDP deficiency (Nousiainen et al., 2004), whereas high concentrations (>300-350 mg/l) are associated with low marginal production responses and poor utilisation.

Manipulating the ratio of degradable protein and fermentable energy supply in the rumen provides the best opportunity to enhance the N efficiency of milk production. Excess RDP and ruminal N losses can be reduced by avoiding excessive levels of N fertilisation, and/or by replacing part of a grass silage-based diet with maize or whole-crop silages. Changes in chemical composition during ensilage have a major impact on the quantity of energy supplied to rumen microbes, which, together with changes in the N fraction, can result in substantial changes in microbial protein production and in the net absorption of ammonia from the rumen. Large amounts of supplementary protein are fed with grass silage-based diets, but despite substantial production responses, N utilisation is consistently compromised. However, differences in production responses between protein supplements do offer an opportunity to improve N efficiency. Attempts to balance nutrient supply by using various glucogenic substrates or AA supplements have had variable effects. Further research including both experimental studies and modelling of existing data needs to be targeted towards optimising the balance of absorbed nutrients. In the future it may be expected that plant breeders will develop new genotypes that result in grass varieties with lower CP content, and vegetable protein sources with reduced protein degradability characteristics and modified AA profile, which together may better complement microbial protein.

References

Ahvenjärvi, S., A. Vanhatalo, P. Huhtanen & T. Varvikko, T. (1999). Effects of supplementation of a grass silage and barley diet with urea, rapeseed meal and heat-moisture-treated rapeseed cake on omasal digesta flow and milk production in lactating dairy cows. *Acta Agriculturae Scandinavica, Section A, Animal Science,* 49, 179-189.

Ahvenjärvi, S., A. Vanhatalo & P. Huhtanen (2002). Supplementing barley or rapeseed meal to dairy cows fed grass-red clover silage: I. Rumen degradability and microbial flow. *Journal of Animal Science,* 80, 2176-2187.

ARC (1984). The Nutrient Requirements of Ruminant Livestock, Supplement 1. Commonwealth Agricultural Bureaux, Farnham Royal, 45 pp.

Bertilsson, J. (1983). Effects of conservation method and stage of maturity upon the feeding value of forages to dairy cows. Sveriges lantbruksuniversitet, Institutionen för husdjurens utfodring och vård. *Rapport,* 104, 21 p.

Chamberlain, D.G. (1987). The silage fermentation in relation to the utilization of nutrients in the rumen. *Process Biochemistry,* 22, 60-63.

Chamberlain, D.G. & J.-J. Choung (1993). The nutritional value of grass silage. In: P. O'Kiely et al. (eds.). Silage Research 1993. *Proceedings of the 10th International Conference on Silage Research. Dublin, Ireland,* 131-136.

Chamberlain, D.G. & J.-J. Choung (1995). The importance of rate or ruminal fermentation of energy sources in diets for dairy cows. In: P.C. Garnsworthy et al. (eds.) *Recent Advances in Animal Nutrition.* Butterworths, London, 3-27.

Chamberlain, D.G., P.A. Martin & S. Robertson (1989). Optimizing compound feed use in dairy cows with high intakes of silage. In: W. Haresign & D.J.A. Cole (eds.) *Recent Advances in Animal Nutrition.* Butterworths, London, 175-193.

Chamberlain, D.G., P.C. Thomas, W. Wilson, C.J. Newbold & J.C. McDonald (1985). The effects of carbohydrate supplements on ruminal concentrations of ammonia in animals given diets of grass silage. *Journal of Agricultural Science*, 104, 331-340.

Choi, C.W., S. Ahvenjärvi, A.Vanhatalo, V. Toivonen & P. Huhtanen (2002). Quantification of the flow of soluble non-ammonia nitrogen entering the omasal canal of dairy cows fed grass silage based diets. *Animal Feed Science and Technology*, 96, 203-220.

Danfær, A., P. Huhtanen, P. Udén, J. Sveinbjörsson & H. Volden (2004). Karoline - a Nordic cow model for feed evaluation: model description. J. Dijkstra. (ed) Sixth International Workshop of Modelling Nutrient Utilisation in Farm Animals, Abstracts, p. 45.

Dewhurst, R.J., K. Aston, W.J. Fisher, R.T. Evans, M.S. Dhanoa & A.B. McAllan (1999). Comparison of energy and protein sources offered at low levels in grass-silage-based diets for dairy cows. *Animal Science*, 68, 789-799.

Dewhurst, R.J., D.R. Davies & R.J. Merry (2000). Microbial protein supply from the rumen. *Animal Feed Science and Technology*, 85, 1-21.

Egan, A.R. (1970). Nutritional status and intake regulation in sheep. VI. Evidence for variations in setting of an intake regulatory mechanism relating to the digesta content of the reticulorumen. *Australian Journal of Agricultural Research*, 21, 735-746.

Givens, D.I. & H. Rulquin (2004). Utilisation by ruminants of nitrogen compounds in silage-based diets. *Animal Feed Science and Technology*, 114, 1-18.

Hristov, A.N. & G.A. Broderick (1996). Synthesis of microbial in ruminally cannulated cows fed alfalfa silage, alfalfa hay, or corn silage. *Journal of Dairy Science,* 79, 1627-1637.

Huhtanen, P. (1998). Supply of nutrients and productive responses in dairy cows given diets based on restrictively fermented silage. *Agricultural and Food Science in Finland,* 7, 219-250.

Huhtanen, P. (2005). A review of the 2001 dairy cattle NRC protein and amino acid model – A European Perspective. *Journal of Dairy Science,* 88, *Supplement* (In press)

Huhtanen, P., S. Ahvenjärvi, M.R. Weisbjerg & P. Nørgaard (2005). Digestion and passage of fibre in ruminants. *Proceedings of the 10th International Symposium in Ruminant Physiology* (In press).

Huhtanen, P., J.I. Nousiainen, H. Khalili, S. Jaakkola & T. Heikkilä (2003). Relationships between silage fermentation characteristics and milk production parameters: analyses of literature data. *Livestock Production Science,* 81, 57-73.

Huhtanen, P., A. Vanhatalo & T. Varvikko (2002). Effects of abomasal infusions of histidine, glucose and leucine on milk production and plasma metabolites of dairy cows fed grass silage diets. *Journal of Dairy Science*, 85, 1, 204-216.

Hvelplund, T. & M.R. Weisbjerg (2000). *In situ* techniques for the estimation of protein degradability and postrumen availability. In: D.I. Givens, E. Owen, R.F.E. Axford & H.M. Omed (eds.) Forage Evaluation in Ruminant Nutrition. CABI Publishing, Wallingford, 233-258.

Jaakkola, S. & P. Huhtanen (1992). Rumen fermentation and microbial protein synthesis in cattle given intraruminal infusions of lactic acid with a grass silage based diet. *Journal of Agricultural Science, Cambridge,* 119, 411-418.

Jaakkola, S. & P. Huhtanen (1993). The effects of the forage preservation method and the proportion of concentrate on nitrogen digestion and rumen fermentation in cattle. *Grass and Forage Science*, 48, 146-154.

Jaakkola, S., P. Huhtanen & V. Kaunisto (1993). VFA proportions and microbial protein synthesis in the rumen of cattle receiving grass silage ensiled with different rates of formic acid. In: P. O'Kiely *et al.* (eds.) Silage research 1993. *Proceedings of the 10th International Conference on Silage Research, Dublin, Ireland*, p.139-140.

Keady, T.W.J., C.S. Mayne & M. Marsden (1996). The effects of concentrate energy source on silage intake and animal performance with lactating dairy cows offered a range of grass silages. *Animal Science*, 66, 21-34.

Keady, T.W.J., C.S. Mayne, D.A. Fitzpatrick & M. Marsden (1999). The effects of energy source and level of digestible undegradable protein in concentrates on silage intake and performance of lactating dairy cows offered a range of grass silages. *Animal Science*, 68, 763-777.

Khalili, H. & P. Huhtanen (1991). Sucrose supplements in cattle given grass silage-based diet. 1. Digestion of organic matter and nitrogen. *Animal Feed Science and Technology*, 33, 247-261.

Khalili, H. & P. Huhtanen (2002). Effect of casein infusion in the rumen, duodenum or both sites on factors affecting forage intake and animal performance of dairy cows fed red clover-grass silage. *Journal of Dairy Science*, 4, 909-918.

Khalili, H., A. Sairanen, K. Hissa & P. Huhtanen (2001). Effects of type and treatment of grain and protein source on dairy cow performance. *Animal Science*, 72, 573-584.

Kim, C.H., T.G. Kim, J.J. Choung & D.G. Chamberlain (1999). Determination of the first limiting amino acid for milk production in dairy cows consuming a diet of grass silage and a cereal-based supplement ontaining feather meal. *Journal of the Science of Food and Agriculture*, 79, 1703-1708.

Kim, C.H., T.G. Kim, J.J. Choung & D.G. Chamberlain (2000). Variability in the ranking of the three most-limiting amino acids for milk protein production in dairy cows consuming grass silage and a cereal-based supplement containing feather meal. *Journal of the Science of Food and Agriculture*, 80, 1386-1392.

Kim, C.H., T.G. Kim, J.J. Choung & D.G. Chamberlain (2001a). Estimates of the efficiency of transfer of L-histidine from blood to milk when it is the first-limiting amino acid for secretion of milk protein in the dairy cow. *Journal of the Science of Food and Agriculture*, 81, 1150-1155.

Kim, C.H., J.J. Choung & D.G. Chamberlain (2001b). The effects of intravenous infusion of amino aicds in dairy cows given diets of grass silage and cereal-based concentrates. *Journal of Animal Physiology and Nutrition*, 85, 293-300.

Korhonen, M. (2003). Amino acid supply and metabolism in relation to lactational performance of dairy cows fed grass silage based diets. Dissertation, 45 p. + encl. University of Helsinki.

Korhonen, M., A. Vanhatalo, & P. Huhtanen, P. (2002a) Effect of protein source on amino acid supply, milk production and metabolism of plasma nutrients in dairy cows fed grass silage. *Journal of Dairy Science* 85, 3336-3351

Korhonen, M., A. Vanhatalo, T. Varvikko & P. Huhtanen (2000). Responses to graded postruminal doses of histidine in dairy cows fed grass silage diets. *Journal of Dairy Science*, 83, 2596-2608.

Korhonen, M., A. Vanhatalo, T. Varvikko & P. Huhtanen. (2002b). Evaluation of isoleucine, leucine, and valine as a second-limiting amino for milk production in dairy cows fed grass silage diet. *Journal of Dairy Science,* 85, 1533-1545.

Lock, A.L & K. Shingfield (2004). Optimisation milk composition. In: E. Kebreab, J. Mills & D. Beever (eds.) UK Dairying: using science to meet consumers' needs. Nottingham University Press, Nottingham, UK. pp. 107-188.

Miettinen, H. & P. Huhtanen (1997). Effects of silage fermentation and postruminal casein supplementation in lactating dairy cows. 2. Blood metabolites and amino acids. *Journal of the Science of Food and Agriculture,* 74, 459-468.

Nocek, J.E. & S. Tammiga (1991). Site of digestion of starch in the gastrointestinal tract of dairy cows and its effect on milk yield and composition. *Journal of Dairy Science*, 74, 3598-3629.

Nousiainen, J., K.J. Shingfield & P. Huhtanen (2004). Evaluation of milk urea concentration as a diagnostic of protein feeding. *Journal of Dairy Science*, 87, 386-398.

NRC (2001). Nutrient requirements of dairy cattle. Seventh Revised Edition. National Research Council. 381 p.

Nsereko, V.L. & J.A. Rooke (1999). Effects of peptide inhibitors and other additives on fermentation and nitrogen distribution in perennial ryegrass silage. *Journal of the Science of Food and Agriculture*, 79, 679-686.

Oldham, J.D. (1984). Protein-energy interrelationships in dairy cows. *Journal of Dairy Science*, 67, 1090-1114.

Ørskov, E.R. & I. McDonald (1979). The estimation of protein degradability in the rumen from incubation measurements weighted according to rate of passage. *Journal of Agricultural Science*, 92, 499-503.

Peltekova, V.D. & G.A. Broderick (1996). *In vitro* ruminal degradation and synthesis of protein from fractions extracted from alfalfa hay and silage. *Journal of Dairy Science*, 79, 612-619.

Reynolds, C.K., S.B. Cammell, D.J. Humphries, D.E. Beever, J.D. Sutton & J.R. Newbold (2001). Effects of postrumen starch infusion on milk production and energy metabolism in dairy cows. *Journal of Dairy Science*, 84, 2250-2259.

Rigout, S., C. Hurtaud, S. Lemosquet, A. Bach & H. Rulquin (2003). Lactational effect of propionic acid and duodenal glucose in cows. *Journal of Dairy Science*, 86, 243-253.

Rinne, M., P. Huhtanen & S. Jaakkola (2002). Digestive processes dairy cows fed silages harvested at four stages of maturity. *Journal of Animal Science*, 80, 1986-1998.

Rinne, M., S. Jaakkola & P. Huhtanen (1997). Grass maturity effects on cattle fed silage-based diets. 1. Organic matter digestion, rumen fermentation and nitrogen utilization. *Animal Feed Science and Technology,* 67, 1-17.

Rinne, M., S. Jaakkola, K. Kaustell, T. Heikkilä & P. Huhtanen. (1999a). Silage harvested at different stages of grass growth versus concentrate foods as energy and protein sources in milk production. *Animal Science,* 69, 251-263.

Rinne, M., S. Jaakkola, T. Varvikko & P. Huhtanen. (1999b). Effects of the type and amount of rapeseed feed on milk production. *Acta Agriculturae Scandinavica, Section A, Animal Science,* 49, 137-148.

Satter, L.D., H.G. Jung, A.M van Vuuren & F.M. Engels (1999). Challenges in the nutrition of high-producing ruminants. In: H.J.G. Jung & G.C. Fahey, Jr. (eds.) Nutritional Ecology of Herbivores. *Proceedings of the Vth International Symposium on the Nutrition of Herbivores*, pp. 609-646.

Santos, F.A.P., J.E.P. Santos, C.B. Theurer & J.T. Huber (1998). Effects of rumen undegradable protein on dairy cow performance. *Journal of Dairy Science*, 81, 3182-3213.

Shingfield, K.J. (2000). Estimation of microbial protein supply in ruminant animals based on renal and mammary purine metabolite excretion. A. Review. *Journal of Animal and Feed Sciences, 9,* 169-212.

Shingfield, K.J., S. Jaakkola & P. Huhtanen (2001). Effects of level of nitrogen fertilizer application and various nitrogenous supplements on milk production and nitogen utilization of dairy cows given grass silage-based diets. *Animal Science, 73,* 541-554.

Shingfield, K.J., A. Vanhatalo & P. Huhtanen (2003). Comparison of heat-treated rapeseed expeller and solvent-extracted soya-bean meal protein supplements for dairy cows given grass silage-based diets. *Animal Science,* 77, 305-317.

Schwab, C.G., P. Huhtanen, C.W. Hunt & T. Hvelplund (2005). Nitrogen Requirements of Cattle. In: E. Pfeffer & A.N Hristov (eds), Interactions between Cattle and the Environment. CABI publishing (In Press).

Tamminga, S. (1992). Nutrition management of dairy cows as a contribution to pollution control. *Journal of Dairy Science*, 75, 345-357

Vanhatalo, A., E. Pahkala, P. Salo-Väänänen, H. Korhonen, V. Piironen & P. Huhtanen (2003). Rapeseed and soyabean as protein supplements of dairy cows fed grass silage based diets. In: Hilmer Sorensen et al. (eds.) *Proceedings of the 11th International Rapeseed Congress. The Royal Veterinary and Agricultural University, Copenhagen,* 1238-1240.

Vanhatalo, A., P. Huhtanen, V. Toivonen & T. Varvikko (1999) Response of dairy cows fed grass silage diets to abomasal infusions of histidine alone or in combinations with methionine and lysine. *Journal of Dairy Science*, 82, 2674-2685

Vanhatalo, A., T. Varvikko & P. Huhtanen (2003a). Effects of casein and glucose on responses of cows fed diets based on restrictively fermented grass silage. *Journal of Dairy Science,* 86, 3260-3270.

Vanhatalo, A., T. Varvikko & P. Huhtanen (2003b). Effect of various glucogenic sources on production and metabolic responses in dairy cows fed grass silage-based diets. *Journal of Dairy Science,* 86, 3249-3259.

Van Bruchem, J., H. Schiere & H. Van Keulen (1999). Dairy farming in the Netherlands in transition towards more efficient use of nutrients. *Livestock Production Science*, 61, 145-153.

Van Vuuren, A.M., P. Huhtanen & J.P. Dulphy 1995. Improving the feeding and health value of ensiled forages. In: Journet, M. *et al.* (eds.). *Recent Developments in the Nutrition of Herbivores.* INRA, Paris. p. 297-307.

Verbic, J., E.R. Ørskov, J. Zgajnar, X.B. Chen & V. Znidarsic-Pongrac (1999). The effect of method of forage preservation on the protein degradability and microbial protein synthesis in the rumen. *Animal Feed Science and Technology,* 82, 195-212.

Volden, H., L.T. Mydland & V. Olaisen (2002). Apparent ruminal degradation and rumen escape of soluble nitrogen fractions in grass and grass silage administered intraruminally to lactating dairy cows. *Journal of Animal Science*, 80, 2704-2716.

Wahlroos, M.T. (2004). Modification of histidine content in transgenic *Brassica rapa* plants. Dissertation, 49 p. + encl. University of Turku.

Weiss, W.P., D.G. Chamberlain & C.W. Hunt (2003). Feeding silages. In: D.R Buxton, R.E. Muck & J.H. Harrison (eds.) Silage Science and Technilogy. *American Society of Agronomy,* Madison, USA, 469-504.

Recent developments in feeding beef cattle on grass silage-based diets

M. McGee

Teagasc, Grange Research Centre, Dunsany, Co. Meath, Ireland.
Email: mmcgee@grange.teagasc.ie

Key points:

1. High digestibility grass silage with moderate concentrate supplementation can sustain a large proportion of the cattle performance achieved on high concentrate diets.
2. Increasing concentrate supplementation reduces the importance of grass silage nutritional value.
3. Subsequent compensatory growth diminishes the advantage of concentrate supplementation of young cattle.
4. Meat quality and fatty acid composition can be influenced by grass silage-based diets.

Keywords: grass silage, beef cattle, concentrate supplementation

Introduction

Grass silage is a basic component of many beef production systems worldwide, particularly in those countries with a temperate climate such as in Northern and Western Europe. In addition to providing forage, usually for the indoor period/winter feeding, grass silage is often an integral part of grassland management through the removal of seasonal surpluses associated with typical grass growth curves. Furthermore, the grass silage land area permits recycling of organic manure and is also a means of reducing internal parasite challenge to grazing cattle. As a ruminant feedstuff, grass silage can be less attractive than alternative forage crops or high concentrate diets due to constraints of often variable and unpredictable weather, the scarcity of silage-cutting contractors, coupled with ever-increasing costs of production and effluent management. In addition to unfavourable relative costs of nutrient supply, changing beef production systems and consumer market demands can necessitate higher levels of concentrate feeding for beef cattle.

Relative to the dairy cow, there is much less research carried out on the nutrition and feeding of beef cattle on grass silage-based diets and consequently the science and technology is much less developed.

Nutrient supply from grass silage

Nutrient supply to the ruminant from grass silage is primarily influenced by altering the cutting date of the grass crop (i.e. digestibility) and by modifying and restricting fermentation through wilting or the use of additives (Thomas & Thomas, 1985). The effects of wilting (Ingvartsen, 1992; Wright *et al.*, 2000) and silage additives (O'Kiely, 2001; Kung *et al.*, 2003) on the performance of cattle has been the subject of previous reviews. Most of the variation in net energy content of grass silage is associated with its digestibility, which in turn is mainly determined by the stage of growth of the grass plant at harvest (Rinne, 2000). It is well established that the performance of beef cattle increases with increasing grass silage digestibility (Flynn, 1981; Randby, 2001; Steen *et al.*, 2002). For example, Steen (1988) calculated from the literature that where silages were offered as the sole feed, carcass gain was increased by 33 g/d per 10 g/kg increase in silage digestibility. The corresponding value where silages were supplemented with concentrates at 0.20 to 0.37 of total dry matter (DM) intake was 28 g/d. The latter value is in good agreement with the value of 29 g per 0.01 increase in digestible organic matter (DOM) reported by Steen *et al.* (2002). As forage DM intake decreases with increasing levels of supplementary concentrates, the effect of forage digestibility diminishes (Van Vuuren *et al.*, 1995) to the extent that at high (0.80 of the diet) concentrate feeding levels silage digestibility had no effect on carcass gain (Steen *et al.*, 2002). Steen (1998) calculated a response in daily carcass gain of 26, 21, 12 and 1 g / 10 g/kg increase in silage digestibility when silages were supplemented with 2.25, 4.5, 6.75 and 9.0 kg concentrates/head daily (~0.20 to 0.80 of total DM intake), respectively. Clearly, this has implications for the cost of producing silage in that silage of lower digestibility may be produced at a lower cost per unit DM (higher yield) where high concentrate feeding levels are practiced.

Intake

It is widely believed that the intake potential of herbage is reduced as a result of ensiling although the variation in the magnitude of the reduction is large (Mayne & Cushnahan, 1995). Mayne & Cushnahan (1995) suggested that factors such as stage of maturity at harvest explained at least some of the reduced intake (and animal performance) with grass silage relative to grass at the stage it is grazed. They attributed the direct effect of ensilage to progressive changes in the nitrogenous components. However, when good ensiling techniques are used and conservation quality is excellent, ensilage *per se* has relatively little effect on the intake of cattle (Mayne & Cushnahan, 1995; Dulphy & Van Os, 1996). Keady & Murphy (1993) presented data from the literature showing that intake of growing heifers and finishing steers ranged from proportionately –0.08 to +0.06 for well preserved grass silage relative to the parent herbage.

Many attempts have been made to estimate the effect of silage composition, digestibility, and fermentation quality on grass silage intake by beef cattle. Steen *et al.* (1998) concluded that the intake of grass silage offered as the sole feed to beef steers is closely related to factors which influence the extent of digestion and rate of passage of material through the animal including *in vivo* apparent digestibility, rumen degradability and the concentrations of the fibre and nitrogen fractions. Intake was poorly related to more conventional factors such as pH, total acidity, buffering capacity and the concentrations of lactic, acetic and butyric acids. The use of Near Infra Red Spectrometry predicted intake with the highest degree of accuracy (R^2 of relationship = 0.90).

Substitution rate

The conventional method of overcoming the deficiencies in nutrient supply from grass silage is to supplement with concentrates. Increasing the level of supplementary concentrates in the diet of beef cattle reduces grass silage intake (Drennan & Keane, 1987a, b; Dawson *et al.*, 2002; Caplis, 2004) but increases total DM intake. The magnitude of the decrease in silage intake is usually greater with silage of higher digestibility (Drennan & Keane, 1987a; Steen, 1998). Similarly, there is also evidence that the substitution rate (SR) is much greater with silage of restricted fermentation (Dawson *et al.*, 2002) although this is not always so (Agnew & Carson, 2000). The effect of silage preservation on SR may be confounded with changes in digestibility (e.g. Shiels, 1998). Mayne *et al.* (1995) reported that SR in growing cattle was positively correlated to intake characteristics of the silage when offered as the sole feed. For diets containing low to moderate levels of concentrate (<0.47 of dietary DM intake) substitution rates range from 0.29 to 0.64 kg silage DM per kg concentrate DM with high digestibility grass silage (Agnew & Carson, 2000; Steen & Kilpatrick, 2000; Patterson *et al.*, 2000; Dawson *et al.*, 2002; Caplis, 2004). Recent studies have reported a curvilinear increase in total DM intake with increasing concentrate level (Steen, 1998; Keane 2001; Caplis, 2004) implying a progressively decreasing intake of grass silage with increasing concentrate level. From a series of experiments with high digestibility grass silage, Steen (1998) calculated substitution rates of 0.33, 0.64, 0.90 and 1.15 kg silage DM/kg concentrate DM for successive increments of concentrates equating to 0.22, 0.42, 0.62 and 0.85 of total DM intake, respectively. Caplis (2004) found substitution rates for high digestibility silage of 0.29, 0.65 and 1.10 kg silage DM/kg concentrate DM for successive increments of concentrate equating to 0.31, 0.55 and 0.85 of total DM intake. Surprisingly, there are relatively few reports in the literature where a wide range of supplementary concentrate levels in the diet of beef cattle offered grass silage-based diets were examined.

The effect of energy supplement type on the intake of grass silage is unclear. Mayne *et al.* (1995) reported that starch or fibre supplements had no significant effect on mean SR in growing cattle when considered across a range of silage compositions but there were interactions between supplement type and silage type. With extensively fermented silages characterised by high lactic acid concentrations, SR were lower with high starch than with high fibre supplements. It was suggested that this difference may reflect better synchronisation of nitrogen and fermentable energy. Steen (1993a) reported that silage intake was higher for fibre than starch-base concentrates for growing cattle. Silage intake of finishing beef cattle was shown to be not differentially affected by starch, fibre or sugar-based concentrates (Moloney *et al.*, 1993), fibre or starch-based concentrates (Steen, 1995a; O'Kiely & Moloney, 1994) and higher for starch than fat-based concentrates (Steen, 1995a; Moloney, 1996) or fibre-based concentrates (Moloney, 1996).

The effect of protein supplementation on grass silage intake is equivocal. Moloney (1991) found no effect of increasing the crude protein concentration (110 to 515 g/kg DM) of the concentrate on grass silage intake in growing cattle while Mayne *et al.* (1995) reported reduced substitution rates from increasing the supplement crude protein concentration (120 to 260 g/kg fresh weight).

In contrast to expectations from published feed table values, increasing the proportion of concentrate in the diet of beef cattle does not necessarily increase the digestible energy value of the total diet (Patterson *et al.*, 2000; Steen & Kilpatrick, 2000; Caplis, 2004), in particular where grass silage of higher digestibility is fed (Drennan & Keane, 1987a; Steen *et al.*, 2002). This negative associative effect is often attributed to a depression in fibre digestibility in the rumen and in the total digestive tract from the inclusion of rapidly fermentable carbohydrates such as barley-based (starch) concentrates (Huhtanen & Jaakkola, 1993) and sucrose (Khalili & Huhtanen, 1991) in grass silage-based diets. The decrease in the digestibility of cell wall constituents is related to a reduction in ruminal pH. Mulligan *et al.* (2002) concluded that for diets based on grass silage and high fibre concentrate supplements, the depressive effect of feeding level *per se* on diet digestibility was greater for high (0.85 dietary DM) than moderate (0.50 dietary DM) concentrate diets. For moderate concentrate diets the reduction in digestibility was attributed to an increased fractional rumen outflow rate for the concentrate component while for high concentrate diets, decreased rumen pH and a decreased rate of concentrate and forage digestion were deemed to be important components. Consequently, dietary energy intake often mirrors total DM intake with increasing concentrate feeding level.

Production response to supplementation

Growing cattle

Due to compensatory growth a close inverse relationship (quadratic) has been found between liveweight gain of weanling cattle in winter and subsequent gain at pasture (Keane, 2002). Consequently, there was a highly significant positive curvilinear relationship between supplementary concentrate (range 0-3 kg) feeding of young/weanling cattle offered grass silage *ad libitum* and liveweight gain during the winter and a significant negative curvilinear relationship with liveweight gain of these same animals when subsequently grazing grass during the following summer (Keane, 2002). There was no difference in compensation at pasture for additional winter gain achieved either by increasing grass silage quality or by concentrate feeding (Keane, 2002). A continuing effect of compensation during a subsequent finishing phase indoors after the grazing season (or common diet) is evident in some, but not all studies (Keane, 2002; Keady *et al.*, 2004a). Similarly, the optimum level of concentrate feeding for cattle in the second winter is higher if animals are to be slaughtered at the end of the winter than for animals ("store cattle") destined to spend the following (3rd) summer at pasture (Keane & Drennan, 1994). Drouillard & Kuhl (1999) highlighted the lack of research pertaining to integrated production systems where a more thorough understanding of the interactions among grazing nutrition and management, finishing performance and carcass traits is needed to facilitate greater economic exploitation of these relationships.

Finishing cattle

Although Scollan *et al.* (2003) concluded that good quality silage will support high levels of performance without the need for supplementation, the time taken to achieve the same slaughter live weight was 277 d for silage only compared to 228 and 196 d for a diet containing 0.3 and 0.7 concentrate on a DM basis, respectively. The carcass growth response to concentrate supplementation is generally lower with higher digestibility (Drennan & Keane, 1987a; Steen, 1998; Randby, 2001) or higher intake/restricted fermentation (Agnew & Carson, 2000) grass silage. Drennan & Keane (1987b) feeding grass silage with an *in vitro* DMD of 725 g/kg to steers reported a linear increase in carcass gain of 52 g per kg concentrate DM fed within the range 2.9 to 10.9 kg/head/d. More recently, Caplis (2004) feeding grass silage with an *in vitro* DMD of 758 g/kg obtained a curvilinear increase in the daily liveweight gain of steers (decreasing from 130 to 13 g liveweight/kg additional concentrate DM) with increasing concentrate level over a comparable concentrate range. Steen (1998) calculated that the response in carcass growth rate to concentrate supplementation within the range 2 to 9 kg/head/d was curvilinear (decreasing from 93 to 4 g carcass/kg additional concentrate) and linear (58 g carcass/kg concentrate) for high (733 g DOM/kg DM) and medium (625 g DOM/kg DM) digestibility silage, respectively. Consequently, the optimum input of concentrates is higher with lower digestibility silage.

Due to a progressive decline in the response to concentrates, high digestibility grass silage plus moderate concentrate inputs can achieve a large proportion of the carcass and lean tissue gain achieved with high concentrate diets. In the study of Patterson *et al.* (2000), bulls offered a high digestibility (730 g DOM/kg DM) grass silage and concentrates at 0.52 of dietary DM intake produced proportionately 0.90 of the rate of carcass gain and 0.99 of the lean tissue gain produced by a diet containing 0.75 concentrates. Indeed, 0.95 of the lean tissue gain was achieved at a relatively low concentrate proportion of 0.39. Steen *et al.* (2002) concluded that a diet containing 0.80 of high digestibility (743 g DOM/kg DM) grass silage and 0.20 concentrate sustained proportionately 0.85 of the carcass gain produced by a diet containing 0.80 concentrate, whereas a diet containing 0.60 high digestibility grass silage and 0.40 concentrate sustained the same carcass gain as a diet containing 0.80 concentrate. At similar metabolisable energy (ME) intakes, indications were that the net energy value of high digestibility silage per MJ of ME can be close to that of a high concentrate diet.

Accordingly, in order to determine the optimum or breakeven level of concentrate supplementation *per se*, estimates of carcass efficiency (kg concentrates per kg carcass), silage substituted (kg DM per kg carcass gain) and the true costs of grass silage and concentrates are required (Keane, 2001).

Energy source

The feeding value of wheat is the same as barley as a supplement to grass silage for finishing cattle (Steen, 1993b; Drennan & Moloney, 1998). When compared with barley as a supplement to grass silage-fed steers, molasses was used more efficiently at 210 g/kg total DM intake, but its relative energy value declined as the level of inclusion increased above 330 g/kg DM (Drennan, 1985). Replacement of barley-based with molasses (0.18-0.21 of dietary DM intake)-based concentrates did not affect the performance of finishing cattle on grass silage-based diets (Moloney *et al.*, 1993; Chapple *et al.*, 1996). The decline in the relative nutritive value of molasses with increasing inclusion level may be attributed to a possible reduction in fibre digestion due to sub-optimal ammonia concentration in the rumen together with excessively high butyrate production and possible differences in the site of digestion of carbohydrate within the digestive tract (Moloney *et al.*, 1994). Drennan *et al.* (1994) found that as a supplement to finishing bulls offered grass silage, a low protein, starch-based, concentrate was superior to a low protein starch + sugar-based (0.18 of dietary DM molasses) concentrate but there was no difference between the energy sources at higher protein inclusions.

Replacing starch with digestible fibre in the concentrate increased the liveweight/carcass weight gain of cattle offered grass silage-based diets in some studies (Moloney *et al.*, 1993; O'Kiely & Moloney, 1994), but not in others (Steen, 1995a; Moloney, 1996; Moloney *et al.*, 2001a). However, in these studies, concentrate feeding level and silage quality differed widely. Differences between concentrate energy sources should be more evident at higher concentrate feeding levels. Further research is required in this area.

Fat-based concentrates are inferior to starch or fibre-based concentrates as supplements to grass silage which is attributed to a lower organic matter digestibility for the former (Steen, 1995a; Moloney, 1996). Dietary supplementation with fat, especially polyunsaturated fats (with the possible exception of fish oil), at greater than 50 g added fat/kg concentrates, has an increasingly adverse effect on ruminal digestion of fibre (Doreau & Chillard, 1997).

Protein source

Growing cattle fed grass silage alone respond to supplementation with ruminally undegraded protein with relatively large improvements in gains (Titgemeyer & Loest, 2001). Silage intake (scaled for bodyweight) was not affected by supplementation and responses to ruminally undegraded protein supplementation were observed across different qualities of silage and different levels of intake. Similarly, including rumen undegradable protein increased the liveweight gain of young steers offered grass silage plus low levels of concentrate (Moloney, 1991; 1993; Rouzbehan et al., 1996). Titgemeyer & Loest (2001) suggested that diets based on grass silage are likely to provide inadequate supplies of amino acids for growing cattle due to low microbial protein production and low ruminally undegraded protein content in the silages. This deficit was corrected by the rumen undegraded protein supplement resulting in increased protein deposition in those cattle. However, Sanderson et al. (2001) concluded that although rumen undegradable protein was an effective means by which to enhance protein deposition in young cattle given silage diets, the composition of the carcass remained unaffected. Supplementation of grass silage alone with a rumen degradable source of protein has also increased the performance of growing steers (Veira et al., 1995; Scollan et al., 2001a). Where young growing cattle are offered grass silage ad libitum and a low level of barley-based concentrates the inclusion of rumen degradable protein has increased (Moloney, 1991) or had no significant effect (Keane, 2002) on growth rate. There are indications of interactions between grass silage digestibility and crude protein degradability. Scollan et al. (2001a) found that as the sole supplement, rumen undegradable protein seemed to be superior to rumen degradable protein with lower digestibility silage with little difference between the protein types with higher digestibility silage. In contrast, Moloney (1993) found that decreasing CP degradability in the concentrate increased liveweight gain with high digestibility silage but not with low digestibility silage. However, when taken in the context of production systems much of the advantage to protein supplementation of young cattle was often lost during a subsequent standard feeding phase (Moloney, 1993) or grazing season (Seoane et al., 1993; Scollan et al., 2001a; Keane, 2002) due to compensatory growth.

For finishing cattle offered high digestibility grass silage plus barley-based concentrates, increasing protein intake by using either a rumen undegradable (Drennan et al., 1994) or degradable (Drennan et al., 1994; Steen & Robson, 1995; Steen, 1996a) protein source did not significantly affect animal growth. Furthermore, feeding excess protein increased nitrogen excretion to the environment (Steen, 1996a). Calculations by Titmeyer and Loest (2001) showed that while amino acids were the limiting factor with lighter weight calves offered grass silage, energy availability was the limiting factor with heavier steers. This shift in the most limiting nutrients as the steers become larger is related to greater energy to protein requirement for growth by the heavier animals. However, there is evidence that finishing cattle are likely to respond to supplementary protein in barley-based concentrates when grass silage digestibility is moderate to low (Waterhouse et al., 1985) and in situations where animals are of very high growth potential (Steen, 1996b). Steen (1996b) and (2000) concluded that in silage-fed finishing cattle, responses to protein in addition to that contained in barley (10%) are likely to be obtained in bulls given a wide range of silage types, but only in steers and heifers given silages that are badly preserved and/or with low digestibility and/or low protein contents.

Concentrate feeding strategies

Feeding frequency

Supplementary concentrates have traditionally been fed in two feeds daily in order to reduce the likelihood of digestive problems. Feeding 6 kg (0.52 of total DMI) of rolled barley (4 experiments) or wheat (2 experiments) (Drennan & Moloney, 1998) or 4.5 kg (0.54 of total DMI) of a cereal-based coarse ration (Keady et al., 2004b) in one as opposed to two daily feeds to finishing cattle offered grass silage ad libitum had no significant effect on carcass gain or feed conversion efficiency. This has positive implications in terms of labour efficiency.

Complete diet feeding/total mixed ration (TMR)

There is a limited quantity of published information comparing separate and TMR feeding of beef cattle offered grass silage-based diets with concentrates. Recent studies show that finishing steers offered grass silage plus a barley-based concentrate (0.31 or 0.55 (Caplis, 2004) and 0.40 or 0.80 (Keane, 2003a) of the dietary DM intake) either fed separately in one feed daily (following gradual introduction) or mixed with the silage through a diet feeder had similar carcass growth rates, carcass traits and muscle and fat colour. Furthermore, feeding method had no significant effect on diet digestibility, ruminal pH, ammonia or total volatile fatty acids concentrations (Caplis, 2004). However, the molar proportions of propionate were higher for the mixed diets than the separate diets. While there are many purported benefits for TMR feeding, these results demonstrate no advantage of mixing *per se* of grass silage and barley-based concentrates on animal efficiency or performance.

Co-ensiling

Previous studies with molassed sugar beet pulp (MBP) used as a silage additive demonstrated improved fermentation, reduced effluent production and increased silage nutritive value under good silage management practices (O'Kiely, 1992). Intake, *in vivo* digestibility and performance of finishing steers were similar when offered comparable amounts of MBP either as MBP co-ensiled with grass or as untreated, well-preserved silage supplemented with MBP. An additional benefit is the convenient manner of concentrate supplementation. Recent results suggest that citrus pulp, and to a lesser extent, soya hulls have similar potential (O'Kiely, 2002). However, some feed ingredients seem less promising and where grass undergoes an extensive clostridial fermentation co-ensiling is considerably less attractive (Stacey *et al.*, 2002).

Concentrate distribution pattern

Feeding weanling cattle a fixed total concentrate allowance offered at a flat daily rate or at a higher rate over the first half of the winter gave a better growth response than when offered at a higher rate over the second half of the winter (Keane, 2002; 2003b). Steen & Kilpatrick (2000) reported that finishing cattle given grass silage plus concentrates at 0.12 of dietary DM intake had similar average liveweight and carcass gains and a lower carcass fat score and fat trim than animals that initially received silage only followed by silage plus concentrates at 0.36 of dietary DM intake. Keane (1998) showed that finishing steers offered grass silage plus a fixed total quantity of supplementary concentrates at (i) a flat rate (5 kg), (ii) a stepped increased rate (2.5, 5.0 and 7.5 kg for each consecutive one-third of the 126-d finishing period) or (iii) silage for the first 42 d followed by *ad libitum* concentrate had similar carcass gains and feed energy efficiency over the same length of finishing period. Combining two experiments it was concluded that there were no significant differences in carcass gains, on efficiency of feed energy utilisation between the flat and *ad libitum* feeding treatments but the latter had a significantly lower carcass fat score and weight and proportion of kidney and channel fat.

Restricted feeding

Restricting the feed allowance to cattle under feed-lot or high concentrate feeding situations has been shown to reduce liveweight gain but to improve the efficiency of conversion of dietary dry matter to carcass weight when compared to *ad libitum* feeding of high cereal-based diets (Hicks *et al.*, 1990) and by-product based diets (French & Moloney, 2001). This improved feed conversion is largely attributed to higher rates of fat deposition in animals offered *ad libitum* concentrates but partly due to a lower digestibility at the higher feeding level (French & Moloney, 2001). Increasing the level of feeding of concentrate ingredients to cattle (offered at 0.85 of dietary DM) decreases the nutrient digestibility (Woods *et al.*, 1999). These findings contrast with studies examining the effect of restricting dry matter intake (~0.8 of *ad libitum*) on the efficiency of carcass gain of cattle finished on grass silage-based diets where results have shown reduced (Steen & Kilpatrick, 2000) and no effect (Keane & Drennan, 1980; Moore *et al.*, 1991; Steen *et al.*, 2002) on efficiency. Steen (1995b) found that restricting the intake of a grass silage-based diet significantly reduced the efficiency of conversion of ME to carcass gain in bulls but had little effect in steers or heifers. The differences in efficiency between restricted and *ad libitum* high concentrate versus forage-based diets are partly attributed to effects of compensatory growth, plane of nutrition and carcass fat content (Steen, 1995b). Furthermore, there was no significant effect on digestibility from restricting the DM intake of grass silage-based diets containing low to moderate proportions of concentrates (Steen, 1995b; Steen & Kilpatrick, 2000; Steen *et al.*, 2002). Restricted feeding must also be considered in relation to the delay in time to slaughter whereby the improvement in feed efficiency must offset the increased days on feed.

Carcass composition

There is considerable evidence that grass silage diets may predispose animals to depositing more fat in the gain than dried grass and wilted silages (e.g. McCarrick, 1966). However, most of those studies involved young animals and the results may not apply to animals slaughtered at commercial weights. Recent studies suggest that cattle fed on grass silage are not fatter. Scollan *et al.* (2003) found that from 350 kg empty body weight upwards the carcasses of animals given supplementary concentrate contained more fat than those offered good quality silage alone. Kim *et al.* (2003) feeding high quality grass silage compared with grass silage and concentrates at similar levels of ME intake per kg metabolic live weight found that concentrate supplementation increased the rate of tissue accretion but nutrient partitioning between fat and protein deposition was similar between 250 and 500 kg live weight. Increasing the protein content of grass silage-based finishing diets by inclusion of soyabean meal with barley increased carcass fatness in a number of studies (Steen & Robson, 1995; Steen, 1996a).

Increasing slaughter weight increases carcass weight and all measures of fatness. For cattle finished on grass silage and concentrates, Steen & Kilpatrick (2000) concluded that reducing slaughter weight is likely to be a more effective strategy to controlling carcass fat content than reducing energy intake either by diet restriction or reducing concentrate proportion.

Meat quality

The perception of healthiness and/or safety, tenderness, juiciness and aroma and flavour are important quality criteria that influence the decision of a consumer to purchase beef (Moloney *et al.*, 2001b). Furthermore, the colour or visual appearance of beef (lean meat and carcass fat) is often a key purchasing factor (Moloney *et al.*, 2004).

Many studies comparing the effects of forage and concentrates on meat quality have been confounded with plane of nutrition effects, such that cattle fed the higher energy concentrate diet have been heavier and fatter than those fed the forage-based diets (Muir *et al.*, 1998). Alternatively, the concentrate fed animals may be younger when grown to a specific bodyweight or back-fat thickness (French *et al.*, 2000). These are all factors which, can alter meat quality, and exist in commercial practice.

There are inconsistent effects of grass silage-based diets on subcutaneous fat colour. Moloney *et al.* (2003) reported that fat colour from steers offered grass was more yellow than fat from animals offered restricted fermentation grass silage-based diets and high concentrate starch or fibre-based diets, which were all similar. However, the fat colour from extensively fermented grass silage-based diets was not significantly different from the grass diet. Dunne *et al.* (2002) concluded that while concentrate feeding led to a time dependent decrease in subcutaneous fat yellowness relative to grass, substituting grass silage with concentrates did not lead to a consistent reduction in yellowness. Other studies have shown that at a similar carcass growth rate animals offered a grass silage/concentrate diet have yellower (Moloney *et al.*, 2000) or similar subcutaneous fat colour (French *et al.*, 2000) compared to non-silage high concentrate diets.

Substituting grass silage with concentrates generally has no effect on muscle redness, lightness and yellowness (Moloney *et al.*, 2000; French *et al.*, 2000; Lively *et al.*, 2004).

At similar carcass growth rates, a straw/concentrate-based diet resulted in beef with similar sensory attributes to a grass silage/concentrate diet (Moloney *et al.*, 2000). Equally, French *et al.* (2000) found that there was no difference in eating quality traits between diets of grass silage plus concentrates, high concentrates, grass plus concentrates and grass only. Moloney *et al.* (2003) comparing grass, restricted and extensive fermentation grass silage-based diets and high concentrate starch or fibre-based diets concluded that after ageing for 14 days, the only difference between the diets in eating quality traits was that beef from extensively fermented grass silage was more juicy than all treatments except the fibre-based concentrates. Similarly, Muir *et al.* (1998) concluded from their review that when compared at similar carcass weights or the same degree of fatness, grass-fed and grain-fed beef had no effect *per se* on tenderness, juiciness, lean meat colour, marbling or pH.

Ruminant meats are generally low in polyunsaturated fatty acids (PUFA) and rich in saturated fatty acids (SFA) due to the biohydrogenation action of rumen bacteria on fat consumed by the animal. Following human health guidelines many studies have aimed at increasing the PUFA content and in particular the *n*-3 long chain fatty acids, as well as the conjugated linoleic acid (CLA) content in intramuscular fat of beef (Raes *et al.*, 2004). One of the limitations of increasing PUFA in beef tissue through dietary supplements with oil or oilseed is the extent of rumen biohydrogenation as well as possible adverse effects on flavour, colour and shelf life of beef cuts with some oils (Mir *et al.*, 2003). In the latter case, high levels of dietary antioxidants are required to help stabilise the effects. This contrasts with grass and grass silage-based diets which contain natural anti-oxidants (Vitamin E) (Warren *et al.*, 2003; Scollan *et al.*, 2005). Protecting the lipid source from the hydrogenating action of the rumen micro-organisms results in reductions in SFA and increases in PUFA (Scollan *et al.*, 2002).

The lipid fraction of most grass-based forages contains alpha-linolenic acid as the main fatty acid and has a very low n-6:n-3 ratio (Givens *et al.*, 2000). Ensiling of forages generally results in fatty acid losses (Scollan & Wood, 2000). Dewhurst & King (1998) found that the effects of various silage additives and consequently fermentation had only a small effect on levels and proportions of fatty acids but shading and wilting had large negative effects on total fatty acids and on alpha-linolenic acid. However, Noci *et al.* (2004) reported that wilting did not impact negatively on the overall content of n-3 PUFA in muscle and increased the concentration of CLA. Diets with higher proportions of concentrates than grass silage have a higher n-6:n-3 PUFA ratio in intramuscular fat (Steen *et al.*, 2002; Warren *et al.*, 2003). Many studies have enhanced the fatty acid profile of beef muscle from grass silage-based diets by fortifying the supplementary concentrates with oils or oilseeds including sunflower oil, fish oil and linseed (Choi *et al.*, 2000; Scollan *et al.*, 2001b; Noci *et al.*, 2004, 2005). There is also evidence that longer term feeding of oils, rather than just in the finishing period, further enhances the intramuscular fatty acid composition (Raes *et al.*, 2004). Nevertheless, while dietary manipulation to enhance the fatty acid composition of meat has invariably resulted in highly significant changes the effects are generally small. The challenge is to increase the magnitude of these effects.

References

Agnew, R.E. & M.T. Carson (2000). The effect of a silage additive and level of concentrate supplementation on silage intake, animal performance and carcass characteristics of finishing beef cattle. *Grass and Forage Science*, 55, 114-124.

Caplis, J. (2004). *The effects of concentrate level and feeding a total mixed ration on intake and performance of finishing steers.* M.Agr.Sc. Thesis, National University of Ireland, 138 pp.

Chapple, D.G., H.F. Grundy, K.P.A. Wheeler & S.P. Marsh (1996). The effect of supplementing grass silage with molasses and/or a mineralised fish meal on performance and carcass characteristics of finishing beef cattle. *Animal Science*, 62, 653-654.

Choi, N.J., M. Enser, J.D. Wood & N.D. Scollan (2000). Effect of breed on the deposition in beef muscle and adipose tissue of dietary n-3 polyunsaturated fatty acids. *Animal Science*, 71, 509-519.

Dawson, L.E.R., R.M. Kirkland, C.P. Ferris, R.W.J. Steen, D.J. Kilpatrick & F.J. Gordon (2002). The effect of stage of perennial ryegrass maturity at harvesting, fermentation characteristics and concentrate supplementation, on the quality and intake of grass silage by beef cattle. *Grass and Forage Science*, 57, 255-267.

Dewhurst, R.J. & P.J. King (1998). Effects of extended wilting, shading, and chemical additives on the fatty acids in laboratory grass silages. *Grass and Forage Science*, 53, 219-224.

Doreau, M. & Y. Chilliard (1997). Digestion and metabolism of dietary fat in farm animals. *British Journal of Nutrition*, 78, Suppl. 1, S15-S35.

Drennan, M.J. (1985). Evaluation of molasses and ensiled pressed beet pulp for beef production. In: Ch. V. Boucque (Ed), *Feeding value of by-products and their use by beef cattle. Commission of the European Communities*, Luxembourg, pp. 171-183.

Drennan, M.J. & M.G. Keane (1987a). Responses to supplementary concentrates for finishing steers fed silage. *Irish Journal of Agricultural Research*, 26, 115-127.

Drennan, M.J. & M.G. Keane (1987b). Concentrate feeding levels for unimplanted and implanted finishing steers fed silage. *Irish Journal of Agricultural Research*, 26, 129-137.

Drennan, M.J. & A.P. Moloney (1998). Effect of concentrate type and feeding method on performance of cattle offered grass silage-based diets. *Irish Journal of Agricultural and Food Research*, 37, p.115.

Drennan, M.J., A.P. Moloney & M.G. Keane (1994). Effects of protein and energy supplements on performance of young bulls offered grass silage. *Irish Journal of Agricultural and Food Research*, 33, 1-10.

Drouillard, J.S. & G.L. Kuhl (1999). Effects of previous grazing nutrition and management on feedlot performance of cattle. *Journal of Animal Science*, 77, Suppl. 2, 136-146.

Dulphy, J.P. & M. Van Os (1996). Control of voluntary intake of precision-chopped silages by ruminants: a review. *Reproduction Nutrition Development*, 36, 113-135.

Dunne, P.G., A.P. Moloney, F.J. Monahan & F.P. O'Mara (2002). Subcutaneous adipose tissue colour of heifers fed grass, grass silage or concentrate-based diets. *Irish Journal of Agricultural & Food Research*, 41, 126.

Flynn, A.V. (1981). Factors affecting the feeding value of silage. In: W. Haresign & D.J.A. Cole (eds), *Recent Advances in Animal Nutrition - 1981*, 265-273.

French, P. & A.P. Moloney (2001). A note on diet digestibility, feed conversion efficiency and growth in steers offered *ad libitum* or restricted allowances of concentrates. *Irish Journal of Agricultural and Food Research,* 40, 271-275.

French, P., E.G. O'Riordan, F.J. Monahan, P.J. Caffrey, M. Vidal, M.T. Mooney, D.J. Troy & A.P. Moloney (2000). Meat quality of steers finished on autumn grass, grass silage or concentrate diets. *Meat Science,* 56, 173-180.

Givens, D.I., B.R. Cottrill, M. Davies, P.A. Lee, R.J. Mansbridge & A.R. Moss (2000). Sources of n-3 polyunsaturated fatty acids additional to fish oil for livestock diets- a review. *Nutritional Abstracts and Reviews,* Series B: Livestock Feeds and Feeding, 70, 1-19.

Hicks, R.B., F.N. Owens, D.R. Gill, J.J. Martin & C.A. Strasia (1990). Effects of controlled feed intake on performance and carcass characteristics of feedlot steers and heifers. *Journal of Animal Science,* 68, 233-244.

Huhtanen, P. S. & Jaakkola (1993). The effects of forage preservation method and proportion of concentrate on digestion of cell wall carbohydrates and rumen digesta pool size in cattle. *Grass and Forage Science,* 48, 155-165.

Ingvartsen, K.L. (1992). Effect of conservation and dry matter concentration of forage from grassland on feed intake, daily gain and feed conversion ratio of growing cattle: A review. *Report 711 National Institute of Animal Science, Foulum,* 44 pp.

Keady, T.W.J. & J.J. Murphy (1993). The effects of ensiling on dry matter intake and animal performance. *Irish Grassland and Animal Production Association Journal,* 27, 19-28.

Keady, T.W.J., A.F. Carson & D.J. Kilpatrick (2004b). The effects of two pure dairy breeds and their reciprocal crosses, and concentrate feeding management, on the performance of beef cattle. *Proceedings of the British Society of Animal Science,* p. 187.

Keady, T.W.J., R.M. Kirkland, D.C. Patterson, D.J. Kilpatrick & R.W.J. Steen (2004a). The effect of plane of nutrition during the growing and finishing phases, and gender, on the performance of beef cattle. *Proceedings of the British Society of Animal Science,* p. 188.

Keane, M.G. (1998). Effects of concentrate distribution pattern on the performance of finishing steers fed silage. Beef Production Series No. 3, Teagasc, Grange, Co. Meath, Ireland, 48 pp.

Keane, M.G. (2001). Response in beef cattle to concentrate feeding. Occasional Series No. 3, Teagasc, Grange Research Centre, Co. Meath, Ireland, 40 pp.

Keane, M.G. (2002). Response in weanlings to supplementary concentrates in winter and subsequent performance. Occasional Series No. 4, Teagasc, Grange Research Centre, Co. Meath, Ireland, 36 pp.

Keane, M.G. (2003a). Comparison of separate and mixed feeding of silage and concentrates for finishing cattle, Experiment 2. Beef Production Research Report, Teagasc, Grange Research Centre, Co. Meath, Ireland, 23-26.

Keane, M.G. (2003b). Response of weanlings to different winter feeding levels. Beef Production Research Report, Teagasc, Grange Research Centre, Co. Meath, Ireland, 29-30.

Keane, M.G. & M.J. Drennan (1980). Effects of diet type and feeding level on performance, carcass composition and efficiency of Friesian steers serially slaughtered. *Irish Journal of Agricultural Research,* 19, 53-66.

Keane, M.G. & M.J. Drennan (1994). Effects of winter supplementary concentrate level on the performance of steers slaughtered immediately or following a period at pasture. *Irish Journal of Agricultural and Food Research,* 33, 111-119.

Khalili, H. & P, Huhtanen (1991). Sucrose supplements in cattle given grass silage-based diet. 2. Digestion of cell wall carbohydrates. *Animal Feed Science and Technology,* 33, 263-273.

Kim, E.J., N.D. Scollan, M.S. Dhanoa & P.J. Buttery (2003). Effects of supplementary concentrates on growth and partitioning of nutrients between different body components in steers fed on grass silage at similar levels of metabolisable energy intake. *Journal of Agricultural Science,* 141, 103-112.

Kung, L., M.R. Stokes & C.J. Lin (2003). Silage additives. In: D.R. Buxton, *et al.* (eds.), Silage science and technology - Agronomy monograph No. 42. American Society of Agronomy, Crop Science Society of America, Soil Science Society of America, Madison, Wisconsin, USA, p. 305-360.

Lively, F.O., T.W.J. Keady, B.W. Moss, R.M. Kirkland, D.C. Patterson & D.J. Kilpatrick (2004). The effect of gender and the plane of nutrition during the growing and finishing phases, on carcass characteristics and meat quality. *Proceedings of the British Society of Animal Science,* p. 68.

Mayne, C.S. & A. Cushnahan (1995). The effects of ensilage on animal performance from the grass crop. In: *68th Annual Report, Agricultural Research Institute of Northern Ireland,* pp. 30-41.

Mayne, C.S., R. Agnew, D.C. Patterson, R.W.J. Steen, F.J. Gordon, D.J. Kilpatrick & E.F. Unsworth (1995). An examination of possible interactions between silage type and concentrate composition on the intake characteristics of grass silage offered to growing cattle and dairy cows. *Proceedings of the British Society of Animal Science,* Paper 21.

McCarrick, R.B. (1966) Effect of method of grass conservation and herbage maturity on performance and body composition of beef cattle. In: A.G.G. Hill *et al.* (eds) *Proceedings of the 10th International Grassland Congress,* Helsinki, 575-580.

Mir, P.S., M. Ivan, M.L. He, B. Pink, E. Okine, L. Goonewardene, T.A. McAllister, R. Weselake & Z. Mir (2003). Dietary manipulation to increase conjugated linoleic acids and other desirable fatty acids in beef: A review. *Canadian Journal of Animal Science*, 83, 673-685.

Moloney, A.P. (1991). Growth, digestibility and nitrogen retention in young Friesian steers offered grass silage and concentrates which differed in protein concentration and degradability. In: *Proceedings of the 6th International Symposium on Protein Metabolism and Nutrition*, 9-14 June, Herning, Denmark, EAAP publication No. 59, 342-344.

Moloney, A.P. (1993). The effects of grass silage digestibility and supplementary protein on growth, digestibility and nitrogen retention in young steers. In: P. O'Kiely *et al.* (eds.) *Proceedings of 10th International Silage Conference*, Dublin City University, Dublin, Ireland, p. 236-237.

Moloney, A.P. (1996). Digestion and growth in steers fed grass silage and starch, fibre or fat-based concentrates. *Irish Journal of Agricultural and Food Research*, 35, p. 65.

Moloney, A.P., A.A. Almiladi, M.J. Drennan & P.J. Caffrey (1994). Rumen and blood variables in steers fed grass silage and rolled barley or sugar cane molasses-based supplements. *Animal Feed Science and Technology*, 50, 37-54.

Moloney, A.P., D. McGilloway, M.T. Mooney, M. Vidal & D.J. Troy (2003). Influence of grass silage fermentation and concentrate composition on the appearance and sensory characteristics of bovine *M. longissimus dorsi*. *Proceedings of the British Society of Animal Science*, p. 201.

Moloney, A.P., T.V. McHugh & A. McArthur (1993). Growth and rumen fermentation in steers fed silage and concentrates differing in energy source. *Irish Journal of Agricultural and Food Research*, 32, p. 101.

Moloney, A.P., F.J. Monahan, F.P. O'Mara & P.D. Dunne (2004). Manipulation of beef fat and lean colour. Beef Production Series No. 58, Teagasc, Grange, Co. Meath, Ireland, 28 pp.

Moloney, A.P., M.T. Mooney, M.J. Drennan, B. O'Neill & D.J. Troy (2000). Fat colour and the quality of meat from beef cattle offered grass silage or concentrate-based diets. *Irish Journal of Agricultural and Food Research*, 39, p. 152.

Moloney, A.P., M.T. Mooney, J.P. Kerry & D.J. Troy (2001b). Producing tender and flavoursome beef with enhanced nutritional characteristics. *Proceedings of the Nutrition Society*, 60, 221-229.

Moloney, A.P., P. O'Kiely, M.C. Hickey & L.A. Adams (2001a). Optimisation of nutrient supply for beef cattle fed grass or silage. Beef Production Series No. 26, Teagasc, Grange, Co. Meath, Ireland, 48 pp.

Moore, C.A., K.J. McCracken, E.F. Unsworth, R.W.J. Steen, F.J. Gordon & D.J. Kilpatrick (1991). Effect of plane of nutrition and slaughter weight on growth and food efficiency of Friesian steers during the finishing period. *Animal Production*, 52, p. 572.

Muir, P.D., J.M. Deaker & M.D. Bown (1998). Effects of forage- and grain-based feeding systems on beef quality: A review. *New Zealand Journal of Agricultural Research*, 41, 623-635.

Mulligan, F.J., P.J. Caffrey, M. Rath, J.J. Callan, P.O. Brophy & F.P. O'Mara (2002). An investigation of feeding level effects on digestibility in cattle for diets based on grass silage and high fibre concentrates at two forage:concentrate ratios. *Livestock Production Science*, 77, 311-323.

Noci, F., A.P. Moloney & F.J. Monahan (2004). The effects of fish oil inclusion in the concentrate and method of silage preservation on fatty acid composition of muscle from steers. *Proceedings of the British Society of Animal Science*, p. 86.

Noci, F., P. O'Kiely, F.J. Monahan, C. Stanton & A.P. Moloney (2005). Conjugated linoleic acid concentration in *M. Longissimus dorsi* from heifers offered sunflower oil-based concentrates and conserved forages. *Meat Science*, 69, 509-518.

O'Kiely, P. (1992). The effect of ensiling sugarbeet pulp with grass on silage composition, effluent production and animal performance. *Irish Journal of Agricultural and Food Research*, 31, 115-128.

O'Kiely, P. (2001). Producing grass silage profitably in northern Europe. In: *Production and utilisation of silage, with emphasis on new techniques*. Nordic Association of Agricultural Scientists; Seminar #326 at Lillehammer, Norway, 27-28 Sept. 2001, 18 pp.

O'Kiely, P. (2002). Soya hulls or citrus pulp as alternatives to molassed beet pulp as silage additives. In: L.M. Gechie and C. Thomas (eds.) *Proceedings of XIII International Silage Conference*, Sept.11-13 (2002), SAC, Auchincruive, Scotland. p. 220-221.

O'Kiely, P. & A.P. Moloney (1994). Silage characteristics and performance of cattle offered grass silage made without an additive, with formic acid or with a partially neutralised blend of aliphatic organic acids. *Irish Journal of Agricultural and Food Research*, 33, 25-39.

Patterson, D.C., R.W.J. Steen, C.A. Moore & B.W. Moss (2000). Effects of the ratio of silage to concentrates in the diet on the performance and carcass composition of continental bulls. *Animal Science*, 70, 171-179.

Raes, K., S. De Smet & D. Demeyer (2004). Effect of dietary fatty acids on incorporation of long chain polyunsaturated fatty acids and conjugated linoleic acid in lamb, beef and pork meat: a review. *Animal Feed Science and Technology*, 113, 199-221.

Randby, A.T. (2001). Beef from forage: The potential of high quality grass silage. In: *Production and utilisation of silage, with emphasis on new techniques.* Nordic Association of Agricultural Scientists; Seminar #326 at Lillehammer, Norway, 27-28 Sept. 2001, 46-51.

Rinne, M. (2000). Influence of the timing of the harvest of primary grass growth on herbage quality and subsequent digestion and performance in the ruminant animal. Academic dissertation, University of Helsinki, Department of Animal Science. Publications 54, 42 p. +5 encl.

Rouzbehan, Y., H. Galbraith, J.H. Topps & J. Rooke (1996). Response of growing steers to diets containing big bale silage and supplements of molassed sugar beet pulp with and without white fish meal. *Animal Feed Science and Technology*, 62, 151-162.

Sanderson, R., M.S. Dhanoa, C. Thomas & A.B. McAllan (2001). Fish-meal supplementation of grass silage offered to young steers: Effects on growth, body composition and nutrient efficiency. *Journal of Agricultural Science, Cambridge*, 137, 85-96.

Scollan, N.D. & J.D. Wood (2000). Improving the nutritional value and eating quality of beef. In: D. Pullar (Ed.), Beef from Grass and Forage. *Occasional Symposium No. 35, Proceedings of the British Grassland Society*, 20 & 21 Nov. 2000, 29-42.

Scollan, N.D., N.J. Choi, E. Kurt, A.V. Fisher, M. Enser & J.D. Wood (2001b). Manipulating the fatty acid composition of muscle and adipose tissue in beef cattle. *British Journal of Nutrition*, 85, 115-124.

Scollan, N.D, M.S. Dhanoa, E.J. Kim, J.M. Dawson & P.J. Buttery (2003). Effects of diet and stage of development on partitioning of nutrients between fat and lean deposition in steers. *Animal Science*, 76, 237-249.

Scollan, N.D., S. Gulati, K.G. Hallett, J.D. Wood & M. Enser (2002). The effects of including ruminally protected lipid in the diet of Charolais steers on animal performance, carcass quality and the fatty acid composition of longissimus dorsi muscle. *Proceedings of the British Society of Animal Science*, p. 9.

Scollan, N.D, I. Richardson, S. De Smet, A.P. Moloney, M. Doreau, D. Bauchart & K. Nuernberg (2005). Enhancing the content of beneficial fatty acids in beef and consequences for meat quality. In: J.F. Hocquette & S. Gigli (eds), *Indicators of milk and beef quality.* EAAP publication No. 112, Wageningen Academic Publishers, 151-162.

Scollan, N.D., A. Sargeant, A.B. McAllan & M.S. Dhanoa (2001a). Protein supplementation of grass silages of differing digestibility for growing steers. *Journal of Agricultural Science, Cambridge*, 136, 89-98.

Seoane, J.R., A. Amyot, A-M. Christen & H.V. Petit (1993). Performance of growing steers fed either hay or silage supplemented with canola or fishmeal. *Canadian Journal of Animal Science*, 73, 57-65.

Shiels, P. (1998). *Defining the conditions under which benefits accrue from the use of bacterial inoculants as silage additives.* Ph.D. Thesis, National University of Ireland, 413 pp.

Stacey, P., P. O'Kiely & F.P. O'Mara (2002). Conservation characterisation of dry concentrates co-ensiled with grass. In: L.M. Gechie & C. Thomas (eds.) *Proceedings of XIII International Silage Conference*, Sept. 11-13, SAC, Auchincruive, Scotland. 222-223.

Steen, R.W.J. (1988). Factors affecting the utilisation of grass silage for beef production. In J. Frame (ed.). *Efficient beef production from grass, Occasional Symposium No. 22, British Grassland Society*, 129-139.

Steen, R.W.J. (1993a). A comparison of supplements to grass silage for growing beef cattle. In: P. O'Kiely *et al.* (eds.) *Proceedings of 10th International Silage Conference*, Dublin City University, Dublin, Ireland, 206-207.

Steen, R.W.J. (1993b). A comparison of wheat and barley as supplements to grass silage for finishing beef cattle. *Animal Production*, 56, 61-67.

Steen, R.W.J. (1995a). A comparison of supplements to grass silage for beef cattle. *Proceedings of the British Society of Animal Science*, Paper 166.

Steen, R.W.J. (1995b). The effect of plane of nutrition and slaughter weight on growth and food efficiency in bulls, steers and heifers of three breed crosses. *Livestock Production Science,* 42, 1-11.

Steen, R.W.J. (1996a). Effects of protein supplementation of grass silage on the performance and carcass quality of beef cattle. *Journal of Agricultural Science, Cambridge*, 127, 403-412.

Steen, R.W.J. (1996b). Factors affecting the optimum protein content of concentrates for beef cattle. Research on beef production. *Occasional Publication No. 22, Agricultural Research Institute of Northern Ireland.* 53-59.

Steen, R.W.J. (1998). A comparison of high-forage and high concentrate diets for beef cattle. In: *71st Annual Report, Agricultural Research Institute of Northern Ireland*, pp. 30-41.

Steen, R.W.J. (2000). Finishing systems based on conserved grass. In: D. Pullar (Ed.), Beef from Grass and Forage. *Occasional Symposium No. 35, Proceedings of the British Grassland Society*, 20 & 21 Nov. 2000, 55-64.

Steen, R.W.J. & D.J. Kilpatrick (2000). The effects of the ratio of grass silage to concentrates in the diet and restricted dry matter intake on the performance and carcass composition of beef cattle. *Livestock Production Science*, 62, 181-192.

Steen, R.W.J. & A.E. Robson (1995). Effects of forage to concentrate ratio in the diet and protein intake on the performance and carcass composition of beef heifers. *Journal of Agricultural Science, Cambridge*, 125, 125-135.

Steen, R.W.J., F.J. Gordon, L.E.R. Dawson, R.S. Park, C.S. Mayne, R.E. Agnew, D.J. Kilpatrick & M.G. Porter (1998). Factors affecting the intake of grass silage by cattle and prediction of silage intake. *Animal Science,* 66, 115-127.

Steen, R.W.J., D.J. Kilpatrick & M.G. Porter (2002). Effects of the proportions of high or medium digestibility grass silage and concentrates in the diet of beef cattle on liveweight gain, carcass composition and fatty acid composition of muscle. *Grass and Forage Science*, 57, 279-291.

Titgemeyer, E.C. & C.A. Loest (2001). Amino acid nutrition: Demand and supply in forage-fed ruminants. *Journal of Animal Science*, 79, E Suppl., E180-E189.

Thomas, C. & P.C. Thomas (1985). Factors affecting the nutritive value of grass silages. In: W. Haresign & D.J.A. Cole (eds) *Recent Advances in Animal Nutrition-1985*, 223-256.

Van Vuuren, A.M., P. Huhtanen & J.P. Dulphy (1995). Improving the feeding and health value of ensiled forages. In: M. Journet *et al.* (eds), *Recent Advances in the Nutrition of Herbivores, Proceedings of the IVth International Symposium on the Nutrition of Herbivores*, INRA Editions, Paris, 279-307.

Veira, D.M., H.V. Petit, J.G. Proulx, L. Laflamme & G. Butler (1995). A comparison of five protein sources as supplements for growing steers fed grass silage. *Canadian Journal of Animal Science*, 75, 567-574.

Warren, H.E., M. Enser, I. Richardson, J.D. Wood & N.D. Scollan (2003). Effect of breed and diet on total lipid and selected shelf-life parameters in beef muscle. *Proceedings of the British Society of Animal Science,* p. 43.

Waterhouse, A., R. Laird & D.P. Arnot (1985). Responses to protein supplements in silage-fed finishing steers: Effects of silage quality and supplement type. *Animal Production*, 40, 538.

Woods, V.B., A.P. Moloney, F.J. Mulligan, M.J. Kenny & F.P. O'Mara (1999). The effect of animal species (cattle or sheep) and level of intake by cattle on *in vivo* digestibility of concentrate ingredients. *Animal Feed Science and Technology*, 80, 135-150.

Wright, D.A., F.J. Gordon, R.W.J. Steen & D.C. Patterson (2000). Factors influencing the response in the intake of silage and animal performance after wilting of grass before ensiling: a review. *Grass and Forage Science,* 55, 1-13.

Ensiled maize and whole crop wheat forages for beef and dairy cattle: effects on animal performance

T.W.J. Keady

Agricultural Research Institute of Northern Ireland, Hillsborough, Co Down BT26 6DR, Email: tim.keady@dardni.gov.uk

Key points

1. Maize silage can be produced and fed to beef and dairy cattle at a similar price to grazed grass.
2. Including maize silage in the diet increases feed intake and performance of beef and dairy cattle.
3. The optimum stage of maturity at harvest for increased performance is at a dry matter concentration of approximately 300 g/kg.
4. Including maize silage in grass silage-based diets has a concentrate sparing effect of up to 5 kg/cow/d.
5. There is a negative relationship between stage of maturity at harvest and milk fat concentration.
6. Whole crop wheat can be produced and fed at a similar cost to grass silage.
7. Including whole crop wheat either fermented, urea- or alkalage-treated in grass silage-based diets increases feed intake but does not alter performance of beef or dairy cattle.

Keywords: grass silage, maize silage, whole crop wheat silage, cost of production, carcass gain, milk yield

Introduction

Traditionally in many parts of Europe, Scandinavia, New Zealand, Australia and North America, grass silage was offered to beef and dairy cattle during the winter indoor feeding period. However in recent times, other ensiled forages, such as maize and whole crop wheat have increased in popularity and have partially replaced grass silage in the diet. Major developments, both in plant breeding and in agronomic practices, have enabled consistent production of high yields of these forages in areas in which it was not possible to grow these crops 20-30 years ago. Alongside these developments major improvements have also occurred in the genetic merit/production potential of beef and dairy herds. Given the reduced margins in beef and milk production, producers aim to reduce costs of production by feeding forages with high intake potential and reduce concentrate feed level whilst maintaining or increasing animal performance. The aim of this paper is to evaluate the effect of including maize and whole crop wheat silages in grass silage-based diets on performance of beef and dairy cattle.

Materials and methods

Data from studies in which grass silage was offered as the sole forage and in which grass silage was replaced by either maize of whole crop wheat were collated to evaluate the effects of replacing grass silage with alternative forages.

Grass silage metabolisable energy (ME) concentration, if not determined in the original study, was estimated from digestible organic matter in the dry matter (D-value) using the equations of AFRC (1993). Where not available, grass silage D-value was estimated from dry matter digestibility (DMD) which was calculated using the equation of Keady *et al.* (2001).

Cost of alternative forages

With the lower prices currently being received for farm produce, it is essential to reduce costs and produce animal product as cheaply as possible. To reduce feed costs it is important to have an accurate cost of all feeds, including forages. Keady *et al.* (2002a) determined the cost of producing forages on the basis of dry matter actually consumed by animals, and also included a land charge, enabling direct comparisons to be made with purchased feeds. The herbage yield data, utilisation rates, in-silo losses and feeding losses used in these costings were based on experimental data from Research Centres in Ireland, the UK and the United States of America. Labour and infrastructure costs and utilisation rates representative of forage production systems on an efficient dairy farm in Northern Ireland were also included. Maize grown in the open and whole crop wheat can be produced at a similar cost to grass silage (Table 1).

Under Northern Ireland conditions, Gilliland (2003) reported that the use of the complete cover plastic mulch system (CCPM), which involves covering the crop with a thin clear film (6 microns), increased mean dry matter yield and the concentrations of dry matter and starch by 3.1 t DM/ha, 66 g/kg and 143 g/kg DM respectively when evaluated with 12 different varieties sown on the same date. Most of the increase in yield can be attributed to increased cob production with modern varieties. The response to CCPM is very much variety dependent. Gilliland, T. (personal communication) reported that the response to CCPM varied from 0.1 to 4.7 t DM/ha and 1.7 to 4.4 t DM/ha for early and late maturing varieties during the 2001 and 2002 growing seasons respectively. The results of these studies indicate that the response to CCPM increases with late maturing varieties. Use of the CCPM system enables the crops to be sown up to one month earlier as protection is provided against frost, and facilitates use of later maturing, higher yielding varieties enabling the system to increase forage yield by up to 6 t DM/ha relative to early maturing varieties grown in the open. Consequently if grown under CCPM, forage yields of 18 t DM/ha can be achieved enabling production and feeding of maize to dairy cows for £75/t DM. Consequently maize can be a very cost effective crop to produce and can be produced at similar costs to that of grazed grass.

Effects of maize silage inclusion on animal performance

Due to major improvements in plant breeding and in agronomic practices the production of maize is moving further into the northern latitudes. For example, in the 1960's (McAllister, 1961) and 1970's (Bartholomew & Chestnutt, 1979) crops of maize produced in Northern Ireland only yielded between 4.1 and 4.9 t DM/ha of low dry matter forage. However, by 1996-1999 yield potential had increased by 300% to 12.2 t DM/ha (Easson, 2000) due to improvements in plant breeding. Development in agronomic practices, particularly the CCPM system has further increased yield potential and maturity of crops grown in more temperate climates. The CCPM system was developed by two producers in Co. Wexford and Co. Limerick, Ireland. The system involves covering the crop with a thin clear film of 6 microns. Currently in Northern Ireland over 50% of maize is produced under CCPM. Current machinery sows the seed, applies herbicide and lays the CCPM in one pass.

The effects of including maize silage in grass silage-based diets on feed intake and performance of dairy cows from 34 comparisons, is presented in Table 2. There is a substantial body of evidence to indicate that the inclusion of maize silage in grass silage-based diets significantly increased forage intake (1.5 kg DM/d), the yields of milk (1.4 kg/d) and fat plus protein (0.15 kg/d), milk fat (0.6 g/kg) and protein (0.8 g/kg) concentrations.

The effects of the including maize silage in grass silage-based diets on feed intake, and performance of finishing beef cattle from nine comparisons is presented in Table 3. Including maize silage increased forage intake (1.5 kg DM/d), carcass gain (0.11 kg/d), liveweight gain (0.23 kg/d) and carcass weight (12 kg).

The studies quoted in Tables 2 and 3 show that the inclusion of maize in grass silage-based diets has produced variable effects on animal performance. These different responses may have been due to variations in the feed value of either the grass and maize silages offered in these studies or in the level of inclusion of maize silage in the diet.

Level of forage maize inclusion

A number of studies have been undertaken to evaluate the effect of level of maize silage inclusion in grass-based diets on the performance of beef cattle (O'Kiely & Moloney, 1995, 2000; Browne *et al.*, 2000) and dairy cows (Phipps *et al.*, 1992b; O'Mara *et al.*, 1998). O'Kiely & Moloney (1995) concluded that replacing up to 0.67 of moderate feed value grass silage (DM 208 g/kg; DMD 746 g/kg DM) with low dry matter (DM = 205 g/kg) maize silage did not alter liveweight gain of beef cattle. However these authors noted that replacing either 0, 0.33, 0.66 or 1.00 of grass silage with maize silage had a quadratic effect on feed intake, with the highest intakes being recorded when 0.33 and 0.67 of the forage component consisted of maize. In a subsequent study O'Kiely & Moloney (2000) replaced a similar feed value grass silage as that used by O'Kiely & Moloney (1995) with either 0.50 or 1.00 of maize silages with DM concentrations of either 260, 300 or 380 g/kg and concluded that increasing the proportion of maize silage in the diet increased carcass gain. Similarly Browne *et al.* (2000) concluded that replacing either 0, 0.33, 0.67 or 1.00 of a medium feed value grass silage (DM = 265 g/kg, ME 10.4 MJ/kg DM) with maize silage (DM = 332 g/kg, starch = 301 g/kg DM) increased feed intake and carcass gain.

Changes in the diet of dairy cows can impact on milk yield and composition as well as body weight change. Phipps *et al.* (1992b) replaced either 0, 0.25, 0.50 or 0.75 of low feed value (DM = 260 g/kg, D-value = 620 g/kg DM) or average feed value (DM = 266 g/kg, D-value = 660 g/kg DM) grass silages with maize silage (DM = 273 g/kg). These authors concluded that with the low feed value grass silage, the higher the level of maize silage inclusion the higher the feed intake and milk yield. However, with the average grass silage, peak feed intake and milk yield were achieved when 0.50 of the grass silage was replaced with maize silage. Similarly O'Mara *et al.* (1998) replaced 0.33, 0.67 or 1.00 high feed value grass silage (DM = 223 g/kg DMD = 759 g/kg DM) with medium DM maize silage (DM = 257 g/kg, starch = 329 g/kg DM) and noted that the highest feed intake and milk yield occurred when 0.33 of the forage component of the diet consisted of maize silage. However, body weight gain improved with maize silage inclusion up to 0.67.

Table 1 The costs of producing and feeding forages

	Grass	Grass silage			Whole crop wheat[a]		Maize	
		3-cut	4-cut	Baled	Fermented	Urea	Mulch	Open
Yield (t DM/ha)								
Forage	10.6	13.8	12.8	13.8	13.0	13.0	18.0	12.2
Utilised	8.0	11.5	10.9	11.7	11.1	11.8	15.5	10.5
ME (MJ/kg DM)	11.7	11.4	12.1	11.4	9.5	8.2	11.6	10.5
Costs (£/ha)								
Reseeding	24	45	32	45	-	-	-	-
Establishment	-	-	-	-	132	132	215	212
Fertiliser	149	160	176	160	51	51	67	73
Sprays	8	8	8	8	110	110	32	32
Plastic mulch	-	-	-	-	-	-	174	-
Additive	-	75	51	-	46	86	59	67
Silo cover	-	4	4	-	4	7	4	4
Infrastructure	49	75	63	26	70	74	67	68
Contractor	74	298	369	417	202	202	163	173
Feed-out	28	114	108	151	110	117	125	104
Land charge	250	200[b]	225[b]	200[b]	250	250	250	250
Total costs	582	979	1036	1007	975	1029	1156	983
Cost of forage (£/t UDM)	73	85	95	86	88	87	75	94
Relative cost of grazed grass	1.0	1.2	1.3	1.2	1.2	1.2	1.0	1.3

(After Keady et al., 2002a)

[a] Winter sown crop

[b] Land charge reduced to allow for grazing after herbage is ensiled

Table 2 The effects of including maize silage in grass silage-based diets on the performance of dairy cows

Reference	Grass silage (GS) DM (g/kg)	ME (MJ/kg DM)	Maize silage (MS) DM (g/kg)	Starch (g/kg DM)	GS:MS ratio	Forage intake (kg DM/d) GS	Maize	Milk (kg/d) GS	Maize	Fat (g/kg) (F) GS	Maize	Protein (g/kg) (P) GS	Maize	F+P (kg/d) GS	Maize
Phipps et al. (1995)	213	10.9	354	339	67:33	9.3	10.6	23.0	26.4	41.7	41.8	29.9	31.2	1.62	1.93
Hameleers (1998)	248	11.4	275	256	60:40	10.6	11.4	27.4	26.3	48.9	46.9	34.1	33.6	2.25	2.11
Patterson et al. (2004)	187	11.2	297	225	60:40	7.9	9.8	25.8	27.4	39.8	39.0	30.9	31.5	1.82	1.94
Murphy (2004)	240	10.5	221	140	33:67	8.8	11.2	27.6	29.6	39.6	40.4	30.7	31.2	1.93	2.11
	231	10.9	302	324	33:67	8.7	13.2	30.8	33.7	36.7	38.5	29.7	31.5	2.05	2.32
Patterson & Kilpatrick (2005)	185	10.7	305	359	50:50	10.4	13.3	27.9	29.8	40.7	42.2	30.8	31.9	1.96	2.20
	234	12.5	305	359	50:50	12.5	13.9	31.5	32.1	38.5	38.5	32.8	32.5	2.23	2.28
Phipps et al. (1992b)	260	9.9	273		50:50	6.9	8.1	23.8	25.2	38.8	38.7	30.2	30.6	1.64	1.75
	266	10.6	273		50:50	8.0	8.5	24.5	25.7	39.4	38.3	30.9	30.8	1.72	1.78
O'Mara et al. (1998)	223	11.5	257	219	33:67	8.8	10.3	21.4	22.9	37.7	37.2	30.6	31.2	1.46	1.56
Phipps et al. (2000)	296	11.1	226	114	75:25	9.2	10.9	28.0	29.4	45.0	45.8	30.6	32.4	2.14	2.31
	296	11.1	290	274	75:25	9.2	13.3	28.0	32.7	45.0	43.4	30.6	32.7	2.14	2.48
	296	11.1	302	309	75:25	9.2	13.1	28.0	33.0	45.0	41.8	30.6	31.9	2.14	2.44
	296	11.1	390	354	75:25	9.2	12.8	28.0	30.8	45.0	44.6	30.6	31.9	2.14	2.37
Keady et al. (2002b)	193	9.8	202	100	60:40	8.7	10.8	24.8	25.8	40.3	42.1	30.4	31.6	1.75	1.89
	193	9.8	280	273	60:40	8.7	11.2	24.8	26.4	40.3	41.6	30.4	31.8	1.75	1.94
	193	9.8	298	270	60:40	8.7	10.9	24.8	25.8	40.3	41.2	30.4	31.2	1.75	1.85
	193	9.8	384	332	60:40	8.7	11.1	24.8	25.5	40.3	41.0	30.4	32.0	1.75	1.86
	326	11.8	202	100	60:40	13.1	13.7	28.8	28.8	39.4	42.6	32.7	33.0	2.03	2.16
	326	11.8	280	273	60:40	13.1	13.8	28.8	29.5	39.4	41.9	32.7	33.0	2.03	2.22
	326	11.8	298	270	60:40	13.1	13.4	28.8	29.3	39.4	41.6	32.7	33.0	2.03	2.19
	326	11.8	384	332	60:40	13.1	13.6	28.8	30.1	39.4	39.4	32.7	33.1	2.03	2.14
Keady et al. (2003)	218	10.2	189	15	60:40	9.8	11.4	25.9	27.3	38.6	39.2	31.8	32.5	1.79	1.93
	218	10.2	249	161	60:40	9.8	11.6	25.9	26.9	38.6	40.6	31.8	32.9	1.79	1.96
	218	10.2	362	270	60:40	9.8	11.9	25.9	27.4	38.6	39.0	31.8	32.7	1.79	1.93
	218	10.2	429	320	60:40	9.8	11.7	25.9	27.5	38.6	40.4	31.8	33.1	1.79	2.00
	234	11.0	189	15	60:40	10.2	11.6	27.3	28.5	38.9	40.3	31.5	32.2	1.93	2.07
	234	11.0	249	161	60:40	10.2	11.2	27.3	28.2	38.9	40.7	31.5	33.4	1.93	2.06
	234	11.0	362	270	60:40	10.2	12.4	27.3	29.0	38.9	39.1	31.5	32.9	1.93	2.08
	234	11.0	429	320	60:40	10.2	11.4	27.3	28.0	38.9	38.9	31.5	32.7	1.93	2.01
	307	12.0	189	15	60:40	14.3	13.6	30.4	30.1	41.5	44.1	34.6	35.0	2.30	2.37
	307	12.0	249	161	60:40	14.3	13.7	30.4	30.8	41.5	43.5	34.6	34.4	2.30	2.39
	307	12.0	362	270	60:40	14.3	14.9	30.4	31.1	41.5	41.3	34.6	34.7	2.30	2.35
	307	12.0	429	320	60:40	14.3	14.7	30.4	31.1	41.5	41.3	34.6	35.0	2.30	2.35
Mean (n = 34)	252	11.0	297	235		10.5	12.0	27.2	28.6	40.5	41.1	31.7	32.5	1.95	2.10
SED						0.227		0.211		0.24		0.12		0.016	
Significance						***		***		*		***		***	

Table 3 The effects of including maize silage in grass silage-based diets on the performance of beef cattle

Reference	Grass silage (GS) DM (g/kg)	Grass silage (GS) ME (MJ/kg DM)	Maize silage (MS) DM (g/kg)	Maize silage (MS) Starch (g/kg DM)	GS:MS ratio	Forage intake (kg DM/d) GS	Forage intake (kg DM/d) Maize	Carcass gain (kg/d) GS	Carcass gain (kg/d) Maize	Liveweight gain (kg/d) GS	Liveweight gain (kg/d) Maize	Carcass weight (kg) GS	Carcass weight (kg) Maize
Keady & Kilpatrick (2004)	192	10.6	276	225	60:40	5.1	5.8	0.51	0.60	0.86	1.10	326	334
Walsh et al. (2005)	161		303		100:0	4.5	6.8	0.48	0.78	0.80	1.20	290	335
O'Kiely & Moloney (2000)	180	11.3	375	446	25:75	5.1	7.2	0.65	0.73	0.85	0.98	324	336
	180	11.3	297	379	25:75	5.1	6.7	0.65	0.72	0.85	0.96	324	334
	180	11.3	256	332	25:75	5.1	6.5	0.65	0.71	0.85	0.95	324	333
Brown et al. (2000)	265	10.4	332	301	33:67	6.3	7.3	0.58	0.79	0.92	1.17	311	321
O'Kiely & Moloney (1995)	208	11.2	205		33:67	6.1	6.8	0.87	0.74	1.39	1.27	316	304
Gorman et al. (1998)	244	10.3	232	93	0:100	6.7	8.6	0.43	0.54	0.73	0.82	369	380
	244	10.3	336	265	0:100	6.7	8.2	0.43	0.61	0.73	0.83	369	387
Mean (n = 9)	206	10.8	290	292		5.6	7.1	0.58	0.69	0.89	1.02	328	340
SED							0.19		0.04		0.045		4.91
Significance							***		*		*		*

Milk composition, namely fat and protein concentrations, have a major impact on milk value, the relative value of each component depending on the prevailing market conditions. Whilst Phipps *et al.* (1992b) and O'Mara *et al.* (1998) reported positive effects on milk yield to varying levels of inclusion of maize silage in grass silage-based diets, they concluded that increasing the level of maize in the diet did not alter milk fat or protein concentration.

The optimum level of inclusion of maize silage in grass silage-based diets depends on the quality of both grass silage and maize silage. If offered low feed value grass silage, then increasing the inclusion rate of maize silage increases animal performance. However, with average quality grass silage, most of the benefit to inclusion of maize silage is obtained from replacing approximately 0.50 of the forage component of the diet with maize silage.

Stage of maturity

Whilst the optimum stage of harvesting grass silage for feeding to finishing beef cattle and lactating dairy cows is at the leafy immature stage (Keady *et al.*, 1999; 2000; 2002b; 2003), for maize silage the intention is to increase starch content and consequently harvest as a mature crop. Major changes occur in the composition of the maize plant as it matures. Neutral detergent fibre, acid detergent fibre and crude protein concentrations decrease (Phipps *et al.*, 2000; Keady *et al.*, 2002b; 2003) whilst starch concentrations increase (Phipps *et al.*, 2000, Keady *et al.*, 2002b; 2003) due to the cob accounting for a larger proportion of plant weight. A number of studies have evaluated the impact of stage of maturity of maize at harvest on performance of dairy cattle (Harrison *et al.*, 1996; Bal *et al.*, 1997, Phipps *et al.*, 2000; Keady *et al.*, 2002b; 2003). Harrison *et al.* (1996) concluded that increasing maize silage DM from 357 to 451 g/kg in mixed maize silage/alfalfa hay-based diets decreased milk yield by 1.4 kg/d. Bal *et al.* (1997) increased maize silage DM from 301 to 420 g/kg in maize silage/alfalfa silage-based diets and observed that the highest milk yields were obtained from diets that included maize silage DM of 351 g/kg. However inclusion of maize silage did not alter milk composition.

Phipps *et al.* (2000) increased maize silage DM content from 226 to 390 g/kg with maize included in grass silage diets at 0.75 of the forage component and concluded that maize silage with DM of 290 and 302 g/kg increased milk yield by up to 3.6 kg/cow/d relative to maize silage of 226 g/kg DM. However, Phipps *et al.* (2000) also observed that as maize silage DM content increased further to 390 g/kg, milk yield was decreased by 2.2 kg/head relative to maize at 302 g/kg. Neither milk fat or milk protein concentrations were affected by maize silage DM content. More recently, Keady *et al.* (2002b) offered maize silages with DM concentrations varying from 202 to 384 g/kg in grass silage-based diets to lactating dairy cows. These authors concluded that the highest yields of milk and fat plus protein were obtained from maize silage of 280 g/kg. Furthermore, Keady *et al.* (2002b) concluded that maize silage with a dry matter of 384 g/kg did not significantly decrease milk solid output relative to maize silage with a dry matter of 280 g/kg.

In the study of Keady *et al.* (2003) in which maize silage dry matter concentrations varied from 189 to 429 g/kg there were no maize silages with a dry matter content between the range of 249 to 362 g/kg DM, possibly explaining the absence of an effect of maize maturity on animal performance. In contrast to the results of Phipps *et al.* (2000), Keady *et al.* (2002b; 2003) observed no negative impacts of high DM maize silage on animal performance. The negative effect of high dry matter maize silage on animal performance as reported by Phipps *et al.* (2000) may be due to silo management, as Keady *et al.* (2002b; 2003) ensiled in narrow silos and used an additive to improve aerobic stability at the point of feed-out.

Using the data of Harrison *et al.* (1996), Bal *et al.* (1997), Phipps *et al.* (2000) and Keady *et al.* (2002b; 2003), the effect of maize silage dry matter content on subsequent milk yield is best described by the following relationship:

MY = 21.86 (s.e. = 2.94***) + 0.0408 (s.e. = 0.01900*) MDM − 0.0000615 (s.e. = 0.0000295*) MDM^2 $R^2 = 0.24, P = 0.12$
where MY = milk yield (kg/d), MDM = maize silage dry matter content (g/kg).

The relationship between maize silage DM and milk yield is presented in Figure 1.

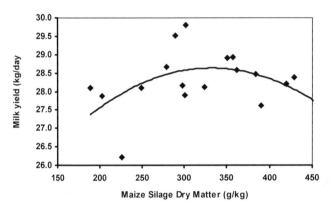

Figure 1 Effect of maize silage dry mater content on milk yield

In contrast to Bal *et al.* (1997) and Phipps *et al.* (2000), Keady *et al.* (2002b; 2003) observed that increasing maize silage DM content decreased milk fat concentration. Keady (2003) using the data from Keady *et al.* (2002b; 2003) presented the following relationship between maize silage dry matter content and milk fat concentration:

Milk fat (g/kg) = 43.61 (s.e. 0.739) − 0.0085 MDM (s.e. 0.00239) $R^2 = 0.68$
where MDM = maize dry matter content (g/kg).

Using the data of Bal *et al.* (1997), Phipps *et al.* (2000) and Keady *et al.* (2002b; 2003) the effect of maize silage dry matter content on milk fat concentration is best described by the following relationship:

MF = 42.42 (s.e. 1.01) − 0.00674 MDM (s.e. 0.00314) $R^2 = 0.25*$
where MF = milk fat concentration (g/kg) MDM = maize silage dry matter (g/kg).

The relationship between silage DM and milk fat concentration is presented in Figure 2.

The negative relationship between milk fat concentration and maize silage dry matter concentration may be attributed to the increased intake of digestible fibre and the decreased intake of starch associated with low dry matter content maize. Previous studies (Keady *et al.*,

1998; 1999) have shown negative relationships between concentrate starch intake and milk fat concentration.

In summary, in order to achieve optimum levels of performance in beef and dairy cattle offered grass silage-based diets, forage maize should be ensiled at a dry matter concentration of approximately 300 g/kg.

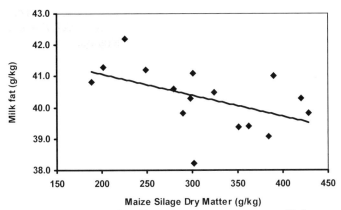

Figure 2 Effect of maize silage dry mater content on milk fat concentration

Interaction between grass silage feed value and maturity of maize at harvest
Whilst many studies have reported differences in the response to the inclusion of maize silage with grass silage-based diets offered to beef and dairy cattle, few have evaluated if the response depends on feed value of maize silage or that of the grass silage. Keady *et al.* (2002b; 2003) evaluated the effects of including four maize silages, differing in stage of maturity at harvest, on performance of lactating dairy cattle offered either two or three grass silages respectively, differing in feed value. Keady *et al.* (2002b, 2003) concluded that the response to maize silage inclusion in the diet in terms of the yields of fat and protein corrected milk (Figure 3), fat plus protein yield and the concentrations of fat and protein were similar irrespective of grass silage feed value or maize silage DM content.

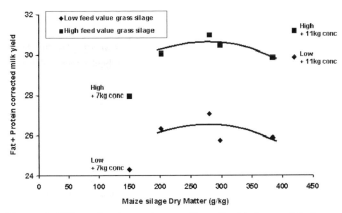

Figure 3 Effect of maize and grass silage feed value on fat plus protein corrected milk yield

However, whilst there were no interactions between maize silage and grass silage feed value for milk yield or composition, it was noted that the response to maize silage inclusion tended to decrease as grass silage feed value increased (Table 4).

Table 4 The effects of the addition of maize silage (40:60 maize silage:grass silage) to diets of differing grass silage feed value on performance of dairy cows

	Grass silage feed value		
	Low	Medium	High
Keady *et al.* (2002)			
Grass silage ME (MJ/kg DM)	9.80		11.84
Dry matter intake (kg/d)	2.25		0.54
Milk yield (kg/d)	+1.08		+0.58
Fat (g/kg)	+1.23		+1.97
Protein (g/kg)	+1.20		+0.37
Keady *et al.* (2003)			
Grass silage ME (MJ/kg DM)	10.17	10.96	11.99
Dry matter intake (kg/d)	+1.83	+1.44	-0.15
Milk yield (kg/d)	+1.38	+1.13	+0.38
Fat (g/kg)	+1.21	+0.84	+1.06
Protein (g/kg)	+1.01	+1.29	+0.16

Potential concentrate sparing effect of including maize silage

One of the potential benefits of including maize silage in the diet of lactating dairy cattle is the ability to maintain animal performance whilst reducing the level of concentrate supplementation required, consequently reducing the costs of production. Keady *et al.* (2002b; 2003) evaluated the potential concentrate sparing effects (reduction in the quantity of concentrates required due to the inclusion of maize silage in the diet to maintain animal performance) of maize silages, differing in stage of maturity at harvest, when offered with grass silages of contrasting feed value to lactating dairy cows. Grass silages offered as the sole forage were supplemented with either 7 or 11 kg concentrates/cow/d and it was assumed that any response in animal performance expressed as combined yield of fat plus protein to concentrate supplementation was linear between these two points. No interactions were observed between stage of maturity of maize silage at harvest or grass silage feed value on the concentrate sparing effect of maize silage. In the study of Keady *et al.* (2002b) the mean concentrate sparing effects for the maize silages, included as 0.40 of the forage proportion of the diet, with dry matter concentrations of 202, 280, 298 and 398 g/kg when determined for fat plus protein yield are presented in Figure 4, and were 2.1, 3.4, 2.1 and 2.0 kg fresh weight respectively. Similarly, Keady *et al.* (2003) concluded the mean concentrate sparing effects for maize silage, included as 0.4 of the forage proportion in grass silage-based diets, with dry matter concentrations of 189, 249, 362 and 429 g/kg when determined for fat plus protein yield were 2.8, 3.1, 2.6 and 2.6 kg fresh weight respectively. In this study a greater concentrate sparing effect was observed as grass silage feed value declined, with concentrate sparing effects of 3.3, 2.6 and 2.5 kg fresh weight for low, medium and high feed value grass silages respectively. More recently, Patterson & Kilpatrick (2005) reported a potential concentrate sparing effect of 5.0 kg fresh weight for maize silage (DM = 305 g/kg) included as 0.50 of the forage proportion of high feed value (ME = 11.6 MJ/kg DM) grass silage-based diets offered to lactating dairy cows.

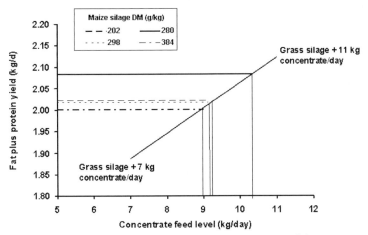

Figure 4 Effect of maize silage dry matter content on potential concentrate sparing effect

With beef cattle, Keady & Kilpatrick (2004) concluded that replacing 0.4 of the forage component of grass silage-based diets with maize silage had a concentrate sparing effect of greater than 2 kg/d.

In production systems where animal product output is limited due to quotas, or concentrate prices are excessively expensive relative to forage, maize silage inclusion in the diet can enable animal performance to be maintained whilst considerably reducing the level of concentrate input.

Effect of whole crop wheat inclusion on animal performance

There has been an increased interest in the production of whole crop cereal silage for feeding to beef and dairy cattle in recent years. The increased interest in this crop is due primarily to the similar cost of production relative to grass silage (Table 1) and the perceived potential benefits in forage intake and subsequently animal performance. Whole crop wheat is predominantly ensiled and fermented at dry matter concentrations ranging from 250 to 450 g/kg. However, whole crop wheat can also be ensiled at high dry matter concentrations ranging from 550 to 800 g/kg and treated with either urea or a urea-based additive to encourage an alkaline environment. Recent developments in the ensiling of whole crop cereals involves the ensiling of crops at high dry matter concentrations (700-800 g/kg), harvested through a forage harvester fitted with a grain processor and ensiled with a urea-based additive. Including whole crop wheat significantly increased the quantity of forage by 1.1 kg DM/kg fat plus protein yield.

The effects of including of whole crop wheat in grass silage-based diets on feed intake and performance of lactating dairy cows from 20 comparisons are presented in Table 5. Including whole crop wheat in dairy cow diets did not significantly alter the yields of milk or fat plus protein, or the concentrations of milk fat or protein. However whole crop wheat inclusion increased feed intake by 2.3 kg DM/d to produce milk solid output. When the 20 comparisons were subdivided into datasets depending on whether the whole crop wheat was either fermented or urea/alkalage treated whole crop wheat inclusion resulted in an increased feed intake of up to 2-3 kg DM/cow/d but had no beneficial effects on milk yield or fat plus protein yield or milk fat and protein concentrations. Furthermore, when including fermented whole crop wheat in grass silage-based diets, milk fat concentration was significantly decreased.

The effects of including whole crop wheat in grass silage-based diets on feed intake and performance of finishing beef cattle based on seven comparisons, is presented in Table 6. Including whole crop wheat in grass silage-based diets increased feed intake by 1.4 kg DM/d but did not alter carcass gain or liveweight gain of finishing beef cattle. When the seven comparisons were sub-divided into datasets depending on whether the whole crop wheat was fermented or urea treated/alkalage treated, whole crop wheat inclusion increased forage intake by 1.2 and 1.6 kg DM/d respectively but did not affect animal performance.

The results of the data presented in Tables 5 and 6 indicate that whilst whole crop wheat inclusion in the diet increased feed intake, it had no beneficial effects on animal performance. A number of studies (Leaver & Hill, 1995; Sutton et al., 2002; Keady & Kilpatrick, 2004) have shown that whole crop wheat has high intake characteristics, but lower ME concentration relative to most grass silages.

Levels of inclusion

When including an alternative forage in the diets of beef or dairy cattle, level of inclusion may impact on the optimum level of animal performance achieved. A number of studies have been undertaken (Hill & Leaver, 1990; Phipps et al., 1992a) to evaluate the effect of level of inclusion of whole crop wheat in grass silage-based diets on the performance of dairy cows.

Phipps et al. (1992a) replaced low feed value grass silage (organic matter digestibility = 658 g/kg DM) with either 0.25, 0.50, 0.75 or 1.0 whole crop wheat (DM = 515 g/kg) and concluded that whilst the inclusion of whole crop wheat increased feed intake it did not alter milk yield or protein concentration. Hill & Leaver (1990) replaced whole crop wheat with either 0.22 or 0.44 low feed value grass silage (ME = 10 MJ/kg DM) and concluded that increasing the proportion of grass silage in the diet increased milk yield but did not alter milk composition. These authors concluded that with mixed forage-based diets, inclusion of whole crop wheat did not improve animal performance, but if included in the diet, it should account for a low proportion of the forage DM component. Sutton et al. (1997) concluded that replacing medium feed value grass silage (DMD = 695 g/kg DM) with either 0.33 or 0.66 urea-treated whole crop wheat, had no effect on milk yield or protein concentration whilst feed intake was increased. However, inclusion of 0.33 whole crop wheat improved butterfat content. Sutton et al. (1997) concluded that the lack of response in milk yield with whole crop inclusion in the diet was due to reduced diet digestibility, particularly starch resulting from the egestion of whole wheat grains in the faeces. However, Jackson et al. (2004) ensiled whole crop wheat at a dry matter concentration of 700 g/kg either after the crop was passed through a harvester with or without a grain cracker and concluded that use of the grain cracker did not affect milk yield or composition of lactating dairy cows.

In studies with beef cattle, O'Kiely & Moloney (2002) replaced medium feed value grass silage (DMD = 732 g/kg DM) with 0.33, 0.67 or 1.00 of either a medium (380 g/kg) or high (519 g/kg) DM whole crop wheat silage. These authors reported that replacing 0.33 grass silage with either whole crop wheat increased carcass gain. However further inclusions of whole crop wheat did not significantly alter carcass gain.

Table 5 The effects of including whole crop wheat (WCW) with grass silage-based diets on the performance of dairy cows

Reference	Type	Grass silage (GS) DM (g/kg)	Grass silage (GS) ME (MJ/kg DM)	Whole crop (WCW) DM (g/kg)	Whole crop (WCW) Starch (g/kg DM)	GS:WCW	Forage intake (kg DM/d) GS	Forage intake (kg DM/d) WCW	Milk (kg/d) GS	Milk (kg/d) WCW	Fat (g/kg) (F) GS	Fat (g/kg) (F) WCW	Protein (g/kg) (P) GS	Protein (g/kg) (P) WCW	F+P (kg/d) GS	F+P (kg/d) WCW
Sinclair et al. (2003)	F	252	11.7	296	52	33:67		11.1	29.8	30.1	40.9	40.9	32.1	30.7	2.15	2.14
	F	252	11.7	371	221	33:67		11.6	29.8	30.7	40.9	39.7	32.1	31.6	2.15	2.19
Leaver & Hill (1995)	F	243	11.3	353	20	60:40	9.4	12.1	28.0	28.7	40.4	40.5	31.3	31.3	2.01	2.06
	U	243	11.3	549	218	60:40	9.4	12.5	28.0	29.6	40.4	39.8	31.3	31.2	2.01	2.10
	F	237	11.4	372	23	67:33	11.7	12.5	30.0	29.1	41.9	41.0	32.5	31.9	2.24	2.12
	U	237	11.4	555	258	67:33	11.7	10.6	30.0	29.4	41.9	40.7	32.5	32.2	2.24	2.13
	U	237	11.4	549	258	67:33	11.7	10.2	30.0	29.9	41.9	41.4	32.5	32.5	2.24	2.21
Phipps et al. (1995)	F	213	10.9	323	11	67:33	9.3	12.2	23.0	24.2	41.7	41.7	29.9	30.8	1.62	1.73
	U	213	10.9	495	280	67:33	9.3	13.1	23.0	24.0	41.7	42.1	29.9	30.8	1.62	1.73
Hameleers (1998)	F	248	11.4	557	359	60:40	10.6	9.6	27.4	27.1	48.9	49.0	34.1	34.0	2.25	2.24
	U	248	11.4	791	373	60:40	10.6	11.2	27.4	26.9	48.9	48.1	34.1	34.3	2.25	2.21
Patterson et al. (2004)	F	187	11.2	316	209	60:40	7.9	13.3	25.8	26.5	39.8	39.0	30.9	31.3	1.82	1.86
Patterson & Kilpatrick (2005)	F	185	10.7	459	350	50:50	10.4	11.3	27.9	28.3	40.7	40.1	30.8	31.1	1.96	1.96
	F	234	12.5	459	350	50:50	12.5	13.8	31.5	31.0	38.5	37.8	32.8	32.9	2.23	2.14
	A	185	10.7	751	420	50:50	10.4	12.8	27.9	26.5	40.7	39.3	30.8	30.8	1.96	1.85
	A	234	12.5	751	420	50:50	12.5	14.8	31.5	30.9	38.5	39.8	32.8	33.2	2.23	2.26
Murphy et al. (2004)	F	240	10.5	406	282	33:67	8.8	14.3	27.6	29.7	39.6	38.4	30.7	31.7	1.93	2.07
	A	240	10.5	733	324	33:67	8.8	16.4	27.6	29.4	39.6	37.8	30.7	31.7	1.93	2.03
	F	231	10.9	370	323	33:67	8.7		30.8	32.8	36.7	36.2	29.7	30.8	2.05	2.19
	A	231	10.9	763	341	33:67	8.7		30.8	31.2	36.7	40.2	29.7	31.9	2.05	2.26
Mean (n = 20)		230	11.3	511	255		10.1	12.4	28.4	28.8	41.0	40.7	31.6	31.8	2.05	2.07
SED							0.50		0.23		0.256		0.17		0.021	
Significance							***		NS		NS		NS		NS	
Comparisons involving fermented WCW																
Mean (n = 11)							9.9	11.9	28.3	28.9	40.9	40.4	31.5	31.6	2.04	2.06
SED							0.57		0.29		0.15		0.23		0.025	
Significance							**		NS		**		NS		NS	
Comparisons involving urea treatment and alkalage																
Mean (n = 9)							10.3	12.9	28.5	28.6	41.1	41.0	31.6	32.1	2.06	2.09
SED							0.85		0.36		0.55		0.26		0.036	
Significance							*		NS		NS		NS		NS	

Table 6 The effects of including whole crop wheat with grass silage-based diets on the performance of beef cattle

Reference	Type	Grass silage (GS) DM (g/kg)	Whole crop (WCW) DM (g/kg)	Whole crop (WCW) Starch (g/kg DM)	GS:WCW	Forage intake (kg DM/d) GS	Forage intake (kg DM/d) WCW	Carcass gain (kg/d) GS	Carcass gain (kg/d) WCW	Liveweight gain (kg/d) GS	Liveweight gain (kg/d) WCW
Keady & Kilpatrick (2004)	Fer	192	319	209	60:40	5.1	5.8	0.51	0.50	0.86	1.01
O'Kiely & Moloney (1995)	Fer	188	371		0:100	5.0	5.2	0.75	0.58	1.05	0.89
	Urea	188	456		0:100	5.0	5.5	0.75	0.53	1.05	0.89
O'Kiely & Moloney (2002)	Fer	191	381		33:67	4.8	6.0	0.64	0.70	0.87	0.98
	Urea	191	519		33:67	4.8	6.1	0.64	0.69	0.87	0.96
Walsh et al. (2005)	Fer	161	391		100:0	4.5	7.1	0.48	0.72	0.80	1.15
	Alk	161	705		100:0	4.5	7.6	0.48	0.69	0.80	1.13
Mean (n = 7)		182	449	209		4.8	6.2	0.61	0.63	0.90	1.00
SED							0.396		0.07		0.077
Significance							**		NS		NS
Comparisons involving fermented WCW											
Mean (n = 4)						4.9	6.1	0.58	0.62	0.89	1.01
SED							0.50		0.088		0.105
Significance							NS		NS		NS
Comparisons involving urea treatment and alkalage											
Mean (n = 3)						4.8	6.4	0.61	0.63	0.91	0.99
SED							0.75		0.127		0.141
Significance							NS		NS		NS

The majority of studies in the literature indicate that even with low feed value grass silage-based diets, inclusion of whole crop wheat either fermented or urea-treated, has little effect on animal performance.

Effect of stage of maturity

As with maize silage, the objective with whole crop wheat is to increase starch content and subsequently harvest as a mature crop. Whilst starch concentration of a cereal crop increases as it matures due to increased grain fill, the fibre concentration does not decrease significantly as a result of lignification of the straw and the accumulation of chaff. Delaying harvesting of the crop increased (O'Kiely & Moloney, 2002), had no effect (Leaver & Hill, 1995; Sutton *et al.*, 2002) or reduced (Phipps *et al.*, 1995; Sinclair *et al.*, 2003) acid detergent fibre concentration but increased (Leaver & Hill, 1995; Phipps *et al.*, 1995; Sinclair *et al.*, 2003; Sutton *et al.*, 2002) starch concentration.

A number of studies have been undertaken to evaluate the effects of stage of maturity and preservation method of whole crop cereals on performance of beef and dairy cattle. Sinclair *et al.* (2003) replaced 0.66 grass silage with fermented whole crop wheat with a dry matter concentration of either 296 or 371 g/kg dry matter content. These authors concluded that neither inclusion of whole crop wheat or stage of maturity at harvest affected milk yield or composition of dairy cows relative to a grass silage-based diet. Leaver & Hill (1995) in two studies replaced either 0.33 or 0.40 of grass silage with whole crop wheat with dry matter concentrations of 370 and 572 g/kg and 346 and 577 g/kg respectively and concluded that neither inclusion or stage of maturity of whole crop wheat affected milk yield or composition of lactating dairy cows.

Concentrate sparing effect

A number of studies (Keady & Kilpatrick, 2004; Patterson *et al.*, 2004; Patterson & Kilpatrick, 2005) have been undertaken to determine the potential concentrate sparing effect of including whole crop wheat in grass silage-based diets. Keady & Kilpatrick (2004); Patterson *et al.* (2004) and Patterson & Kilpatrick (2005) reported non-significant potential concentrate sparing effects/head/d of 0 kg, 0.5 kg and 1.3 kg for whole crop wheat. These studies clearly indicate that if whole crop wheat replaces grass silage as the forage component of the diet, concentrate feed level should not be reduced if animal performance is to be maintained.

Comparison of maize and fermented whole crop wheat silages

A number of studies (Phipps *et al.*, 1995; Hameleers, 1998; Patterson *et al.*, 2004; Murphy *et al.*, 2004 and Patterson & Kilpatrick, 2005) have been undertaken to compare whole crop wheat and maize silages when included in grass silage-based diets of dairy cows. All of these authors concluded that including either maize or whole crop wheat increased feed intake. Phipps *et al.* (1995); Murphy *et al.* (2004 - in two comparisons), Patterson *et al.* (2004) and Patterson & Kilpatrick (2005) concluded that maize silage inclusion increased milk yield. Furthermore, Phipps *et al.* (1995) and Murphy *et al.* (2004) in one comparison reported increased milk protein concentration. However, when whole crop wheat was included, only Murphy *et al.* (2004) in one comparison reported increased milk yield and protein concentration.

The effect of replacing grass silage with either maize or fermented whole crop wheat silages on dairy cow performance is presented in Table 7. Including either maize or whole crop wheat in grass silage-based diets increased feed intake and forage conversion rate. However, while maize silage inclusion increased milk yield by 2.2 kg/cow/d whole crop wheat inclusion had no significant effect.

Table 7 Direct comparisons of replacing grass silage (GS) with either fermented whole crop wheat (WCW) or maize (MS) on dairy cow performance

Reference	Forage intake (kg DM/d)			Milk yield (kg/d)[1]		
	GS	WCW	MS	GS	WCW	MS
Phipps et al. (1995)	9.3	10.6	10.6	22.9	24.4	26.8
Murphy et al. (2004)	8.8	12.8	11.2	27.0	29.0	29.5
	8.7	14.3	13.2	28.5	30.6	32.9
Patterson & Kilpatrick (2005)	9.8	12.3	13.6	28.0	29.3	31.2
Patterson et al. (2004)	7.9	9.6	9.8	25.4	25.9	26.9
Hameleers (1998)	10.6	12.2	11.4	31.7	31.3	29.5
Mean (n = 6)	9.2[a]	12.0[b]	11.6[b]	27.3[a]	28.4[ab]	29.5[b]
SED		0.58			0.71	
Significance		***			*	

[1]Fat plus protein corrected milk yield

In studies with beef cattle Keady & Kilpatrick (2004) observed that whilst maize inclusion increased carcass gain, whole crop wheat had no effect. However, Walsh et al. (2005) observed increased carcass gain from either maize or whole crop wheat silage-based diets.

In deciding which alternative forage to produce, the literature indicates that maize silage will increase animal product output. In contrast, whole crop wheat silage inclusion has not significantly affected either milk yield or carcass gain, whilst increasing forage intake and therefore potentially increasing the cost of production.

Conclusions

From an extensive review of the literature on the effect of including maize and whole crop wheat silages with grass silage-based diets it is concluded that:
1. High yields of maize can be achieved in marginal areas due to improvements in plant breeding and use of the complete cover plastic mulch (CCPM) system.
2. Maize silage can be produced and fed at a similar cost to grazed grass.
3. Including maize silage in the diet of dairy and beef cattle increases animal performance.
4. The optimum stage of maturity to harvest maize for ensiling is at a dry matter content of approximately 300 g/kg.
5. There is a negative relationship between maize silage dry matter content and milk fat concentration.
6. Including maize silage in the diet has a concentrate sparing effect of up to 5 kg/cow/d.
7. Whole crop wheat silage can be produced and fed to dairy cows at a similar cost to grass silage.

8. Including whole crop wheat silage in the diet increases feed intake but has no beneficial effect on animal performance. However additional forage intake is required to produce each kg of carcass from beef cattle or fat plus protein yield from dairy cows.
9. Ensiling whole crop wheat either fermented, urea-treated or as alkalage does not alter animal performance.

References

Agricultural and Food Research Council (1993). Energy and Protein Requirements of Ruminants. CAB International, Wallingford.

Bal M.A., J.G. Coors & R.D. Shaver (1997). Impact of the maturity of corn for use as silage in the diets of dairy cows on intake, digestion, and milk production. *Journal of Dairy Science*, 80, 2497-2503.

Bartholomew P.W. & D.M.B. Chestnutt (1979). An assessment of the feasibility of forage maize production in Northern Ireland. *Record of Agriculture Research, Department of Agriculture for Northern Ireland*, 25, 17-23.

Browne, E.M., M.J. Bryant, D.E. Beever & A.V. Fisher (2000). Intake, growth rate and carcass quality of beef cattle fed forage mixtures of grass silage and maize silage. *Proceedings of the British Society of Animal Science*, p. 72.

Easson D.L. (2000). The effects of plastic mulch on the growth and development of forage maize in Northern Ireland. *73rd Annual Report, Agricultural Research Institute of Northern Ireland*, p. 41-49.

Gilliland T. (2003). Forage maize : Recommended Varieties for Northern Ireland 2003. *Department of Agriculture and Rural Development*.

Hameleers A. (1998). The effects of the inclusion of either maize silage, fermented whole crop wheat or urea-treated whole crop wheat in a diet based on a high-quality grass silage on the performance of dairy cows. *Grass and Forage Science*, 53, 157-163.

Harrison, J.H., L. Johnson, R. Riley, S. Xu, K. Loney, C.W. Hunt & D. Sapienza (1996). Effect of harvest maturity of whole plant corn silage on milk production and component yield, and passage of corn grain and starch into faeces. *Journal of Dairy Science*, 75 (Suppl. 1), 149.

Hill, J. & J.D. Leaver (1990). Urea-treated whole crop wheat for dairy cattle. *Proceedings of the British Society of Animal Production*, p. 113.

Jackson, M.A., R.J. Readman, J.A. Huntington & L.A. Sinclair (2004). The effects of processing at harvest and cutting height of urea-treated whole-crop wheat on performance and digestibility in dairy cows. *Animal Science*, 78, 467-476.

Keady, T.W.J. (2003). Maize silage in the diet of beef and dairy cattle – the influence of maturity at harvest and grass silage feed value, and feeding value relative to whole crop wheat. *76th Annual Report, Agricultural Research Institute of Northern Ireland*, p. 43-54.

Keady T.W.J. & D.J. Kilpatrick (2004). The effect of the inclusion of maize and whole crop wheat silages in grass silage-based diets on the performance of beef cattle offered two levels of concentrate. *Proceedings of the British Society of Animal Science*, p. 65.

Keady T.W.J., C.M. Kilpatrick, A. Cushnahan & J.A. Murphy (2002a). The cost of producing and feeding forages. *Proceedings of the XIIIth International Silage Conference, Auchincruive, Scotland*, pp. 322-323.

Keady T.W.J., C.S. Mayne & D.A. Fitzpatrick (2000). Prediction of silage feeding value from the analysis of the herbage at ensiling and effects of nitrogen fertiliser, date of harvest and additive treatment on grass silage composition. *Journal of Agricultural Science, Cambridge*, 134, 353-368.

Keady T.W.J., C.S. Mayne, D.A. Fitzpatrick & M. Marsden (1999). The effects of energy source and level of digestible undegradable protein in concentrates on silage intake and performance of lactating dairy cows offered a range of grass silages. *Animal Science*, 68, 763-777.

Keady, T.W.J., C.S. Mayne & D.J. Kilpatrick (2001). Prediction of silage dry matter digestibility from digestible organic matter digestibility. *Proceedings of the British Society of Animal Science*, p. 93.

Keady T.W.J., C.S. Mayne & D.J. Kilpatrick (2002b). The effect of maturity of maize silage at harvest on the performance of lactating dairy cows offered two contrasting grass silages. *Proceedings of the British Society of Animal Science*, p. 16.

Keady T.W.J., C.S. Mayne & D.J. Kilpatrick (2003). The effect of maturity of maize silage at harvest on the performance of lactating dairy cows offered three contrasting grass silages. *Proceedings of the British Society of Animal Science*, p. 126.

Keady, T.W.J., C.S. Mayne & M. Marsden (1998). The effects of concentrate energy source on silage intake and animal performance with lactating dairy cows offered a range of grass silages. *Animal Science*, 66, 21-33.

Leaver J.D. & J. Hill (1995). The performance of dairy cows offered ensiled whole-crop wheat, urea-treated whole-crop wheat or sodium hydroxide-treated wheat grain and wheat straw in a mixture with grass silage. *Animal Science*, 61, 481-490.

McAllister J.S.V. (1961). Nitrogen top-dressing for maize. *35th Annual Report, Agricultural Research Institute of Northern Ireland 1960-1961*, p. 36-37.

Murphy, J.J., S. Kavanagh and J.J. Fitzgerald (2004). Comparative evaluation of grass silage, fermented whole crop wheat silage, urea-treated processed whole crop wheat silage and maize silage in the diet of early lactation cows. *Proceedings of the 55th Annual Meeting of the EAAP*, Paper N 4.4, p. 95.

O'Gorman, C., P.J. Caffrey & G.P. Keane (1998). The feeding value of maize silage for finishing beef cattle. *Proceedings of the Agricultural Research Forum*, pp. 99-100.

O'Kiely P. & A.P. Moloney (1995). Performance of beef cattle offered different ratios of grass and maize silage. *Irish Journal of Agricultural and Food Research*, 34, 76.

O'Kiely P. & A.P. Moloney (1999). Whole crop wheat silage for finishing beef heifers. *Irish Journal of Agricultural and Food Research*, 38, 296.

O'Kiely P. & A.P. Moloney (2000). Nutritive value of maize and grass silage for beef cattle when offered alone or in mixtures. *Proceedings of the Agricultural Research Forum*, pp. 99-100.

O'Kiely P. & A.P. Moloney (2002). Nutritive value of whole crop wheat and grass silage for beef cattle when offered alone or in mixtures. *Agricultural Research Forum*, pp. 42.

O'Mara F.P., J.J. Fitzgerald, J.J. Murphy & M. Rath (1998). The effect on milk production of replacing grass silage with maize silage in the diet of dairy cows. *Livestock Production Science*, 55, 79-87.

Patterson, D.C. & D.J. Kilpatrick (2005). The effects of maize and whole crop wheat silages and quality of grass silage on the performance of lactating dairy cows. *Proceedings of the XVth International Silage Conference*, Belfast, Northern Ireland (in press).

Patterson D.C., D.J. Kilpatrick & T.W.J. Keady (2004). The effects of maize and whole crop wheat silages on the performance of lactating dairy cows offered two levels of concentrates differing in protein concentration. *Proceedings of the British Society of Animal Science,* p. 4.

Phipps R.H., J.D. Sutton & B.A. Jones (1995). Forage mixtures for dairy cows : the effect on dry matter intake and milk production of incorporating either fermented or urea treated whole-crop wheat, brewers grains, fodder beet or maize silage into diets based on grass silage. *Animal Science*, 61, 491-496.

Phipps R.H., J.D. Sutton, D.E. Beever & A.K. Jones (2000). The effect of crop maturity on the nutritional value of maize silage for lactating dairy cattle. 3. Food intake and milk production. *Animal Science*, 71, 401-409.

Phipps R.H., R.F. Weller & A.J. Rook (1992b). Forage mixtures for dairy cows : the effect on dry matter intake and milk production of incorporating different proportions of maize silages into diets based on grass silage of differing energy value. *Journal of Agricultural Science (Cambridge)*, 118, 379-382.

Phipps, R.H., R.F. Weller & J.W. Siviter (1992a). Whole crop cereals for dairy cows. In: Whole crop cereals, (Ed. B.A. Stark & J.M. Wilkinson), Chalcombe Publications, pp. 51-58.

Sinclair, L.A., R.G. Wilkinson & D.M.R. Ferguson (2003). Effects of crop maturity and cutting height on the nutritive value of fermented whole crop wheat and milk production in dairy cows. *Livestock Production Science*, 81, 257-269.

Sutton, J.D., A.L. Abdalla, R.H. Phipps, S.B. Cammell & D.J. Humphries (1997). The effect of the replacement of grass silage by increasing proportions of urea-treated whole-crop wheat on feed intake and apparent digestibility and milk production by dairy cows. *Animal Science*, 65, 343-351.

Sutton, J.D., R.H. Phipps, E.R. Deaville, A.K. Jones & D.J. Humphries (2002). Whole-crop wheat for dairy cows : effects of crop maturity, a silage inoculant and an enzyme added before feeding on food intake and digestibility and milk production. *Animal Science*, 74, 307-318.

Walsh, K., P. O'Kiely & F. O'Mara (2005). The feeding value of conserved whole-crop wheat and forage maize relative to grass silage and *ad libitum* concentrates for beef cattle. *Proceedings of the XVth International Silage Conference,* Belfast, Northern Ireland (In press).

Update on technologies for producing and feeding silage

P.D. Forristal[1] and P. O'Kiely[2]

Teagasc, Crop Production and Engineering Department, Oak Park, Carlow, Ireland
Teagasc, Beef Research Centre, Grange, Dunsany, Co. Meath, Ireland

Key Points

1. Mechanisation and engineering inputs are key factors which contribute to silage costs.
2. Sensing technologies will improve management precision in many areas of ensilage.
3. While current harvesting machines have high output capacity reflecting mature design, there is a need to revisit the area of energy efficiency.
4. Baled silage technology, particularly in the areas of covering film and wrapping technology, needs further research.
5. Feeding systems are well researched but there is a need for systems research to underpin farmers' decisions concerning housing design and feeding system.

Keywords: silage, engineering, mechanisation, costs

Introduction

Silage making on farms requires technologies that ensure efficient conservation of forage. While the term technology could encompass all aspects of the application of science to ensilage, in the context of this paper it is restricted to the engineering- or mechanisation-related components of the ensiling and feeding processes. These technologies can reduce costs and labour requirements and improve work/process quality and timeliness. This paper has two main objectives:

1. Review recent developments in engineering-related technologies involved in the production, storage and utilisation of silage.
2. Identify engineering related areas that require further research and development in order to improve the ensiling process.

The scope of the paper is limited to those crops for ensilage that are normally grown in northern European climates. For a broad overview of ensiling technologies see Muck and O'Kiely (2002).

Costs and cost components

The cost or value of silage as a feed can be estimated as a cost per unit dry matter (DM) or energy consumed by an animal and this enables comparison with other feeds. Examples of cost breakdowns for a grass silage crop stored in a conventional silo, or stored as bales, and for a whole crop wheat crop using a crop costing model (O'Kiely *et al.*, 1997) with 2005 data, are given in Table 1. The yields and costs attributed are those appropriate for Ireland. The importance of mechanisation as a cost component is evident with between 52% and 64% of the production costs being attributable to machinery. These costs are averages around which there is considerable variation. They do not reflect the contribution that mechanisation technology can make to silage quality and value, or to labour-saving and convenience.

The science of silage is often thought of as the non-mechanisation processes directly involved with the production and preservation of forage, the effects of which are measured by various analyses of the forage and animal performance. While these are considered they are not the only factors that will influence a farmer's decision concerning choice or specification of forage conservation system. Machinery cost and logistical considerations in the supply of labour and mechanisation for the production, harvesting, storing and feeding of forage have a significant influence on forage conservation practised on farms.

Table 1 Forage production costs[1] and partitioning of cost components

	Conventional silage		Baled silage		Whole crop wheat	
Crop DM yield (t/ha)	6.64		6.64		14.13	
Crop DM (g/kg)	200		280		440	
Silage DM yield (t/ha)	5.31		5.64		12.44	
[2]Total cost/ha (€)	532.08		607.17		1193.67	
[2]Total cost/t DM (€)	100.20		107.65		95.95	
[3]Total cost/t DDM (€)	138.79		149.10		141.10	
Cost categories	Machinery	Materials	Machinery	Materials	Machinery	Materials
Production (%)	7	22	6	19	22	45
Harvest (%)	45	0	53	0	23	0
Additive (%)	0	13	0	0	0	0
Storage (%)	0	8	0	17	0	3
Feed (%)	5	0	5	0	7	0
Category total (%)	57	43	64	36	52	48
Total (%)	100		100		100	

[1]Excluding land charges [2]Cost of feed consumed [3]DDM: Digestible dry matter

Crop production

Mechanisation technology involved in the production of the forage crop can influence crop yield, quality and production costs.

Crop establishment

Grass establishment is expensive. Seed and mechanisation are the primary costs with sward longevity determining annual costs. While lower-cost establishment/improvement using minimal cultivation and direct drilling techniques continue to be developed, the techniques used, their success and application, depend on local soil, climate and machine availability factors.

The establishment cost for annual forage crops such as maize, cereals or fodder beet is significant. Rapid and successful establishment is essential. Over the last decade the use of polythene mulch systems with forage maize has been shown to give improved yields and higher dry matter and starch levels (Crowley, 1998; Keane *et al.,* 2003; Easson & Fearnehough, 2003). While traditional mulch systems required the seeds to be sown through a hole punched in the plastic mulch, an alternative complete cover system where the growing plant must push through a total polythene cover has been developed. While this system has given advanced early growth, the effect on performance of the harvested crop has been inconsistent (Keane *et al.*, 2003) (Crowley, 2005).

Nutrient application

The application of nutrients can affect forage costs and quality. Even application of nutrients aids the achievement of optimum yield, consistent quality and ensilability. Modern fertiliser spreaders are capable of very even spreading (coefficient of variation <0.05) at wide (>15 m) bout widths (Anon, 1999). These spreaders ensure more even application than that achieved by most spreaders marketed 20 years ago (Sogard & Kierkegaard, 1994). However, fertiliser physical quality has not improved over this time period, with farmers often purchasing product with physical characteristics that have a deleterious effect on spread patterns (Hofstee & Huisman, 1990). In addition to the improved technical performance of the spreaders, accuracy is today aided by the use of GPS technology which guides the driver along equidistant bout widths.

Newer organic nutrient application techniques can also play a role in improving the efficiency of nutrient use. The use of a trailing-shoe slurry applicator which places the slurry directly on the soil surface, increased grass yield by 23% compared to using a splash-plate spreader (Binnie & Frost, 2003). This was equivalent to 48% of the N in the slurry being available as fertiliser N compared to just 12% for the splash plate system. When slurry is applied at higher rates these systems are also less likely to adversely affect silage quality than splash plate spreading (O'Kiely *et al.*, 1994).

Crop sensing

The scope for the application of sensing technology to forage crops is vast, with the possibility of generating data to aid management and ultimately lead to more accurate application of inputs and improved forage quality.

Multi-spectral crop sensing

The development of precision farming technology, which allows spatial variability within fields to be assessed and managed, has increased research activity in the area of crop and soil multi-spectral sensing. Multi-spectral crop sensing can be carried out with sensors close to the crop on field machines, by aircraft mounted sensors, or from satellites. The prediction of required crop nitrogen input based on spectral imaging has been evaluated and has led to the marketing of a commercial sensing system (Wollring *et al.*, 1998). These technologies have limitations but developments in this area are proceeding. Development of complimentary soil and water status sensing technologies coupled with the use of historic spatial yield patterns should help improve the usefulness of crop sensing.

Yield sensing

The estimation of crop yield using real-time sensing technology offers potential benefits including improved feed budgeting, accurate additive application, a better base for contractor charging, and opportunities to utilise precision agricultural management techniques in forage crops. Various on-harvester yield-sensing techniques have been assessed mainly on metered chop type machines, but also on mowers, balers and silage trailers (Savoie *et al.*, 2002; Demmel *et al.*, 2002; Wild & Auernhammer, 1999). The accuracies of these yield sensors vary. Correlations of meter readings with actual mass-flow rates range from 0.60 to 0.98 (Savoie *et al.*, 2002). The uneven flow of forage through a forage harvester compared to grain flowing through a combine harvester, and the importance of grass dry matter content are factors which limit the usefulness of the sensed information. Fresh-weight yield data is adequate for silage additive application control and as a base for contractor charging (Forristal & Keppel, 2001a). Additive application accuracy can be improved by automating control and varying application rate in proportion to grass or crop throughput. Evaluation of a prototype system showed a decrease in coefficient of variation from 0.41 where fixed rate application was used to 0.12 where a throughput-based control was used (Forristal & Keppel, 2001b). Currently at least one forage harvester manufacturer is marketing a forage yield sensor.

Dry matter sensing

To provide yield data for feed management and precision-agriculture field applications, dry matter sensing technology is necessary. This will probably be in the form of a separate sensor to the mass flow sensors outlined above. The application of NIR sensing technology to real-time harvesting has been evaluated with some success (Paul *et al.*, 2000). The technology is expensive. Alternative less expensive microwave and capacitance systems have been evaluated but their accuracy has been much less than that of NIR techniques (Kormann & Auernhammer, 2002). The latter authors suggest that the ability of NIR techniques to predict various qualitative parameters during harvest may make the system economic in the future.

Harvesting

Mowing and wilting

Engineering developments in mowing and crop tedding to increase work rate have resulted in wider and faster machines. Tractor mounted, trailed and self-propelled mowers with working widths of up to 14 m are available. Almost all mowers used to cut silage crops are high cutting speed (knife speed >70m/s) rotary mowers which rely on the impact speed of the cutting flail to cut the crop stems. The cutting technology is simple, robust and fast, but requires high power inputs (O'Dogherty & Gale, 1986). The need to conserve energy for environmental and/or cost reasons will require shear mowing systems which are proven to be less power demanding (Copeland, 1993) to be developed. The concept of a light-weight saw-band mower, which can optionally be developed as a shear cutting concept, is at an early development stage (Ehlert & Kraatz, 2004).

Mowing research since the 1970s has focused on the development of conditioning systems to accelerate the wilting process. Early conditioning developments increased the wilting potential of the crop, but often narrow swath structures failed to exploit this potential. The development of rapid wilting systems using tedders led to successful and practical wilting systems. (Bosma & Verkaik 1987). Intensive conditioning using close-coupled differentially speeded rollers has been investigated since the 1980s. The drying rate of lucerne in good drying conditions can be dramatically increased by maceration (Hintz *et al.*, 1999). In north western Europe, high grass yields and variable weather have affected the performance of the technique. Research in Northern Ireland showed that rained-on macerated swaths dried more slowly than unconditioned grass and suffered greater losses (81.2 g/kg compared to 42.3 g/kg) (Binnie & Frost, 1996). While this research did show improved drying rates, it concluded that the expenditure on the equipment was not justified.

The success of the spreading and tedding wilting systems led to the development of mowers with semi-intensive conditioners coupled with spreading attachments (Bosma, 1995). From this research, European mower manufacturers have chosen to adopt simple spreading devices which spread the cut grass over 80 to 100% of the cut area thus increasing exposure to the natural drying elements. Spreading systems can speed up the crop drying process with high yielding crops in temperate climates (Forristal, 1996), achieving satisfactory wilts within 32 hours in all of the harvesting season provided it does not rain (Frost, 1988). However wilting continues to be difficult with high-yielding crops in unpredictable weather conditions. In these areas, extra harvesting capacity, in addition to specific wilting machines, would be necessary to consistently improve wilting performance. In practice wilting is carried out opportunistically in these areas, when weather conditions allow, with minimal investment in equipment. Anecdotal evidence suggests that the inability of swath tedding and raking equipment, working in high yielding crops, to present even swaths for harvesting results in a slower harvesting rate. There is scope to develop improved harvester and baler pick-up systems that incorporate active crop presentation components to improve the evenness of crop feeding to the chopping or baling mechanism.

Conventional harvesting

The principles of forage chopping on chopper-harvesters have changed little, with development concentrating on increased output self-propelled machines of up to 500 kW power output. These machines are now technologically mature and are capable of substantial annual and lifetime throughputs. However, the high-speed chopping and largely pneumatic grass delivery systems used are not power efficient. Alternative systems capable of working with 50% of the power input have been researched (Knight, 1984). Low speed chopping mechanisms and mechanical grass delivery systems such as those used in pick-up wagons are more power efficient (Tremblay *et al.*, 1991). The popularity of the pick-up wagon system has been in decline as contractors preferred continuous harvesting systems of more robust construction. However there is renewed interest in the system with high-output, more robust pick-up wagons teamed with high-powered tractors. In a recent trial in Northern Ireland, a wagon system was shown to be twice as fuel-efficient and more labour efficient than a conventional self-propelled harvesting system (Frost & Binnie, 2005). If energy prices continue to increase and if CO_2 reduction policies are implemented, the need to adopt fuel-efficient systems will increase.

The pick-up wagon harvesting system involves two separate components: crop pick-up/chop and transport. The crop pick-up and chopping mechanism is idle during the transport part of the cycle. There would be benefit in developing a system with the low power requirement of the wagon coupled with the operation efficiency of a system with separate and continuously working harvesting and transport elements. An experimental system developed in the 1980s which used slow speed slicing and mechanical grass delivery, reduced power requirement by up to 62% compared to a conventional harvester (Knight, 1984). This concept is worthy of further development.

Baled silage

Baled silage, which is now a well-established conservation system, is characterised by its unique individual-package storage system. The characteristics of typical bales and bale wrapping practice in Ireland are outlined in Table 2. Until recently baling and wrapping were carried out by separate machines, thereby allowing wrapping to be carried out either in the field or following transport, at the storage site. Recently combined baler/wrapper units that can only wrap in the field, have become common.

Compared to clamp silage, baled silage often has a more restricted fermentation (e.g. Jones & Fychan, 2002), but the feeding value of the resultant silage is usually similar (O'Kiely et al., 1998). Research on baled silage has focused on bale wrapping as the implications of an imperfect seal are more far-reaching than with conventional silage for a number of reasons:

- For a given quantity of silage stored, the surface area in contact with film in baled silage is typically 6 to 8 times that of conventional silage.
- With bales, 50% of the silage stored is within 12 cm of the polythene film whereas with conventional silage less than 10% of the volume would be within this distance.
- The normal thickness of stretched film on baled silage is 70 μm (4 layers), compared to 250 μm for a double-sheeted silage clamp.
- The porosity of baled silage is usually greater than clamp silage.
- Bales are allowed to drop to the ground and are handled after wrapping, exposing the polythene to a greater risk of damage.
- The grass ensiled in bales can often be stemmy, dry and is not lacerated and consequently presents a greater polythene puncture challenge.

Failure to maintain anaerobic conditions will result in aerobic deterioration. Survey results indicate that most Irish farms (87%) have bales with some mould growth (O'Kiely et al., 1998; O'Brien et al., 2004). In Norway the principal moulds found in baled silage were *Penicillium, Aspergillus, Mucor, Rhizopus, Geotrichum* and *Byssochlamys* (Skaar, 1996). The fungal growths found in Ireland were predominantly *Penicillium, yeasts, Geotrichum* and *Schizophyllum* species (O'Brien et al., 2004). Mould growth causes deterioration in the nutritive value of the silage, as well as being a direct source of potential health challenges through the production of spores and mycotoxins.

Table 2 Characteristics of round[1] bales and wrapping of grass for silage

	Typical	Range
Nominal bale size (m)	1.25 x 1.25	-
Bale weight (kg)	650	350 – 1000
Dry matter content (g/kg)	300	160 – 700
Porosity (% pore space)	60	50 – 80
Density (dry matter kg/m^3)	130	90-200
Film width (mm)	750	250[2]-750
Film thickness (μm)	25	12-30
Pre-stretch on wrapper (%)	70	25-70
No. of layers	4	4 – 8
Film weight per bale (g)	850	500 -1700

[1]Various sizes of large rectangular (e.g. 0.8 m x 0.8 m x 1.2 m) and small rectangular bales are also wrapped
[2] 250 mm only used on small rectangular bale wrappers

Research on baled silage has sensibly been focused on the creation and maintenance of anaerobic conditions. In particular the stretch film applied to bales and its effect on the preservation of silage has been the subject of much study.

Film cover

The quantity of film applied to bales, usually expressed as the number of layers, has a marked effect on the cost of silage production and is consequently the subject of evaluation. In Scandinavia, the application of six layers of 25 µm polythene film resulted in less mould than where four layers were applied (Lingvall, 1995). In the same climatic region, a significant reduction in mould growth on high dry matter forage (564 g/kg) was obtained when eight layers of film were used compared to six (Jacobsson *et al.*, 2002). Similarly Heikkila *et al.* (2002) showed less mould growth with six layers of film compared to four.

In more temperate conditions, such as in Ireland, increasing the level of cover from two to four to six layers of film progressively reduced mould levels from 21.5 to 1.7 to 0.7% respectively of the surface area (Forristal *et al.*, 1999). Carbon dioxide profiles in the bale are a good index of the integrity of seal achieved and this trial indicated that the use of more layers of film resulted in better retention of CO_2 levels. These results indicate that in these conditions a minimum of four layers of film were required. A recent study in the UK by Harrison *et al.* (2004) with four, six and eight layers showed that mould was reduced by increasing the number of covering layers.

Film colour

Black film is the least expensive to manufacture as the UV inhibitor (carbon black) does not interfere with the other properties of the film. The polythene film is technically permeable to gases – but at very low gas transmission rates. Dark films absorb heat and the consequent rise in film temperature increases the permeability of the individual film layers (Möller *et al.*, 1999). In temperate regions, film colour was not shown to influence baled silage quality or mould development (Forristal *et al.*, 1999; Harrison *et al.*, 2004). In mini silos where white, green and black films were used in natural and artificial light conditions, silage quality was not affected even where film temperature differences of up to 16°C (black: 37.6°C, white: 21.6°C) were recorded (Snell *et al.*, 2003). Countries with high sunshine levels generally use white or light-coloured films which reduce film temperature and heat transfer.

Stretch level

The tension created by stretching film during the application process, typically to 1.7 or 1.5 times its original length, is essential to ensure that the film remains tight on the bale during the storage period. A trial that assessed the effect of three different stretch levels (1.4, 1.7 and 2.1 times original length) on silage composition, mould growth and gas composition showed no significant effect of stretch level (Forristal *et al.*, 2000). The effect of stretching on the films performance is complex. In laboratory trials, stretching has been shown to decrease the permeability coefficient (i.e. less permeable per unit of thickness), however it may also adversely affect the mechanical properties of the film (Laffin *et al.*, 2005).

Film type

A thinner (12-14 µm) polythene film that is pre-stretched in production has been evaluated (Forristal *et al.*, 2002). There was relatively little difference in performance between the thin film and a conventional film, however at certain points in the storage period, the thin film had a poorer gas profile indicating possible air entry. A stronger stretch film has been shown to give better sealing and less fungal growth in the silage compared to conventional film (Jacobsson *et al.*, 2002).

Film damage

Polythene film can be damaged during wrapping in the field, or while the wrapped bale is being transported to, or in, storage. Machines, stubble and wildlife are sources of damage. The effect of relatively small holes in the film on silage preservation has been shown to be quite significant. A study of the effects of damage (McNamara *et al.*, 2002a) showed damaged film to allow greater levels of surface mould, more rotted silage, and more inedible silage.

A detailed study of the prevention of damage by birds to the film surrounding bales highlighted practical control strategies (McNamara *et al.*, 2002b). In the field, the most effective strategy was to remove bales from the mown field before wrapping. While the use of painted eye designs and red or transparent films had some deterrent properties, they were not completely effective. Chemical repellents had little effect. For season-long storage, the use of nets or closely spaced (0.5 m) monofilament lines placed 1 m above and to the side of the bales, were the only truly effective protection strategies.

Polythene film requirements and research needs

Because of the high incidence of mould, the current systems used for baled silage on farms can only be considered partly satisfactory. There is a need to devise improved wrapping methods which give better sealing. The move towards film standards similar to the 'P' mark in Sweden should remove inferior films from the market. However, there is a need to develop stretch film with improved qualities and to devise wrapping methodologies that give improved and more robust sealing compared to what is commonly used today.

Harvesting logistics

The logistics of harvesting influences the efficiency of utilisation of labour and machinery. Logistics include the selection of machine types and capacities and the organisation of the field operation to optimise the utilisation of all resources. Logistics can be examined on a single farm, or on a group of farms harvested by a single contractor, or on all farms in a region where many harvesting units operate. The aim should be to determine the optimum mechanisation supply to ensure satisfactory supply of forage. There has been relatively little formal research in this area. Ward *et al.* (1986) examined the seasonal capacity of harvester systems and calculated the opportunity cost for crop digestibility losses. Bernhardt *et al.* (2004) showed the importance of matching transport capacity to harvesting capacity for a range of dairy farm sizes in two regions of Germany. The effect of field size and transport distance on work rate, labour requirement and costs were calculated for baled silage (Wagner & Seufert, 2000). More information is needed to support mechanisation selection decisions including:
- comprehensive machine performance data for all crops and conditions, and support information such as trailer packing density for various forages
- crop growth models indicating changes with time in yield, ensilability and feed value
- regional information such as field size, farm harvest area, transport distance, feed quantity and quality requirement and weather patterns

A more formal approach to the subject would determine optimal logistical strategies at farm and at regional level.

Mechanisation and soil effects

As harvesting output has increased, the weight of the field machinery used has also increased with many of today's machines exerting individual axle loads in excess of 6 tonnes. Increased axle loading increases the risk of soil damage with possible effects on crop yield. Much of the early work studying the impact of traffic on crop performance was with cultivated crops. Early work on grassland assessed the impact of wheel traffic on grass, with first harvest yield reductions of 13 to 33% recorded directly in the wheeled area (Frost, 1988). In a complete harvesting system trial where the cumulative effect of traffic was examined, the use of low ground pressure tyres resulted in yield increases of between 9 and 16% in annual grass yield depending on site (Table 3) (Fortune *et al.*, 1995). This research showed that where grass was harvested without applying traffic on a low bearing capacity soil, annual yield increases of up to 32% were recorded. The crop response was accompanied by soil structure changes and reduced uptake of nitrogen.

The mechanisation changes required to achieve a significant reduction in ground pressure are considerable. Research on arable soils would also suggest that very heavy axle loads are capable of causing deep compaction even when low ground pressure tyres are fitted (Hakansson & Petelkau, 1994). Machinery developments through larger, heavier and more labour efficient machines are contributing to the compaction problem. Reducing ground pressure will increase machinery cost and as a result will add to the harvesting costs, and this must be considered. It may be necessary to rethink current transport systems which use three to five trailers in the field and on the road. To equip all of these trailers with low ground pressure tyres would be an expensive option. An alternative would be to couple the harvester with a dedicated single low ground pressure trailer which then transfers the load to a fleet of road-going trailers.

Table 3 Annual grass yield and nitrogen removal following three levels of silage harvesting traffic using a three-cut system on two sites (Fortune *et al.*, 1995)

Traffic system	Wet site		Dry site	
	DM yield (kg/ha)	N removed in crop (kg/ha)	DM yield (kg/ha)	N removed in crop (kg/ha)
Conventional	9500	211	12500	326
Low ground pressure	11100	263	13600	364
Zero traffic	13000	327	13700	362

Silage feeding

The feeding of silage utilises significant resources in the form of machine, labour and associated building costs. While the costs attributed to feeding (Table 1) are relatively low, costs can vary significantly with herd size and mechanical system used for feeding. Feeding systems used in different regions and for different animal rearing systems are influenced by a combination of tradition, labour demand, cost and suitability for the feeds being offered.

Mechanisation-based research has focused mainly on the requirements for processing in terms of chop length and grain cracking along with method of presentation including mixing of feeds.

Early work on chop length focused on the need for short chop. A range of chop lengths evaluated by Gordon (1982) with dairy cows and research by O'Kiely and Flynn (1991) with beef cattle showed little benefit from short chopping. As precision chop harvesters produced shorter chop lengths to make the harvested crop easier to handle, concerns about the effects of overly fine chopping were raised. For example, work on lucerne silage indicated that very fine chopping could negatively affect fibre digestion in the rumen (Grant et al., 1990). However, the importance of chop length in rumen function is influenced by forage species with Mertens (1997) showing that grass silage particle size had less impact on chewing than with lucerne.

Where maize silage is harvested with well-developed cobs, the need for grain processing with corn-cracker rollers arises. Animal trials comparing processed with unprocessed maize silage have shown varied results. Processing can increase intake, starch digestion and performance (Bal et al., 2000). However in many trials where processing is compared with different varieties (Moreira et al., 2000) or different crops (Pressinger et al., 1998) the response to processing is variable although generally showing some benefit.

Where forage alone is fed, the method of presentation (e.g. self-feeding or easy feeding) has no significant effect on animal performance provided the feed on offer remains fresh and sufficient feeding space is available. The benefit from mixing forage and concentrate components has been the subject of research for a considerable period with variable responses reported. The reported responses in animal production to total mixed ration feeding (TMR) have been variable. Trials in Northern Ireland have shown a positive response in milk production to TMR in two trials (Gordon et al., 1995; Yan et al., 1998) with no effect in another trial (Agnew et al., 1996). In a beef trial where finishing steers were fed 3.5 kg or 7.0 kg supplementary concentrates daily, mixed or separate feeding gave the same animal performance (Caplis et al., 2003). Overall, other than where very high levels of concentrate supplementation are fed (>50% of diet), mixed feeding of forage and concentrates is unlikely to give an animal performance benefit compared to careful separate feeding of the same feedstuffs.

For most farmers, the attractions of TMR feeding systems are associated with management and labour saving aspects. The ability to weigh, mix and accurately dispense feed-stuffs is particularly beneficial where a variety of feeds are to be fed including combinations of forages (e.g. maize and grass silage) or a number of concentrate feed components. The choice of feeding system coupled with building design can have a significant impact on the time and labour required to feed animals. For example in a study in Ireland, feeding with a mixer wagon required the same time and labour input as separate mechanical feeding of the diet components, but the task was easier with no manual handling (Forristal, 1992).

The labour associated with feeding tasks is dependent on mechanisation, feeding system and housing design and consequently differs among geographical regions being studied. For example in Ireland, a recent study showed that the time taken to feed livestock on beef farms with an average of 93 livestock units exceeded 2 hours per day over the winter period (Leahy et al., 2004). In this work silage feeding accounted for 72% of that time.

There is renewed interest in developing low-cost animal wintering systems based on outdoor systems, or simple housing. Animal performance on these units is good (Hickey *et al.*, 2002) and the selection of appropriate feeding systems is now being considered.

There is a need for a comprehensive evaluation of feeding systems to include costs, labour and integration with house design which would allow farmers to make optimal choices concerning feeding mechanisation on their farm. More research is needed to underpin such evaluations.

Environmental constraints

Silage storage systems present a number of environmental challenges. While effluent production and control systems are well understood, the post-use fate of the large quantities of polythene used with baled and conventional silage is of concern. The baled silage system uses considerably more polythene at 23.5 kg/ha than clamp silage at 4.7 kg/ha (Hamilton *et al.*, 2005). Collection and recycling systems are effective but the current systems may not be the most effective means of handling used polythene. The use of biodegradable polythenes has been considered. However, development for this application is difficult since polythenes must maintain anaerobic conditions in silage up to the time of use (Keller, 2000).

Ensilage for non forage uses

The concept of using crops for energy purposes is not new. Anaerobic digestion of carbon-rich forages is always possible (Plochl & Heiermann, 2004). If a biogas plant using agricultural crops is to be effective, then the issue of crop storage prior to use must be considered. Ensilage is the obvious choice for many low dry-matter content crops. Efficient ensiling technologies for crops destined for biogas production may need to be developed.

Conclusions

Silage research has largely concentrated on the biological and biochemical aspects of the ensiling process. Engineering and mechanisation technologies are significant inputs which influence the cost and value of silage and which significantly impact on choices made by farmers. Research is needed to underpin the decisions concerning mechanisation and engineering that must be made at farm level, and to exploit new technologies which are becoming available. Developments in the areas of baled silage research and sensing technologies are continuing and offer scope to improve conservation efficiency. Energy efficiency must again become a topic in forage conservation research. In a broad context, energy efficiency applies to all aspects of silage conservation from production to feeding.

References

Agnew, K.W., C.S. Mayne, & J.G. Doherty (1996). An examination of the effect of method and level of concentrate feeding on milk production in dairy cows offered a grass silage-based diet. *Animal Science*, 63, 21-31.
Anon, (1999) Anbau-Zweischeiben-Dungerstreur (Fertiliser spreader test) Bogballe EX trend. Danish Institute of Agricultural Sciences, Bygholm, Denmark. 35 pp.
Auernhammer, H., M. Demmel & P.J.M. Pirro (1996). Lokale ertragsermittlung mit dem feldhacksler (Local field mapping with a forage harvester). *Landtechnik*, 51(3), 152-153.
Bal, M.A., R.D. Shaver, A.G. Jirovec, K.J. Shinners & J.G. Coors (2000). Crop processing and chop length of corn silage: effects on intake, digestion and milk production by dairy cows. *Journal of Dairy Science*, 83, 1264-1273.

Bernhardt, H., M. Kilian & H. Seufert (2004). Silage harvesting operations for growing dairy farms. *Landtechnik,* 59 (1), 40-41.

Binnie, R.C. & J.P. Frost (1996). The effect of rainfall on the drying rate and dry matter loss from intensively conditioned grass. In: *Proceedings of the Irish Grassland and Animal Production Association 22nd meeting,* University College Dublin, 79-80.

Binnie, R.C. & J.P. Frost (2003). Effect of method of slurry application on productivity of grass in silage swards. In: *Proceedings of the Agricultural Research Forum,* 14-15th March 2003, Tullamore, Ireland, p73.

Bosma, A.H. & A.P. Verkaik (1987). Pre-wilted silage within 24 hours. *Landbouwmechanisatie,* 38 (4), 262-265.

Bosma, A.H. (1995). New systems for wilting grass. In: *Proceedings of the 50th Anniversary meeting of the British Grassland Society,* Harrogate, UK, 247-249.

Caplis, J., M.G. Keane & F.P. O'Mara (2003). Comparison of separate and mixed feeding of silage and concentrates for finishing cattle. In: *Proceedings of the Agricultural Research Forum,* 3-4th March 2003, Tullamore, Ireland, p38.

Copeland, T. (1993). Developments in cutting of grass. *Agricultural Engineer,* 48 (2), 38-41.

Crowley, J.G. (1998). Improving the yield and quality of forage maize. End of project report. Teagasc, Oak Park, Carlow, Ireland. 10 pp.

Crowley, J.G. (2005). Effects of variety, sowing date and photodegradable plastic cover on yield and quality of maize silage. End of project report. Teagasc, Oak Park, Carlow, Ireland (in press).

Demmel, M., T. Schwenke, H. Heuwinkel, F. Locher & J. Rottmeier (2002). Yield mapping on pasture – first results. *Landtechnik,* 57(3), 146-147.

Easson, D.L. & W. Fearnehough (2003). The ability of the Ontario heat unit system to model the growth and development of forage maize sown under plastic mulch. *Grass and Forage Science,* 58, 372-384.

Ehlert, D. & S. Kraatz (2004). Mowing by steel band. *Landtechnik,* 59 (2), pp 80-81.

Forristal, P.D. (1992). The performance of mechanised feeding systems on farms. Teagasc, Oak Park Research centre, Carlow Ireland. 15 p.

Forristal, P.D. (1996). The effect of mechanical swath treatments on grass wilting rates. In: *Proceedings of the Irish Grassland and Animal Production Association 22nd meeting,* University College Dublin, 135-136.

Forristal, P.D. & D. Keppel (2001a). The application of harvester mounted forage yield sensing devices. End of project report. Teagasc, Oak Park, Carlow, Ireland. 22pp.

Forristal, P.D. & D. Keppel (2001b). The use of a harvester-mounted forage yield sensing device to control additive application on a forage harvester. In: *Proceedings of the Agricultural Research Forum,* 3-4th September 2001, Tullamore, Ireland, p. 53.

Forristal, P.D., P. O'Kiely & J.J. Lenehan (1999). The influence of the number of layers of film cover and film colour on silage preservation, gas composition and mould growth on big bale silage. In: *Proceedings of XII International Silage Conference,* Uppsala, Sweden, 251-252.

Forristal, P.D., P. O'Kiely & J.J. Lenehan (2000). The influence of the number of layers of film cover and stretch level on silage preservation, gas composition and mould growth on big bale silage. *Irish Journal of Agricultural and Food Research,* 39(3): 467.

Forristal, P.D., P. O'Kiely & J.J. Lenehan (2002). The influence of polythene film type and level of cover on ensiling conditions in baled silage. *Proceedings of the Agricultural Research Forum,* 11-12th March 2002, Tullamore, Ireland, p. 82.

Fortune, R.A., P.D. Forristal & J.T. Douglas (1995). Development and assessment of grass harvesting systems to reduce soil compaction, improve trafficability and slurry disposal in high rainfall areas. EU RTD contract no 8001-CT91-101 final report, 134 p.

Frost, J.P. (1988). Effects on crop yields of machinery traffic and soil loosening: part 1, effects on grass yield of traffic frequency and date of loosening. *Journal of Agricultural Engineering Research,* 39, 301-312.

Frost, J.P. (1995). Mechanisation to achieve rapid wilting. In: Rapid Wilt Silage. DARD Northern Ireland, 2 p.

Frost, J.P. & R.C. Binnie (2005). A comparison of two systems for harvesting herbage for silage. In: *Proceedings of the Agricultural Research Forum,* 14-15th March 2005, Tullamore, Ireland, p. 24.

Gordon, F.J. (1982) The effects of degree of chopping grass for silage and method of concentrate allocation on the performance of dairy cows. *Grass and Forage Science,* 37, (1) 59-65.

Gordon, F.J., D.C. Patterson, T. Yan, M.G. Porter, C.S. Mayne & E.F. Unsworth (1995). The influence of genetic index for milk production on the response to complete diet feeding and the utilization of energy and nitrogen. *Animal Science,* 61, 199-210.

Grant, R.J., V.F. Colenbrander & J.L. Albright (1990). Effect of particle size of forage and rumen cannulation upon chewing activity and laterality in dairy cows. *Journal of Dairy Science,* 73 (11) 3158-3164.

Hakansson, I. & H. Petelkau (1994). Benefits of limited axle load. In: Soane, B.D., Van Ouwerkerk, C. (eds.) Soil Compaction in Crop Production. Elsevier Sciences, 479-499.

Hamilton, W.J., P. O'Kiely & P.D. Forristal (2005). Plastic film use on Irish farms. In: *Proceedings of the Agricultural Research Forum,* 14-15th March 2005, Tullamore, Ireland, p. 77.

Harrison, S., A.K. Phipps, A.K. Jones & J. Siviter (2004). An evaluation of the effect of colour and number of layers of plastic used for big bale grass silage on the quality of grass silage. Report No. 220. Centre for Dairy Research, The University of Reading, Reading, UK, 15 pp.

Heikkila, T., S. Jaakkola, A. Saarisalo, A. Suokannas & J. Helminen (2002). Effects of wilting time, silage additive, and plastic layers on the quality of round bale silage. In: *Proceedings of the XIIIth International Silage Conference,* 11-13 September 2002, Auchincruive, Scotland. SAC. 158-160.

Hickey, M.C., P. French & J. Grant (2002). Out-wintering pads for finishing beef cattle: animal production and welfare. *Animal Science,* 75 (3), 447-458.

Hintz, R.W., R.G. Koegel, T.J. Kraus & D.R. Mertens (1999). Mechanical maceration of alfalfa. *Journal of Animal Science,* 77 (1), 187-193.

Hofstee, J.W. & W. Huisman (1990). Handling and spreading of fertilisers part 1: Physical properties of fertilisers in relation to particle motion. *Journal of Agricultural Engineering Research,* 47, 213-234.

Jacobsson, F., P. Lingvall & S.O. Jacobsson (2002). The influence of film stretch quality, number of layers and type of baler on bale density, silage preservation, mould growth and nutrient losses on big bale silage. In: *Proceedings of the XIIIth International Silage Conference,* 11-13 September 2002, Auchincruive, Scotland. SAC, 164-166.

Jones, R. & A.R. Fychan (2002). Effect of ensiling method on the quality of red clover and lucerne silage. In: *Proceedings of the XIIIth International Silage Conference,* 11-13 September 2002, Auchincruive, Scotland. SAC, 104-105.

Keane, G.P., J. Kelly, S. Lordan & K. Kelly (2003). Agronomic factors affecting the yield and quality of forage maize in Ireland: effect of plastic film system and seeding rate. *Grass and Forage Science,* 58, 362-371.

Keller, A. (2000). Biodegradable films for silage bales: basically possible. *Agrarforschung,* 7 (4), 164-169.

Knight, A.C. (1984). Forage harvester design including developments for low energy use. In: *Proceedings of British Grassland Society Occasional Symposium No. 17,* 43-50.

Kormann, G. & H. Auernhammer (2002). Continuous moisture measurements in self-propelled forage harvesters. *Landtechnik,* 57 (5), 264-265.

Laffin, C., G.M. McNally, P.D. Forristal, P. O'Kiely & C.M. Small (2005). Uni-axially stretching of LDPE/LLDPE silage wrap films: gas permeation properties. In: *Proceedings of the Agricultural Research Forum,* 14-15th March 2005, Tullamore, Ireland, p. 78.

Leahy, H., E.G. Riordan & D.J. Ruane (2004). Labour use on Irish Suckler farms over winter. In: *Proceedings of the Agricultural Research Forum,* 1-2nd March 2004, Tullamore, Ireland, p. 95.

Lingvall, P. (1995). The Balewrapping Handbook. Trioplast AB, Sweden. 52 pp.

Mertens, D.R. (1997). Creating a system for meeting the fibre requirements of dairy cows. *Journal of Dairy Science,* 80 (7), 1463-1481.

Möller, K., T. Klaesson & P. Lingval (1999). Correlation between colour and temperature of LDPE stretch film used in silage bales. *Proceedings XII International Silage Conference,* Uppsala, Sweden 251-252.

McNamara, K., P. O'Kiely, J. Whelan, P.D. Forristal & J.J. Lenehan (2002a). Simulated bird damage to the plastic stretch-film surrounding baled silage and its effects on conservation characteristics. *Irish Journal of Agricultural and Food Research,* 41, 29-41.

McNamara, K., P. O'Kiely, J. Whelan, P.D. Forristal & J.J. Lenehan (2002b). Preventing bird damage to wrapped baled silage during short- and long-term storage. *Wildlife Society Bulletin,* 30 (3), 809-815.

Moreira, V.R., L.D. Satter & M.I. Endres (2000). Effect of two corn hybrids with or without kernel processing on milk production. *Journal of Animal Science,* 78, supplement 1, p. 112.

Muck, R.E. & O. O'Kiely (2002). New technologies for ensiling. In: *Proceedings of the XIIIth International Silage Conference,* 11-13 September 2002, Auchincruive, Scotland, SAC, 334-342.

O'Brien, M., P. O'Kiely, P.D. Forristal & H. Fuller (2004). Pilot survey to establish the extent and the identity of visible fungi on baled silage. In: *Proceedings of the Agricultural Research Forum,* 1-2nd March 2004, Tullamore, Ireland, p. 50.

O'Dogherty, M.J. & G.E. Gale (1986). Laboratory studies of the cutting of grass stems. *Journal of Agricultural Engineering Research,* 35 (2), 115-129.

O'Kiely, P. & A.V. Flynn (1991). Comparison of unwilted grass silages made using the pick-up wagon and the double-chop harvester systems. *Canadian Agricultural Engineering,* 33 (1), 119-125.

O'Kiely, P., O.T. Carton & J.J. Lenehan (1994). Effect of time, method and rate of slurry application to grassland grown for silage. FAO network on animal waste utilization, 7th consultation. Bad Zwischenahn, Germany. 11pp.

O'Kiely, P., P.D. Forristal & J.J. Lenehan (1998). Big bale and precision chop silage systems: conservation characteristics and silage nutritive feed value for beef cattle. In: *Proceedings of the Agricultural Research Forum,* 28-29th March 1996, University College Dublin, 33-34.

O'Kiely, P., A. Moloney, L. Killen & A. Shannon (1997). A computer programme to calculate the cost of providing ruminants with home-produced feedstuffs. *Computers and electronics in agriculture,* 19(1), 23-26.

Paul, C., M. Rode & U. Feuersteib (2000). From laboratory to harvester: forage analysis by NIRS doide array instrumentation. *Grassland Science in Europe*, 5, 259-261.

Plochl, M. & M. Heiermann (2004). From field to fuel cell: a strategy for biogas farming. Agricultural Engineering 2004, Leuven Belgium, 12-16 September 2004, 8pp.

Preissinger, W., F.J. Schwarz & M. Kirchgessner (1998). The effect of size reduction of corn silage on feed intake, milk production and milk composition of dairy cows. Arch Tierernahr. 51 (4) 327-329.

Savoie, P., P. Lemire & R. Theriault (2002). Evaluation of five sensors to estimate mass flow rate and moisture of grass in a forage harvester. *Applied Engineering in Agriculture*, 18 (4), 389-397.

Skaar, I. (1996). *Mycological survey and characterisation of the mycobiota of big bale grass silage in Norway.* PhD thesis. Norwegian College of Veterinary Medicine, Norway, 101 pp.

Snell, H.G.J., C. Oberndorfer, W. Lucke & H.F.A. van den Weghe (2003). Effects of polyethylene colour and thickness on grass silage quality. *Grass and Forage Science,* 58 (3), 239-248.

Sogard, H.T. & P. Kierkegaard (1994). Yield reduction from uneven fertiliser distribution. Transactions of the ASAE, 37 (6), 1749-1752.

Tremblay, D., P. Savoie & R. Theriault (1991). Self loading wagon power requirements for coarse chopping forage. *Canadian Agricultural Engineering,* 33, 31-38.

Wagner, A. & H. Seufert (2000). The proportion of transport and transport costs in grassland farming in the middle-German uplands. *Landtechnik,* 55 (5), 342-343.

Ward, S.M., P.B. McNulty & M.B. Cunney(1986). Determining least cost silage mechanisation systems. *Irish Journal of Agricultural Research,* 25 (2), 191-196.

Wild, K. & H. Auernhammer (1999). A weighing system for local yield monitoring of forage crops in round balers. *Computers and Electronics in Agriculture,* 23 (2), 119-132.

Wollring, J., S. Reusch & C. Karlsson (1998). Variable nitrogen application based on crop sensing. In: *Proceedings of the Fertiliser Society*. Fertiliser Society, UK. no. 423, 28 pp.

Yan, T., D.C. Patterson & F.J. Gordon (1998). The effect of two methods of feeding the concentrate supplement to dairy cows of high genetic merit. *Animal Science,* 67, 395-403.

Silage production from tropical forages

L.G. Nussio
Universidade de São Paulo, Escola Superior de Agricultura "Luiz de Queiroz",
Depto. de Zootecnia, Av. Pádua Dias, 11. 13418-900. Piracicaba, São Paulo, Brazil.
Email: nussio@esalq.usp.br

Key Points

1. The determination of overall DM recovery is important in tropical grass silage systems
2. Silage fermentation profile and aerobic stability: trends and additional effects
3. Combinations of chemical and microbial additives might be useful for control of losses
4. Fermentation products should be considered in order to better predict animal performance
5. A critical points database would be a helpful management tool for the development of a set of HACCP principles for the production and utilisation of tropical grass silages

Keywords: tropical grass silage, effluent, gases, silo losses, aerobic stability

Introduction

In the tropics, silage production supplies feed for the dry (winter) season, when forage growth rates do not match the nutritional demands of the animal. In the warm season, the yield of tropical forages (mainly grasses) is high but the high moisture content and low soluble carbohydrate pool may limit the uptake of ensilage as a conservation technique for such forages. The success of the ensiling process is dependent on many factors, including some associated with forage quality and feed safety. Uncontrolled growth of microorganisms leads to heating of silage with consequent nutritional losses and potential risks to animal health. Occasionally, recommended management practices related to animal health issues, such as the application of silage additives or mechanical processing, are used in the field (Chin, 2002).

Silages made from tropical forages are prone to increased losses across different stages of the ensiling process. Such losses decrease the net output of edible silage and are more marked in legume silages. To maximise the overall efficiency of silage production and utilisation, all sources of loss should be identified and their consequences quantified. Historically however, fermentation patterns and in-silo losses have been more extensively studied than field losses (Nussio *et al.*, 2000; Balsalobre *et al.*, 2001a; Reis & Coan, 2001). This review suggests a rationale for integrated systems of tropical forage silage production, mainly focused on the use of the warm-season C4 grasses specially studied in South America.

Harvesting related losses

The initial losses in the ensiling process occur during forage harvesting and chopping. Igarasi (2002) used a pull type double-chop forage harvester fitted with flail knives and cutter head to ensile unwilted "Tanzania" guineagrass (*Panicum maximum* Jacq cv Tanzania) and found harvesting losses of 3.2% and 5.3% of the total available forage in winter and summer, respectively. Wilting the forage for 5 hours increased harvesting losses to 12.2% and 20% during winter and summer respectively thereby calling into question the benefits of forage wilting. To evaluate ensiling losses with Tifton 85 (*Cynodon dactylon*), Castro (2002) used a self propelled precision-chop harvester provided with multi-rotary drum mower. The forage was harvested at a dry matter of 25% and was allowed to wilt to 35, 45, 55 and 65% DM. Harvesting losses were reduced from 6 to 3% with the increase in forage DM content.

Opposing trends between harvesting losses and wilting losses might be explained by forage species, but are mainly due to differences in the design of the harvesting equipment.

In silo losses

Moisture control and substrate availability for fermentation

Techniques that reduce water activity, such as wilting or the use of absorbent substrates as additives, may promote absorption of free water. Since tropical grasses have a high moisture content and a low soluble carbohydrate pool at harvesting time (Vilela, 1998), the addition of soluble carbohydrate sources may reduce losses from undesired fermentation. Recent data from Igarasi (2002) indicated lower water activity in tropical grasses than in temperate grasses at the same moisture level, possibly as a result of a higher ionic charge in the cell content. Accordingly, it might be possible to successfully control undesirable microorganisms, such as Clostridium, even when DM contents are slightly below 30%.

Some absorbents, such as finely ground grain or citrus pulp, may simultaneously increase both the soluble sugar concentrations and dry matter content of the ensiled product. However, these absorbents act differently to control the free water content of the product. Even though the use of absorbents with higher NDF content reduces silage bulk density, the water retention capacity of the silage may be increased (Jones & Jones, 1996; Giger-Reverdin, 2000).

Aguiar *et al*. (2001) cited by Sollenberger *et al*. (2004) observed lower levels of ammonia-N with a 10% addition of pelleted citrus pulp to ensiled "Tanzania" guineagrass, when compared to the control silage. Surplus soluble carbohydrate associated with the higher DM content probably lowered the activity of proteolytic enzymes due to the rapid pH drop in the silage. A lower moisture content, resulting from citrus pulp addition, also improves the fermentation pattern, as observed by Balsalobre *et al*. (2001a). However, in silage wilted to a DM content similar to that observed with the addition of citrus pulp, the pH was higher, suggesting that the addition of soluble sugars might have promoted the additional drop in pH. Igarasi (2002) ensiled "Tanzania" guineagrass with citrus pulp included at between 5 and 10% and noted better fermentative characteristics (pH, ammonia-N), higher DM and improved TDN (digestible energy) recovery as a result of lower losses of gas and effluent. These quality characteristics contrasted with a lower aerobic stability after opening of the silo and with an increase in TDN relative cost, mainly for summer harvested forage.

Evangelista *et al*. (2001) included wheat meal or pelleted citrus pulp at 0, 5, 10 or 15% when ensiling Coastal Bermuda grass (*Cynodon spp*) at 7 or 9 weeks of vegetative re-growth and observed no additive effects on silage pH values at 7 weeks of growth. However, all wheat meal and citrus pulp inclusion levels resulted in a more rapid pH drop in forages harvested at 9 weeks re-growth. As expected, silage DM content increased in line with additive inclusion rate. This was more noticeable in forage harvested at 9 weeks re-growth. With both harvesting intervals, increasing additions of wilted sugarcane (sacharina) or finely ground maize meal were associated with an increase in silage ammonia-N content, indicating more protein degradation with both additives (Lima *et al*., 2001).

Pedreira *et al*. (2001) observed higher ammonia-N concentrations in silages with unaltered moisture contents probably due to a greater degree of proteolysis by plant enzymes or through *Clostridium* activity. Addition of pelleted citrus pulp affected the pH and led to lower ammonia-N values, presumably because of lower *Clostridium* activity, even though the crude protein content and the other cell wall constituents were reduced as a result of the low NDF, ADF and CP concentrations in citrus pulp. Vilela *et al.* (2001) wilted hybrid Paraíso (Elephantgrass and Pearl millet) for 0, 6 and 12 hours, and found lower ammonia-N contents in the wilted silages. However, wilting may reduce soluble carbohydrates and consequently lactic acid, because of cell respiratory activity. Wilting may also increase silage ash content due to an increase in soil contamination from raking. These observations suggest a combined effect of a loss of cell contents through silage effluent losses and carbohydrate disappearance during the wilting process, resulting in lower energy availability and increased ash concentration (Balsalobre *et al.*, 2001a).

Souza *et al.* (2001) evaluated elephantgrass silage of 14.5% DM ensiled with ground coffee hulls as an absorbent and observed a lower effluent yield and pH values close to 3.9. Another benefit was the maintenance of silage CP content close to that of the fresh forage. Many additives reduce the CP content through a dilution effect. Quadros *et al.* (2003) added increased amounts of coffee hulls to elephantgrass (up to 20% on a fresh weight basis) and noticed better silage fermentation profiles and silage DM digestibility when coffee hulls were added at rates from 5% to 10%. Ferrari & Lavezzo (2001) used cassava meal as an additive in elephantgrass silages and found increased DM and total soluble carbohydrate contents. However, ammonia-N and butyric acid concentrations suggested that these silages were inferior. Fermentation of "Tanzania" guineagrass silage was changed by citrus pulp addition and there was an interaction with particle size reduction (Balsalobre *et al.*, 2001b).

Bulk density: particle size and packing

Among the factors that affect silage bulk density are: weight and pressure applied at packing, packing duration, layer thickness between loads, filling rate, forage DM content and mean particle size (Ruppel *et al.*, 1995; Mayne, 1999; Holmes & Muck, 1999; Balsalobre *et al.*, 2001a). Silage bulk density determines the amount of residual gas in voids in the forage mass. In situations where the reduction of particle size is limited by equipment design, it represents the main restrictive factor to an increase in silage bulk density. The study of Ruppel (1992) cited by Holmes & Muck (1999), showed DM losses of 202 and 100 g/kg for silage bulk densities of 160 and 360 kg/MS/m³, respectively, demonstrating the benefits in terms of loss reduction that can be achieved with increased bulk density. In a field survey by Igarasi (2002), the average bulk densities of *Panicum* and *Brachiaria* silages on farms was 141.9 kg DM/m³ (from 86.7 to 230 kg MS/m³) with 93% of the samples below 200 kg DM/m³ and 21% below 100 kg DM/m³. Evaluating "Tanzania" guineagrass silages, Igarasi (2002) observed mean silage densities around 150 kg DM/m³ in samples of about 25% DM and with satisfactory fermentation. The maximum estimated bulk density of 159.5 kg DM/m³ was observed with a wilted forage containing 33.3% DM. Particle size reduction may improve fermentation due to better packing and increased surface contact area between substrate and microorganism, with a resultant greater access to cellular contents. McDonald *et al.* (1991) pointed out that when particle size is smaller than 20-30 mm, positive effects on the availability of soluble carbohydrates may be noticed and, consequently, lactic acid bacteria may be stimulated. However, the literature is still unclear about the benefits of reducing particle size on grass silage fermentation. According to Mayne (1999), the positive effects of particle size reduction on the fermentation process were more generally observed in higher DM content forages.

Forage chopping can alter silage fermentation patterns through altering the extent of plant tissue damage. Electrical conductivity is a useful indicator of the extent of mechanical processing (Kraus *et al.*, 1997) and can help assess the degree of cell disruption and variations in cell content exchange due to chopping and shredding of forage by equipment of different design. Cell wall rupture may improve homogeneity, creating a liquid film surrounding the grass particles and can lead to more uniform growth conditions for the lactic acid producing bacteria (Pauly, 1999). In spite of this, Pauly (1999) questioned the benefits of mechanical processing for the loss of cell contents in effluent and in facilitating fermentation. The author implied that possible advantages might be due to the establishment of anaerobic conditions in a shorter period of time. Additionally, larger particle sizes could lead to a slower pH drop and higher DM losses, mainly of water soluble carbohydrates and protein (Woolford, 1972). Mari (2003) reported a higher bulk density with smaller particles in "palisadegrass" (*Brachiaria brizantha* A. Rich, Stapf) silage. Igarasi (2002) studied "Tanzania" guineagrass silages and observed that electrical conductivity increased from 1694 to 1823 mS/cm with a reduction in particle size, the effect being more noticeable in wilted forages (from 1774 to 1985 mS/cm). Smaller particle size increased silage bulk density, improved aerobic stability after unloading and produced a faster pH drop and gave higher DM recovery. A reduction in particle size did not alter gaseous or effluent yield and left the recovery and relative cost of TDN unchanged.

Forage particle size reduction may be an alternative way in which to minimise Clostridium fermentation, by promoting greater packing density and closer substrate contact with the fermenting bacteria, leading to a higher lactate yield and a faster pH drop. However, in silages with low DM content, particle size reduction may increase water activity and effluent losses. In this way, it could result in the same overall DM loss, but through a different mechanism, demonstrating mutual, negative, relationships between effluent and gaseous losses. However, in higher DM content silage, total losses are diminished as a result of the higher osmotic pressure associated with the significant reduction in water activity. In this situation, by promoting particle size reduction, effluent yield is minimised (Balsalobre *et al.*, 2001b).

Effluent

There are some models that attempt to quantify silage effluent yield (Haigh, 1999). However, they use only forage DM as a factor in their prediction and do not consider other factors such as silo type and size, packing density, chopping type and use of additives. According to these models, DM values of between 28.5 and 30% would be necessary to eliminate effluent yield. From Figure 1, the exponential regression equation to predict effluent yield from "Tanzania" guineagrass silage revealed a trend for lower effluent (38.3 to 9.3 l/t) when DM increased from 20 to 30% (Igarasi, 2002). Even though the model proposed by Haigh (1999) predicts effluent yield based on DM content in a negative correlation, only in few cases the trend was significant. . As a result, it can be assumed that other parameters besides forage DM content may be correlated with effluent losses. More intensive packing (for higher silage bulk density) might increase effluent yield, dependent on plant DM. Although effluent losses are frequently measured, they are not always the most important source of loss. The composition of the effluent allows a more realistic evaluation and varies with DM content and with the type of additive used. In perennial ryegrass silages with a DM content between 16 and 19.5%, effluent DM contents were between 7 and 8.5% (Jones *et al.*, 1990). Evaluating elephantgrass (*Pennisetum purpureum*) silage with low DM content (13%) and made under different packing densities (356 to 791 kg/m^3), Loures (2000) observed that the effluent yield and DM content were both increased with greater silage packing pressure.

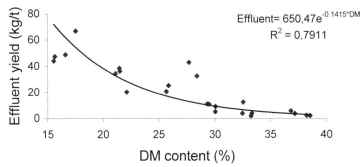

$$\text{Effluent} = 650{,}47e^{-0.1415 \cdot DM}$$
$$R^2 = 0{,}7911$$

Figure 1 Prediction equation for effluent yield (kg/t of fresh forage) based on DM content in Tanzania grass silage; Source: IGARASI (2002)

Both wilting and addition of citrus pulp to "Tanzania" guineagrass, harvested and re-chopped at three different particle sizes, resulted in lower effluent losses compared to a control silage. However, the effect due to wilting was greater than that of citrus pulp addition, suggesting that even under similar DM, the water activity and probably the location of this water in the plant tissue is important for effluent yield (Balsalobre, 2001b). According to Aguiar *et al.* (2001) cited by Sollenberger *et al.* (2004), the addition of citrus pulp to tropical grass prior to ensilage or wilting of the forage reduced effluent yield from 40-50 l/t to less than 10 l/t.

In general, as particle size is reduced, a better fermentation pattern is obtained, and there are lower total DM losses, despite a trend towards greater effluent production. The ratio of losses through effluent in relation to total losses is increased, as particle size becomes smaller.

Gases

Wilted silages made from tropical grasses require careful control of packing and gaseous losses in order to obtain a satisfactory fermentation profile which is itself strongly affected by the wilting period. In general, increasing DM content increases silage pH and decreases acetic and propionic acid yield and produces lower ammonia-N (Castro, 2002). For wet silages, excessive moisture at ensiling may lead to even higher DM losses in the form of gases, often arising from undesirable fermentation by *Clostridium*. In this scenario, energy losses are also high (McDonald *et al.*, 1991). Heterolactic bacteria also produce CO_2 and alcohols and these also contribute significantly to the greater DM losses from wet silages.

In "Tanzania" guineagrass silages of low DM content (20%), there was a significant reduction in gaseous losses as the particle size was reduced (as evaluated by the percentage retained on a 1.9 cm sieve). However, a compensatory increase in effluent DM losses was also observed, resulting in unchanged overall losses averaging 27% (Balsalobre *et al.*, 2001b). In such silages, made with grass of similar mean chop length, addition of 5 and 10% of pelleted citrus pulp raised the DM contents to 23 and 28%, respectively, but still allowed effluent losses. However, in both cases there were significant reductions in gaseous losses, as well in the overall DM. In the 28% DM silage (containing 10% citrus pulp), mean particle size fell from 80 to 10% of material retained by an 1.9 cm sieve and resulted in a drop in overall DM losses, from 18.18% to 13.75% (Balsalobre *et al.*, 2001b). The most dramatic effects of smaller particle size were observed in low moisture silages. Nevertheless, the silage DM increase associated with an increase in soluble carbohydrate supply (through addition of pelleted citrus pulp) showed a greater effect on the control of losses than did particle size reduction alone.

Although overall DM loss was not significantly reduced with smaller chop length in wet silages, a reduction in mean particle size should still be aimed for in tropical grass silages because smaller particle size allows greater bulk density and lower unit cost and minimises haulage and storage costs. Additionally, smaller particle size promotes smaller physical losses during unloading and feeding. However, the most important effects of smaller particle size are related to the higher potential intake by animals (Balsalobre *et al.*, 2001a, Paziani, 2004).

Additives and inoculants

Acids, salts, fermentable carbohydrates, lactic acid bacterial cultures and enzymes can all be used to control silage fermentation. Recent reviews of additives (McDonald *et al.*, 1991; Lindgren, 1999; Weinberg & Muck, 1996; Vilela, 1998) have characterised their different mode of actions and emphasised that different forages require different types of additive.

Coan *et al.* (2001) observed that the use of enzymatic-bacterial inoculant did not improve the quality, fermentation and nutritional characteristics of guineagrass (*Pannicum maximum)* silage, regardless of forage variety (Tanzania or Mombaça) or regrowth stage (45 or 60 days), and they therefore questioned the benefits of using inoculants. Nussio *et al.* (2001) evaluated the use of enzymatic-bacterial inoculant in Tifton 85 (*Cynodon dactylon*) silage of increasing DM content (25 to 65%) and observed positive results from the use of inoculant only in higher DM silage (>45%), mainly due to a more rapid fall in pH. Temperatures of inoculated silages were lower up to 8 days from silo sealing. There was no control of proteolysis with the use of inoculants in low DM silage (25%), as indicated by higher ammonia-N concentrations, but there was improved aerobic stability in inoculated silages of intermediate DM content (45%).

Enzymes that degrade plant cell walls have been suggested as silage additives based on the proposition of supplying additional soluble carbohydrate for lactic acid bacteria to promote a more rapid fall in pH and also as a way to increase forage organic matter digestibility (Lavezzo *et al.*, 1983; Henderson, 1993). Castro (2002) concluded that lower electrical conductivity and an increase in water activity associated with the use of enzymatic-bacterial inoculant, was consistent with cell membrane disruption and leaching of cell contents, thus providing additional substrate for microorganisms and leading to a pronounced pH drop. Additionally, inoculant decreased ruminal N degradation and increased milk fat and protein yield. According to Loures (2004), addition of fibrolytic enzymes to "guineagrass" prior to ensilage - alone or combined with *L. plantarum* - led to lower NDF, ADF, cellulose and hemicellulose contents in wilted silage. However, no improvement was seen in either *in vitro* digestibility or *in vivo* digestibility. The lack of a response might be related to a lower digestibility of the residual cell wall fraction after enzyme degradation or to depletion of soluble sugars during silage fermentation resulting in negative effects on animal performance.

The effects of homolactic bacteria addition to tropical grass silages are not conclusive. Much of the data currently available in the literature suggests a more rapid drop in pH and a lower ammonia-N content and lower temperature but other data (Rodrigues *et al.*, 2002a; Rodrigues *et al.*, 2002b) contradicts such claims. Paziani (2004) noted lower DM recovery and no change in ammonia-N in "guineagrass" silage made with added *L. plantarum*. When using large scale pressed bag silos, addition of *L. plantarum* increased unloading losses by 57%

(P<0.10) but DM intake and animal weight gains were not different to those obtained with control silages.

Recent data also showed that inoculation with *Lactobacillus buchneri*, which produces acetic acid, may improve silage stability during storage (Lindgren, 1999). This effect is observed mainly when applied to high DM forage (Davies & Hall, 1999), and seems to be related to the inhibition of yeast development and a reduction of aerobic losses at the silo face during unloading. Pedroso (2003) observed remarkable effects of *L. buchneri* addition to sugarcane silages with lower ethanol levels, gaseous losses reduced by 25% and increased DM recovery (15%). Dairy heifers fed sugarcane silage-based diets of 46% DM inoculated with *L. buchneri* gained 32% more than those fed the control diet. During this trial, an important integrated effect was noticed in terms of silage fermentation, aerobic stability and animal performance. Among other treatments, the use of *L. buchneri*, urea, sodium benzoate and potassium sorbate were identified as being especially effective in terms of the overall efficiency of silage use.

Losses related to unloading of silos

Aerobic instability is an important source of DM losses and energy losses after opening a silo and few additives are able to prevent it. Considerations of the efficiency of conservation of energy in specific management practices must take account of such losses. Even though the effects of inoculants and chemical preservatives on silage fermentation are widespread, it is uncommon to find data on the effects of aerobic instability at unloading and on animal performance in temperate (Lindgren, 1999) or tropical grass silages. The benefits of any inoculant must consider fermentation efficiency and animal performance (Flores *et al.*, 1999).

Exposure of silage to air, after unloading and during feeding, may lead to aerobic losses with higher costs due to DM and energy losses. The ability to predict the extent of tropical grass silage deterioration, based on fermentation pattern, is still uncertain, even though it would be a valuable farm tool. Some studies have concluded that the desired fermentation pattern does not always prevent unloading related losses. Such losses are not only associated with silage management, but with factors such as forage source and silo filling management. The greater the unloading rate, the lower are the losses but the extent of silage deterioration during unloading is related to its aerobic stability. Conventionally, aerobic stability is assessed from the time necessary, after silage unloading, to allow the temperature of the silage to rise by 2°C above the ambient temperature pattern (Kung Jr., 2001). Aerobic stability may be also assessed by measurement of the accumulated temperature rise compared to the environmental pattern, during 5 or 10 days after unloading (O'Kiely *et al.*, 1999).

Since aerobic stability is one of the most important parameters in silage management, many studies have been conducted to minimise its effect and prevent undesired fermentation. By preventing butyric acid fermentation or restricting acetic acid yield, the risk of aerobically unstable silage is increased. A high concentration and prevalence of lactic acid in well-fermented silage is not necessarily a positive predictor of aerobic stability (Weinberg & Muck, 1996). In fact, better aerobic stability was observed in silages containing some acetic acid as well as lactic acid (Mayrhuber *et al.*, 1999). The aerobic depletion of lactate by fungi, yeast and Bacillus reduces the potential stability of silage (Lindgren *et al.*, 1985) due to lactate aerobic conversion to acetate or degradation to butyric or acetic acid with a consequent pH rise. Aerobic microorganisms readily degrade lactic acid to CO_2, ethanol and acetic acid after unloading the silo, while also generating heat through exothermic reaction (Kung Jr., 2001).

According to Salawu & Adegbola (1999), tropical grass silages of 30% DM tend to have higher concentrations of lactic acid and are more prone to aerobic instability. The higher content of intact sugars or soluble carbohydrates preserved from fermentation associated with low acetic acid concentration, and the presence of yeast and poor packing practice might explain this trend with high DM silages.

"Tanzania" guineagrass, chopped for silage at three different particle sizes and combined with three levels of citrus pulp addition (0, 5 and 10%) showed a strong trend of lower aerobic stability with reducing particle size. The silages containing citrus pulp also had lower aerobic stability than the control treatment (Balsalobre *et al.*, 2001b). In contrast, Moura *et al.* (2001) observed a gradual increase in CO_2 yield with aerobic exposure up to 8 d in elephantgrass (*Pennisetum purpureum*) silage, but the addition of poultry litter (20%) and molasses (3%) during ensiling slowed the rate of deterioration. These results may indicate that the use of recommended silage practices aided carbohydrate preservation and assisted towards a good fermentation, attaining the ideal pH more rapidly but creating higher susceptibility to aerobic loss after unloading. The controversy surrounding unloading stability is due to the negative correlation between this parameter and the adequacy of the fermentation pattern.

Other factors may play a role in the aerobic stability of tropical grass silages. According to Bernardes (2003), in "palisadegrass" (*Brachiaria brizantha (A. Rich.) Stapf)* silages made with the addition of 5 or 10% pelleted citrus pulp, yeasts and fungal strains detected were more related to aerobic instability and resulted in small temperature increases. However, in wet silages (22% DM) the most important microorganisms growing under aerobic conditions were bacterial strains which lead to higher ammonia-N and pH increases. In both wet and additive-treated silages, the temperature increases were not closely related to DM losses or reductions in nutritive value indicating that temperature rises might not be considered as unique.

Conclusions

Tropical grasses have enormous potential as silage sources because of their biomass yield and their relatively low cost of production. However, production of tropical grass silages can incur overall losses greater than 30-40%, which in turn, may result in an expensive nutrient resource and lead to the wrong conclusions about the value of such silages. In order to assure the value of this system of forage conservation for animal feeding and for nutritive value and nutrient cost effectiveness, the control of losses at every step from the field through to the animal is essential.

References

Balsalobre, M.A.A., L.G. Nussio & G.B. Martha Jr (2001a). Controle de perdas na produção de silagens de gramíneas tropicais (Control of losses in tropical grass silages). In: Simpósio "A produção animal na visão dos brasileiros". Anais da 38ª Reunião Anual da Sociedade Brasileira de Zootecnia, 38. Piracicaba, SP, Brazil, 890-911.

Balsalobre, M.A.A., L.G. Nussio, R.V. Santos, R.F. Crestana, R.N.S. Aguiar & M. Corsi (2001b). Dry matter losses in Tanzania grass (*Panicum maximum* Jacq. cv. Tanzânia*)* silage. In: *Proceedings of XIX International Grassland Congress, São Pedro, SP, Brazil*, 789-790.

Bernardes, T.F. (2003). Características fermentativas, microbiológicas e químicas do capim-Marandú (*Brachiaria brizantha* Hoescht ex. A. Rich) Stapf cv. Marandu) ensilado com polpa cítrica peletizada (Fermentative, microbial and chemical profile of "palisadegrass" ensiled with pelleted citrus pulp). Dissertação de Mestrado – UNESP/FCAVJ, Jaboticabal, SP, Brazil, 108 pp.

Castro, F.G. (2002). Uso de pré-emurchecimento, inoculante bacteriano-enzimático ou ácido propiônico na produção de silagem de Tifton 85 (*Cynodon dactylon* ssp) (Pre-wilting, bacterial/enzymatic inoculant or propionic acid on the production of Tifton 85 silage). Tese de Doutorado– USP/ESALQ, Piracicaba, SP, Brazil, 136p.

Chin, F.Y. (2002). Ensiling of tropical forages with particular reference to South East Asian Systems. In: *Proceedings of the XIIIth International Silage Conference, Auchincruive, Scotland, UK*, 21-36.

Coan, R.M., P.F. Vieira, R.N. Silveira, M.S. Pedreira & R.A. Reis (2001). Efeitos do inoculante enzimático-bacteriano sobre a composição química, digestibilidade e qualidade das silagens dos capins Tanzânia e Mombaça (Effects of the bacterial/enzymatic inoculant on the chemical composition, digestibility and quality of "Tanzânia"and "Mombaça" grasses). In: Anais da 38ª Reunião Anual da Sociedade Brasileira de Zootecnia, 38. Piracicaba, SP, Brazil,124.

Davies, O.D. & P.A. Hall, (1999). The effect of applying an inoculant containing *L. buchneri* to high dry matter ryegrass swards ensiled in wrapped, round bales. In: Workshop E. Determination and control of aerobic instability. *Proceedings of XIIth International Silage Conference, Uppsala, Sweden,* 262-263.

Evangelista, A.R., J.A. Lima, G.R. Siqueira & R.V. Santos (2001). Aditivos na ensilagem de coast-cross (*Cynodon dactylon* (L.) Pers) 1. Farelo de trigo e polpa cítrica (Additives for ensiling Coast-cross grass 1. Wheat meal and pellet citrus pulp). In: Anais da 38ª Reunião Anual da Sociedade Brasileira de Zootecnia, 38. Piracicaba, SP, Brazil, 71.

Ferrari Jr., E. & W. Lavezzo (2001). Qualidade da silagem de capim Elefante (*Pennisetum purpureum* Schum.) emurchecido ou acrescido de farelo de mandioca (Silage quality of elephantgrass wilted or added with dried cassava meal). Revista da Sociedade Brasileira de Zootecnia (Brazilian Journal of Animal Science), Viçosa, 30, 5, 1424-1431.

Flores, G., J. Castro, A.G. Arraez, A. Amil, T. Brea & M.G. Warleta (1999). Effect of a biological additive on silage fermentation, digestibility, ruminal degradability, intake and performance of lactating dairy cattle in Galicia (NW Spain). In: Workshop C. Methods to predict feeding value of silage-based diets. *Proceedings of XIIth International Silage Conference, Uppsala, Sweden,* 181-176.

Giger-Reverdin, S. (2000). Characterisation of feedstuffs for ruminants using some physical parameters. *Animal Feed Science and Technology,* 86, 53-69.

Haigh, P.M. (1999). Effluent yield from grass silages treated with additives and made in large-scale bunker silos. *Grass and Forage Science,* 54, 280-218.

Henderson, N. (1993). Silage additives. *Animal Feed Science and Technology,* 45, 1, 35-56.

Holmes, B.J. & R.E. Muck (1999). Factors affecting bunker silos densities. Madison, University of Wisconsin, USA, 7 p.

Igarasi, M.S. (2002). Controle de perdas na ensilagem de capim Tanzânia (*Panicum maximum* Jacq. cv. Tanzânia) sob os efeitos do teor de matéria seca, do tamanho de partícula, da estação do ano e da presença do inoculante bacteriano (Control of losses in "Tanzânia"grass silages based on the effects of dry matter content, particle size, seasonality and bacterial inoculant). Dissertação de Mestrado – ESALQ/USP, Piracicaba, SP, Brazil,132pp.

Jones, D.I.H., R. Jones, & G. Moseley (1990). Effect of incorporating rolled barley in autumn-cut ryegrass on effluent yield, silage fermentation and cattle performance. *Journal of Agricultural Science*, 115, 399-408.

Jones, R. & D.I.H. Jones (1996). The effect of in-silo effluent absorbents on effluent yield and silage quality. *Journal of Agricultural Engineering Research*, 64, 173-186.

Kraus, T.J., R.G. Koegel, R.J. Straub & K.J. Shinners (1997). Leachate conductivity as an index for quantifying level of forage conditioning. ASAE, Mineapolis, MN, USA.

Kung Jr., L. (2001). Aditivos microbianos e químicos para silagem – Efeitos na fermentação e resposta animal (Microbial and chemical additives for silages – Effects on fermentation and animal performance) In: Anais do 2º Workshop sobre Milho para silagem. Piracicaba, SP, Brazil, 53-74.

Lavezzo, W., L.E. Gutierrez, A.C. Silveira . (1983). Utilização de capim elefante (*Pennisetum purpureum,* Schum.), cultivar Mineiro e Vruckwona, como plantas para ensilagem (Utilization of elephantgrass cv. Mineiro and Vruckwona as fodder plants for silage). Revista da Sociedade Brasileira de Zootecnia (Brazilian Journal of Animal Science), 12, 163-176.

Lima, J.A., A.R. Evangelista, R.V. Santos & G.R. Siqueira (2001). Aditivos na silagem de coast-cross (*Cynodon dactylon* (L.) Pers.) 2. Sacharina e fubá (Additives for ensiling Coast-cross grass 1. Sacharina and corn grain meal) In: Anais da 38ª Reunião Anual da Sociedade Brasileira de Zootecnia, Piracicaba, SP, Brazil, 72.

Lindgren, S. (1999). Can HACCP principles be applied for silage safety? In: *Proceedings of XIIth International Silage Conference, Uppsala, Sweden,* 51-66.

Lindgren, S., A. Jonsson, K. Petterson & A. Kasperson (1985). Microbial dynamics during aerobic deterioration of silage. *Journal of Science and Food Agriculture,* 36, 765-774.

Loures, D.R.S. (2000). Características do efluente e composição químico-bromatológica da silagem sob níveis de compactação e de umidade do capim-elefante (Pennisetum purpureum Schum) cv. Cameroon (Chemical composition of effluent from elephantgrass silages obtained under different packing pressures and moisture). Dissertação de Mestrado.UFV,Viçosa. MG, Brazil, 85 pp.

Loures, D.R.S. (2004). Enzimas fibrolíticas e emurchecimento no controle de perdas da ensilagem e na digestão de nutrientes em bovinos alimentados com rações contendo silagem de capim Tanzânia (Fibrolytic enzymes and wilting on the control of silage losses and nutrient digestion of bovine fed with "guineagrass"silage containing rations) – Tese de Doutorado, USP/ESALQ, Piracicaba, SP, Brazil, 146 pp.

Mari, L.J. (2003). Intervalo entre cortes em capim-Marandu (Brachiaria brizatha (Hochst. Ex A.Rich.) Stapf cv. Marandu): produção, Valor nutritivo e perdas associadas à fermentação da silagem (Cutting intervals in "palisadegrass": yield, nutritive value and losses associated to silage fermentation). Dissertação de Mestrado – USP/ESALQ, Piracicaba, SP, Brazil, 138 pp.

Mayne, C.S. (1999). Post harvest management of grass silage – effects on intake and nutritive value. In: *Proceedings of XVIIIth International Grassland Congress, 1997. Winnipeg/Saskatoon* CFC/CSA/CSAS, (Compact disk).

Lactobacillus strains as silage inoculum to improve aerobic stability In: Workshop E. Determination and control of aerobic instability, *Proceedings of XIIth International Silage Conference, Uppsala, Sweden*, 270-271.

McDonald, P., A.R. Henderson & S.J.E. Heron (1991). The biochemistry of silage. 2 ed. Aberystwyth: Chalcombe Publications, UK, 340 pp.

Moura, M.S.C., F.F.R. Carvalho, A. Guim, D.H.M. Marques & R.C. Ferreira (2001). Efeitos de aditivos sobre a velocidade de deterioração de silagem de capim Elefante (*Pennisetum purpureum* Schum.) (Effects of additives over the rate of deterioration of elephantgrass silages). In: Anais da 38ª Reunião Anual da Sociedade Brasileira de Zootecnia, Piracicaba, SP, Brazil, 363.

Nussio, L.G., F.G. Castro, J.M. Simas, C.M. Haddad, P. Toledo & A.L. Merchan (2001). Effects of dry matter content and microbial additive on Tifton 85 (*Cynodon dactylon sp*) wilted silage fermentation parameters. *Proceedings of XIXth International Grassland Congress, São Pedro, SP, Brazil*, 790-792.

Nussio, L.G., R.P. Manzano, R.N. Aguiar, R.F. Crestana & M.A.A. Balsalobre (2000). Silagem do excedente de produção das pastagens para suplementação na seca (Ensiling the forage biomass surplus in grazing systems to supplement cattle during the winter season). In: Anais do Simpósio sobre Manejo do Gado de Corte, Goiânia, GO, Brazil, 121-138.

O'Kiely, P., A. Moloney, T. Keating & P. Shields (1999). Maximising output of beef within cost efficient, environmentally compatible forage conservation systems. *Beef Yield Series No 10*, Teagasc, Grange Research Centre, Dunsany, Co. Meath, 64 pp.

Pauly, T.M. (1999). *Heterogeneity and hygienic quality of grass silage*. Doctoral Thesis, *Agraria, 157*. Uppsala. Swedish University of Agricultural Sciences.

Paziani, S.F. (2004). Controle de perdas na ensilagem, desempenho e digestão de nutrientes em bovinos de corte alimentados com rações contendo silagens de capim Tanzânia (Control of silage losses, performance and nutrient digestion in beef cattle fed with "Tanzânia" grass silage containing diets). Tese de Doutorado. USP/ESALQ, Piracicaba, SP, Brazil, 208 pp.

Pedreira, M.R., A.L. Moreira, R.A. Reis, N.S. Gimenes & T.T. Berchielli (2001). Características químicas e fermentativas do Tifton 85 (*Cynodon* spp.) ensilado com diferentes conteúdos de matéria seca e níveis de polpa cítrica (Chemical and fermentative parameters of Tifton 85 grass silages with different moisture content and citrus pulp addition) In: Anais da 38ª Reunião Anual da Sociedade Brasileira de Zootecnia, Piracicaba, SP, Brazil, 100.

Pedroso, A.F. (2003). Aditivos químicos e microbianos no controle de perdas e na qualidade de silagem de cana-de-açúcar (*Saccharum officinarum* L.) (Chemical and microbial additives on the control of losses and sugarcane silage quality). Tese de Doutorado –USP/ESALQ, Piracicaba, SP, Brazil, 120pp.

Quadros, D.G., M.P. Figueiredo, L.G. Nussio, N.C. Santos, J.V. Feitosa & J.Q. Ferreira (2003). Fermentative and nutritional traits of elephantgrass silage added with increasing proportions of coffee hulls. *Acta Scientiarum*, 25, 1, 207-214.

Reis, R.A. & R.M. Coan (2001). Produção e utilização de silagens de gramíneas (Production and utilization of grass silages). In: Anais do Simpósio Goiano sobre manejo e nutrição de bovinos. Goiânia, GO, Brazil, 91-120.

Rodrigues, P.H.M., A.L. Senatore, S.J.T. Andrade, J.M. Ruzante, C.S. Lucci & F.R. Lima (2002b). Efeitos da adição de inoculantes microbianos sobre a composição bromatológica e perfil fermentativo da silagem de sorgo produzida em silos experimentais. (Effects of microbial inoculants on the composition and chemical profile of sorghum silage stored in experimental silos). Revista Brasileira de Zootecnia (*Brazilian Journal of Animal Science*), 31, 6, 2373-2379.

Rodrigues, P.H.M., S.J.T. Andrade, J.M. Ruzante, F.R. Lima & L. Meloti (2002a). Valor nutritivo da silagem de milho sob o efeito da inoculação de bactérias ácido-láticas (Nutritive value of corn silage inoculated with lactic acid bactéria). Revista Brasileira de Zootecnia (*Brazilian Journal of Animal Science*), 31, 6, 2380-2385.

Ruppel, K.A., R.E. Pitt, L.E. Chase & D.M. Galton (1995). Bunker silo management and its relationship to forage preservation on dairy farms. *Journal of Dairy Science*, 78, 141-155.

Salawu, M.B. & A.T. Adegbola (1999). Aerobic stability of pea-wheat bi-crop silages treated with different additives. In: Workshop E. Determination and control of aerobic instability. *Proceedings of XIIth International Silage Conference, Uppsala, Sweden,* 282-283.

Sollenberger, L., R.A. Reis, L.G. Nussio, C. Chambliss & W. Kunkle (2004). Conserved Forage. In: Warm-Season C4 Grasses, ASA, CSCA & SSSA (eds), Madison, WI, USA, 1171 p.

Souza, A.L., F.S. Bernardino, R. Garcia, O.G. Pereira & F.C. Rocha (2001). Valor nutritivo da silagem de capim Elefante (*Pennisetum purpureum* Schum. cv. Cameroon) com diferentes níveis de casca de café (Nutritive value of elephantgrass silage added with increasing levels of coffee hulls). In: Anais da 38ª Reunião Anual da Sociedade Brasileira de Zootecnia, Piracicaba, SP, Brazil, 255.

Vilela, D. (1998). Aditivos para silagens de plantas de clima tropical (Silage additives for tropical climate plants) In: Anais da 35ª Reunião Anual da Sociedade Brasileira de Zootecnia, Piracicaba, SP, Brazil, Compact Disk.

Vilela, H., F.A. Barbosa, E.T. Dias, N. Rodriguez & E. Benedetti (2001) Qualidade das silagens de capim Elefante Paraíso (*Pennisetum hybridum* cv. Paraíso) submetidas a três tempos de emurchecimento (Silage quality of the elephantgrass hybrid "Paraíso" submitted to increasing wilting periods). In: Anais da 38ª Reunião Anual da Sociedade Brasileira de Zootecnia, Piracicaba, SP, Brazil, 323.

Weinberg, Z.G. & R.E. Muck (1996). New trends and opportunities in the development and use of inoculants for silage. *FEMS Microbiology Reviews*, 19, 53-68.

Woolford, M.K. (1972). Some aspects of the microbiology and biochemistry of silage making. *Herbage Abstracts*, 42, 105-111.

Recent developments in methods to characterise the chemical and biological parameters of grass silage

R.S. Park, R.E. Agnew and M.G. Porter

Agricultural Research Institute of Northern Ireland, Hillsborough, Co. Down BT26 6DR, U.K., Email: rae.park@dardni.gov.uk

Key points

1. Chemical analysis of forages is expensive, time consuming, environmentally unfriendly and relates poorly to the feed value for production purposes.
2. *In vivo* characterisation of animal feed is not a feasible option in terms of cost and analysis time.
3. NIRS is a rapid, non destructive, environmentally friendly, multi-analytical technique which can estimate the nutritive value of the feed.
4. NIRS predictive equations developed on a master instrument can be transferred to local and international sites.
5. Future assessment of forages necessitates rapid, stable, instrumentation for 'in field' studies.

Keywords: near infrared reflectance spectroscopy, diode array, forages

Introduction

Accurate characterisation of available feeds is a key requirement for cost-effective feeding of farm livestock. It forms the basis of not only making decisions on appropriate feeding levels to meet production targets but also optimising economic combinations for differing feed sources. While traditional approaches to silage characterisation provide some insight into feed quality, they are laborious to carry out and their accuracy in estimating biological parameters is limited. Ruminant production is now facing new challenges in a difficult economic climate, with the need to predict the quality of farm output, for example milk components (fat, protein, and lactose) or meat components (fatty acids profiles) rather than just volume or weight. As a consequence there has been a need to develop more accurate, rapid and cost-effective methods to characterise grass silage.

Previous approaches

The nutritive value of grass silage has traditionally been expressed in terms of organic matter digestibility (OMD) and digestibility of crude protein (DCP). These were derived from prediction equations based on chemical analysis of the materials. Alternatively in some countries more biologically meaningful *in vitro* techniques using rumen micro-organisms or commercial cell-free enzymes (Givens *et al.*, 1995) have been adopted. Table 1 shows the relationships between laboratory measurements, which have been used to predict the OMD of silages and the actual *in vivo* data. The precision of the relationships, as shown by R^2 values of 0.34–0.74, and particularly the modified acid detergent (MAD) fibre relationship previously used in the UK, are relatively poor. Although these wet chemistry approaches were, at the time, a reasonable compromise between simplicity and accuracy of prediction, they were also labour intensive and slow to carry out.

Table 1 Correlations between *in vivo* organic matter digestibility values (mean OMD = 0.71) and a range of laboratory measurements in 122 grass silage samples

	RSD	R^2
MAD fibre	0.051	0.34
Pepsin cellulase digestibility	0.042	0.55
In vitro digestibility	0.032	0.74

(Barber *et al.*, 1990)

Silage feeding value

Silage feeding value depends on the nutritive value of the silage expressed either as the digestibility or the ME concentration and the quantity of silage the animal will consume, that is the intake potential. It must be recognised however that the intake potential is often more important than the nutritive value in determining the feeding value of a silage as it exhibits greater variation. Accurate prediction of both intake potential and ME concentration, or digestibility of a silage are both therefore essential pre-requisites to the effective rationing of dairy cows and beef cattle offered grass silage *ad libitum*. In addition to feeding value and chemical composition, it is becoming increasingly important to be able to predict the rates of digestion of the protein components in silages and the amount of fermentable metabolisable energy. These latter parameters are required in situations where rations must be formulated to produce specific animal end-products. In view of the need to develop a new approach to predicting the feeding value of grass silage a series of studies were initiated at the Agricultural Research Institute of Northern Ireland, Hillsborough (funded by the Department of Agriculture and Rural Development). The studies were designed to ensure that both chemical and near infrared reflectance spectroscopy (NIRS) methods could be explored.

Animal experimentation to provide database

At the onset of the Hillsborough research programme it was recognised that a new approach to feed evaluation could only be developed if the intake, digestibility, chemical composition and rates of digestion of a large number of silages, representing the range of types of silages across the industry, were characterised within a standard animal protocol. The study, involved selecting silages on the basis of their pH, dry matter, ammonia and metabolisable energy contents. A total of 136 silages were selected on this basis from farms across Northern Ireland. Approximately seven tonnes of each silage were brought to the Institute, mixed in a mixer wagon to achieve uniformity and stored in polythene-lined boxes until feeding one to four weeks later. Care was taken to ensure that there was no deterioration of the silages during storage and work by Pippard *et al.* (1996) showed that their chemical composition remained constant throughout the storage period. The silages were offered *ad libitum* to 192 individually fed steers, which were crosses of the continental beef breeds and had a mean initial live weight of 415 kg, in a partially balanced changeover design experiment. Detailed chemical compositions of the silages were also determined. All silages were offered to sheep to determine *in vivo* digestibility. The rates of disappearance and rumen degradabilities of the dry matter, nitrogen and fibre fractions in the 136 silages were determined using the *in sacco* method. Steen *et al.* (1998) have reported the main results of this study, particularly the prediction of silage feeding value from chemical analyses (organic matter digestibility (OMD) was predicted from chemical constituents with a coefficient of determination ($R^2 = 0.30$)).

This paper is concerned with the further development of the data from this study to explore the potential of NIRS as a means of predicting a range of feeding parameters of grass silage.

Near infrared reflectance spectroscopy

Spectroscopy means 'looking at light' and the matter to be analysed interacts with electromagnetic radiation. Near infra red light is defined as the wavelength region from 780 to 2500 nm in the electromagnetic spectrum. When a sample is scanned, light is absorbed selectively according to the specific vibration frequencies of the molecules present and gives rise to a spectrum. All organic bonds have absorption bands in the NIR region, whereas minerals may only be detected in organic complexes. In forages, NIRS is therefore detecting the bonds in the protein, oil and carbohydrate fractions. The optical data stored as the NIRS spectrum is then regressed against known parameters to produce the best correlations and the resulting equation is then used to predict the parameters in unknown samples.

NIRS offers a number of advantages over traditional chemical and *in vitro* analyses. It is a rapid, physical, non-destructive method, requiring minimal or no sample preparation and it's accuracy is high. It is an environmentally friendly technique as no harmful, corrosive chemicals are used and there are no waste products to dispose. NIRS is a multi-analytical technique as many parameters can be predicted simultaneously.

NIRS was first shown to be a reliable, practicable method for predicting the chemical composition of forages by Norris *et al.* (1976) in the USA. Since then numerous workers have explored the use of NIRS for the prediction of both chemical composition and digestibility of grass silages. Attempts have also been made to use NIRS to predict the voluntary intake potential of dried, milled forages (Norris *et al.*, 1976; Shenk *et al.*, 1977; Ward *et al.*, 1982; Coelho *et al.*, 1988; Abreu *et al.*, 1991; Flinn *et al.*, 1992).

Prediction of intake and organic matter digestibility using dried samples

As all previous uses of NIRS to evaluate forages were based on scanning of dried milled samples this procedure was adopted as the first approach in the present study. Samples were air equilibrated after drying and milling and scanned at 2 nm intervals over the visible and near infrared spectral range (400-2500 nm) using a Foss NIRSystems 6500 scanning spectrophotometer. The full details of the methods and results of this work are presented by Park *et al.* (1997a).

The spectral data were subjected to a range of mathematical treatments to develop the optimum prediction methods. These methods included regression and derivatisation techniques and three scatter correction procedures. Appropriate cross validation was performed by removing one sample from the population of 136 in turn and forming a calibration on the remaining 135 samples and predicting the excluded sample. The best three mathematical treatments of the data, within each of these regression techniques, were selected on the basis of the lowest Standard Error of Calibration (SEC) and highest coefficient of determination (R^2). On the basis of the calibration statistics (SEC), the modified partial least squares (MPLS) technique achieved the best performance, having the lowest individual SEC and highest R^2 for intake (SEC 3.4 g/kg $W^{0.75}$, $R^2 = 0.90$) and organic matter digestibility (SEC 1.64 g/kg $W^{0.75}$, $R^2 = 0.94$) respectively.

Prediction of intake and organic matter digestibility using undried silage

The drying of silage is time consuming and leads to the loss of volatile acids, alcohols, esters, amines and ammonia which may influence the accuracy of predicting the components of feeding value (intake and digestibility). The development of more rapid, less expensive and potentially more accurate NIRS methodologies using fresh (undried) silage was therefore considered a desirable objective. However it was recognised that in this approach the sample preparation method could have a major influence on the accuracy of any data obtained particularly where NIRS scanning instruments only scanned small areas of sample. A programme was therefore undertaken to examine a range of sample preparation and scanning methods for undried silages using a Bran & Luebbe InfraAlyzer 500 Spectrophotometer either internally in a closed or viscous cup or externally using a probe attachment on the NIRS instrument.

In this study the 136 undried silages were scanned using eight different methods, involving various combinations of five sample preparation methods (intact with no pre-treatment; coarsely chopped; finely milled following freezing in liquid nitrogen; expressed liquor or eluent) and three scanning techniques (external probe; internal closed or internal viscous sample cup). Liquor was extracted from 100 g intact silage using a manual screw press and the eluent was decanted from 50 g intact silage immersed in 100 ml distilled water overnight. Six spectral scans (log 1/Reflectance, 1/R) were produced for each of the 136 silage samples by each of the 8 methods. The full methods and results of this study are given in Gordon *et al.* (1998) and a brief summary of the OMD data is presented in Table 2. The lowest standard error of cross validation (SECV) for the prediction of intake and digestibility was obtained by using finely milled silage presented to the NIRS instrument in the internal cup, and the highest SECV was obtained using the eluent. This work has demonstrated that in a static scanning-based system enhanced comminution has been an important factor in increasing the accuracy of the prediction of intake.

Table 2 The performance statistics for the prediction of organic matter digestibility (mean value 678 g/kg) using NIRS on undried silage samples (n=136) prepared and scanned by a range of methods

Preparation Method	Calibration and validation statistics for organic matter digestibility		
	SEC	R^2	SECV
(a) Intact tray	25	0.87	31
(b) Intact tube	23	0.89	27
(c) Coarse tube	26	0.86	29
(d) Fine tube	24	0.88	27
(e) Coarse cup	18	0.94	26
(f) Fine cup	18	0.94	26
(g) Liquor	37	0.73	46
(h) Eluent	52	0.44	56

(Gordon *et al.*, 1998)

Reeves and Blosser (1991) have indicated that enhanced comminution of fresh forages improved the accuracy of NIRS for prediction of fibre components but not crude protein.

However no similar data are available on the effects of particle size in fresh forages on the accuracy of biological parameters, such as digestibility and intake. Nevertheless it is likely that due to the linkage of digestibility and intake to the extent and nature of the fibre fraction these also will be influenced by particle size.

Calibrations based on scanning intact undried grass silage

The previous section has shown that when static scanning techniques are adopted it is preferable to finely comminute the undried material before scanning. However, this implies freezing and milling prior to scanning, both of which incur time delays and increased labour costs. Recently NIRS instruments have become available which incorporate a moving sample transport mechanism which may permit irradiation of a much larger sample surface area and minimise the effects of sample heterogeneity and so eliminate the need for sample comminution (e.g. Foss NIRSystems). Therefore, to evaluate this system NIRS calibrations for a range of chemical and biological parameters based on scanning intact undried grass silage were developed using this equipment. Two sub-samples of each silage were wrapped in non PVC cling film before being packed into a rectangular sample cell with internal dimensions of width 4.1 cm, length 17.2 cm and depth 1.4 cm. Each sample was scanned 24 times over the entire area of the sample and then automatically averaged using ISI NIRS3 Version 4.00 (Infrasoft International, Port Matilda, PA, USA) software, to produce one representative spectrum, thus minimising the effects of sample heterogeneity. Scanning takes less than a minute and optical data were recorded as log 1/R at 2 nm intervals. Equations were developed to predict dry matter, pH, total nitrogen, insoluble nitrogen, neutral detergent fibre, ether extract, gross energy and organic matter digestibility (Park *et al.*, 1996). The cross validation statistics are given in Table 3 where $1-VR$ = the coefficient of determination showing the proportion of variation in the reference method values explained by cross validation predicted values.

Table 3 Cross-validation statistics for the equations developed on the Foss NIRSystems 6500 using 136 uncomminuted undried grass silage samples.

Parameter	Mean	SECV	1-VR
Oven dry matter (g/kg)	201	6.38	0.98
pH	4.15	0.14	0.85
Total nitrogen (g/kg fresh)	4.64	0.34	0.92
Insoluble nitrogen (g/kg fresh)	1.98	0.15	0.89
Neutral detergent fibre (g/kg fresh)	118	5.13	0.95
Ether Extract (g/kg fresh)	7.73	0.61	0.88
Gross Energy (MJ/kg fresh)	4.05	0.14	0.96
Organic matter digestibility (%)	71.0	2.60	0.87

(Park *et al.*, 1996)

While the above work was undertaken using a Foss NIRSystems 6500 instrument, another instrument, which involves a rotating (600 rpm) sample cup and also provides a similar scanning area of the sample is also available (Bran & Luebbe InfraAlyzer 500). Work undertaken by Offer *et al.* (1996) had suggested that this latter approach may be a more accurate methodology. A study was therefore undertaken to compare the accuracy of both types of scanning monochromators in the development of calibrations for seven chemical and biological parameters of undried silage. Samples of 136 undried silages were scanned in the two instruments and calibrations were produced using the MPLS regression technique, in conjunction with either first or second order derivatisation and three scatter correction procedures. Optimum equation selection was based on the lowest SECV. The results showed no consistent differences in cross-validation statistics between the two instruments (Table 4).

Table 4 Cross-validation statistics for a range of parameters of undried silage developed on two types of NIRS instruments

Parameter	Foss NIRSystems 6500		B&L Rotating Cup	
	SECV	1-VR	SECV	1-VR
Alcohol corrected toluene dry matter (g/kg)	7.38	0.97	8.61	0.95
pH	0.14	0.85	0.16	0.83
Total nitrogen (g/kg fresh)	0.34	0.92	0.27	0.95
Neutral detergent fibre (g/kg fresh)	6.23	0.92	6.40	0.91
D-value (g/kg alcohol corrected toluene dry matter)	22.4	0.85	21.8	0.86
Voluntary Intake (g/kg $W^{0.75}$)	6.39	0.71	6.84	0.65

(Park *et al.*, 1998b)

Cloning of NIRS instruments

Whilst research at Hillsborough had shown that NIRS is an accurate, rapid method for characterising grass silage, it was important to consider the transfer of this predictive system to other NIRS instruments either within or between laboratories across the wider industry. This has been shown to be particularly difficult for heterogeneous and high moisture samples when differing types of equipment are being used. It would therefore be advantageous if calibrations developed on one NIRS instrument could be successfully transferred to another NIRS instrument, irrespective of manufacturer. Unfortunately spectral differences exist even between instruments of the same make and model (Dardenne & Biston, 1990). These differences in spectra are due to differences in the optical and electronic characteristics of the instruments and even very small differences have considerable effects on the predictions of parameters produced using the same equations.

Transfer of NIRS calibrations across instruments is commonly achieved by adjusting the equations for slope and bias. An alternative approach is to standardise the instruments so that they produce identical spectra (Shenk *et al.*, 1985). This 'cloning' approach has been shown to be successful when using homogeneous materials, such as dried milled samples (Forina *et al.*, 1995; Flinn *et al.*, 1995) or whole grain samples (Dardenne *et al.*, 1992). However difficulties arise when samples, with more than 200 g/kg moisture or high log 1/R values, are used because at high log 1/R values the reflectance data become non linear due to stray light (Shenk & Westerhaus, 1995). Grass silage is a very heterogeneous material consisting of leaves, stems, and dead material from a wide range of plant species and with a dry matter content ranging from approximately 120 to over 500 g/kg.

Two studies (Park *et al.*, 1998b and 1999) were undertaken to develop the technique of successfully transferring undried grass silage calibrations developed from one make of NIR spectrophotometer (i.e. Foss NIRSystems 6500) to another make of NIR spectrophotometer (i.e. Bran & Luebbe InfraAlyzer 500) and also between instruments of the same type (i.e. Foss NIRSystems 6500 to Foss NIRSystems 5000).

In the first study a range of undried grass silage samples was selected to clone two different makes of scanning monochromators using the ISI (Infrasoft International) cloning software. This software produces standardisation files which when applied to spectra from the slave instrument makes them 'look like' spectra from the master instrument and so allows accurate prediction by equations developed on the master instrument. Using an independent set of silages to compare the predictions of the unstandardised and standardised slave spectra to the master spectra predictions, using the same equations, showed that the standard error of prediction (SEP) was greatly reduced and the R^2 increased after standardisation of the spectra. The standardised spectra predictions were highly correlated to the master predictions, thus proving that this method of transferring calibrations worked very successfully. In comparison the alternative and commonly used method of sloping and biasing equations for use on other instruments was examined. Again the SEP values were reduced in line with the cloning method but the resultant slope and bias values were not improved over the cloning method. In addition, the average 'H' values, a measure of the closeness to the original data set, were not reduced by sloping and biasing as this technique matches the equations and not the spectra. Calibrations for biological parameters transferred more successfully when the cloning technique was employed. This study has demonstrated that this method of cloning monochromators of two different types and without the use of sealed sample sets, has proven very successful even with forages of high moisture content.

In the second study the cloning technique of Shenk *et al.* (1985) was used to spectrally match a Foss NIRSystems 6500 and a Foss NIRSystems 5000. A range of grass silage samples were scanned through both instruments using both a coarse transport cell and a natural product cell. These are the two most commonly used cell types with this instrument for scanning fresh forages. Cloning was based on using either 30 samples or one central sample. Standardisation files were produced (using the ISI software) and applied to the spectra of the validation samples, scanned on the Foss NIRSystems 5000 (regarded as the slave instrument). These standardised spectra were predicted by the master equations developed on the Foss NIRSystems 6500 instrument and the results compared to the corresponding master validation spectra scanned in the coarse transport cell and predicted by the master equations (regarded as the reference values). The spectra of a silage sample scanned in the coarse transport cell on the master, and the slave instruments and the standardised slave spectrum are shown in Figure 1.

In all instances the cloning technique (based on either one or thirty samples) proved very successful, clearly indicating that undried grass silage calibrations can be transferred across NIRS instruments of the same type with little loss in accuracy of prediction.

Use of NIRS in the Hillsborough Feeding Information System

The NIRS prediction equations developed on the Foss NIRSystems 6500 (Park *et al.*, 1998b) for chemical composition, digestibility, intake potential and rates of digestion have subsequently been used to establish the Hillsborough Feeding Information System. This is a commercial service available from the Institute for the evaluation of grass silage and the provision of associated feeding information. The system is based around 3 main functions; sample registration, sample analysis, and sample reporting with automation being used to increase efficiency and reduce costs. Silage sample information is logged into computer software and the undried (fresh) silages are scanned via a coarse transport cell mechanism on the NIRS instrument. This provides a NIRS spectrum which is examined through a set of calibration equations which allow the key attributes of the silage to be predicted automatically. This information is then appraised through a computer model to provide feeding strategies for dairy cows, growing cattle, suckler cows, breeding ewes and growing lambs. Wherever possible the System has been automated to keep the cost to the customer at a minimum and produce a final report within four working days for grass silage. Since its inception in 1996, the forage analysis service at Hillsborough has processed in excess of 80,000 samples of grass silage (approximately 12,000 silage samples/year at present).

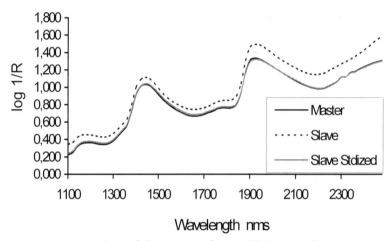

Figure 1 Comparison of the spectra of an undried grass silage sample scanned in a coarse transport cell, on the master (Foss 6500) and the slave (Foss 5000) NIRS, plus the slave spectrum standardised to the master spectrum

In January 2000 a new body known as the Forage Analytical Assurance (FAA) Group was formed. It comprises of the vast majority of silage analysis laboratories in the UK and one in the Republic of Ireland, accounting for approximately 85,000 forage samples per year. The FAA group use a range of equations developed on the Master NIRS instrument at the Agricultural Research Institute of Northern Ireland. This was accomplished by cloning thirteen instruments to the Hillsborough Master. Since it was established, the group has developed common quality control methods, with twice yearly stringent ring tests carried out between all sites plus monthly bias checks performed to maintain forage characterisation accuracy. It is hoped that these robust equations could also be expanded and new parameters developed for use internationally.

Future developments in forage characterisation

The NIRS programme of silage characterisation described here has developed an excellent advisory tool for quantifying the energy and protein components of ruminant diets – including intake, digestibility and rates of degradation etc. The key reasons for this success at the advisory level have been:
- Rapid sample turn round.
- Relatively inexpensive technique.
- Robust predictive relationships.
- Excellent repeatability.

However in the future from both an animal and an environmental perspective there is a need to provide similar accuracy in relation to some of the key cations and anions in ruminant diets, especially dairy cows. For example the relationship between dietary Na, K, Cl and S are the primary drivers of the Dietary Cation-Anion Difference (DCAD) which is important in dry cow management (control of milk fever) and lactation. Even within relatively small geographical areas (such as Northern Ireland) DCAD levels in grass silage can vary widely and this can have important implications for animal performance.

Equally phosphorus (P) levels in ensiled forages (grain and cereals) can also vary widely. With the present move towards on-farm P balances becoming a key environmental factor driving on-farm profitability it will become increasingly important to ensure that minimal levels are purchased onto the farm via concentrates. This can only be accurately achieved, and risks to animal health minimised, if P levels in individual forages are available. While it is fully recognised that NIRS may not be a good direct determinant of some of the key minerals it remains quite possible that indirectly NIRS may provide sufficiently robust relationships for the advisory situation.

To really assess the quality of forages, evaluate the nutrient status (N, P, K etc), detect early signs of disease problems and therefore improve the yield and quality of forage, we need to be able to perform 'in-field' (real time) studies with hand held or machine mounted instrumentation which can rapidly evaluate the crop. This type of evaluation is currently being enhanced by the development of new rapid scan NIR monochromator/detection systems without mechanically moving parts, such as acoustic-optic tuneable filters (AOFT), diode arrays or charge-couples devices. The advantages of new developments would be speed of measurement, mechanical robustness, temperature stability and small instrument size. In-field instruments would have to be robust in order to withstand mechanical shock as vehicles pass over rough ground. Real time data would provide objective rationale for adding fertiliser, fungicides etc and thus reduce over use of these chemicals and unnecessary expense.

Research is also progressing to determine on-line evaluation of total mixed rations in dairy feeding systems, where an NIR device is mounted directly on a feed mixer. At present problems exist in solving sample presentation in front of the scanning window of the feed mixer. When presentation and software problems are resolved, the farmer will have instant access to a complete breakdown of the concentration of all constituents in the diet enabling monitoring and adjustment of the feed for production purposes. Major minerals levels could also be monitored, especially phosphate levels which are an environmental concern in dairy rationing.

There seems to be little doubt that for reasons of speed and costs InGaAs-based diode array spectrometers are top of the instrumental shortlist. However these instruments have a reduced wavelength range and acceptable solutions for calibration transfer among these instruments still require work.

On a much larger scale, at present in America, sophisticated airborne and space-based imagers from NASA are enabling development of precision farming systems. Using hyperspectral, multispectral and infrared imaging, farmers can identify areas that are experiencing water, nitrogen or micronutrient stress and target their relief. The spectral information can be fed into a digital controller and, using the global positioning system for navigation, tractors can be guided to problem areas where they can distribute nutrients or pesticides within 19 mm of where they are needed. This technique has also allowed the farmer to reduce fertiliser application where it is not needed, thus affording production of higher yields at a lower cost. Early identification of limiting plant nutrients could substantially benefit EU farmers who now have to comply with the limits set by the Phosphate and Nitrates Directive.

Conclusions

The Near Infrared Reflectance Spectroscopy Research Programme based at the Agricultural Research Institute of Northern Ireland has developed a rapid, cheap and effective method for predicting a wide range of chemical and biological parameters, including a prediction of intake potential of grass silage for use in dairy cow, beef cattle and sheep rationing systems. These new methods enable cost effective and accurate predictions of a range of nutritionally important parameters of grass silage and are the basis of proven commercial silage analysis systems.

Further research has demonstrated that these calibrations can be successfully transferred to other NIRS instruments of either the same or different type.

With the rapid advances in instrumentation the future looks set to enable real time assessment of forages/crops, allowing in-field decisions to be made regarding the health and nutrient status of the plant. Thus yields should improve with reduced costs due to identification of mineral/nutrient requirements of the plants. These improvements in automation and sensor technology will play a significant role for farmers seeking to compete in today's global markets plus reduce environmental pollution due to overuse of nitrates and phosphates.

References

Abreu, J.M.F., M.T.A. Paco, T. Acamovic & I. Murray (1991). NIR Calibration for nutritional value of a diverse set of Portuguese forage crops. In: Murray I. & Cowe I.A. (eds) *Making Light Work: Advances in Near Infra-red Spectroscopy*, pp. 318-322. Weinham: VCH.

Barber, G.D., D.I. Givens, M.S. Kridis, N.W. Offer & I. Murray (1990. Prediction of the organic matter digestibility of grass silage. *Animal Feed Science and Technology*, 28, 115-128.

Coelho, M., F.G. Hembry, F.E. Barton & A.M. Saxton (1988). A comparison of microbial, enzymatic, chemical and near-infrared reflectance spectroscopy methods of forage evaluation. *Animal Feed Science and Technology*, 20, 219-231.

Dardenne, P. & R. Biston (1990). Standardization procedure and NIR instrument network. In: Biston R. & Bartiaux-Thill N. (eds) *Proceedings 3rd International Conference on Near-Infrared Spectroscopy*. p. 655. Gembloux: Agricultural Research Centre Publishing.

Dardenne, P., R. Biston & G. Sinnaeve (1992). Calibration transferability across NIR instruments. In: Hildrum K.I., Isaksson T., Naes T. & Tandberg A. (eds) *Near Infra-Red Spectroscopy. Bridging the Gap Between Data Analysis and NIR Applications*. p. 453. Ellis Horwood, Chichester.

Flinn, P.C. & L.D. Saunders (1995). Making the old a slave to the new: using existing calibrations on a new NIR instrument. In: Batten G.D., Finn P.C., Welsh L.A. & Blakeney, A.B. (eds) *Leaping Ahead with Near Infrared Spectroscopy*, p. 129. Near Infrared Spectroscopy Group, North Melbourne: RACI,

Flinn, P.C., W.R. Windham & H. Dove (1992). Pasture intake by grazing sheep estimated using natural and dosed *n*-alkanes: A place for NIR? In: Hildrum K.I., Isaksson T. & Tandberg A. (eds). *Near Infra-red Spectroscopy: Bridging the Gap Between Data Analysis and NIR Applications,* pp. 173-178. Chichester: Ellis Horwood.

Forina, M., G. Drava, C. Armanino, R. Boggia, S. Lanteri, R. Leardi, P. Corti, P. Conti, R. Giangiacomo, C. Galliena, R. Bigoni, I. Quartari, C. Serra, D. Ferri, O. Leoni & L. Lazzeri (1995). *Chemometrics and Intelligent Laboratory Systems,* 27, 189.

Givens, D.I., B.G. Cottyn, P.J.S. Dewey & A. Steg (1995). A comparison of the neutral detergent-cellulase method with other laboratory methods for predicting the digestibility *in vivo* of maize silage from three European Countries. *Animal Feed Science and Technology,* 54, 55-64.

Gordon, F.J., K.M. Cooper, R.S. Park & R.W.J. Steen (1998). The prediction of intake potential and organic matter digestibility of grass silages by near infrared spectroscopy of undried samples. *Animal Feed Science and Technology*, 70, 339-351.

Norris, K.H., R.F. Barnes, J.E. Moore & J.S. Shenk (1976). Predicting forage quality by infrared reflectance spectroscopy. *Journal of Animal Science*, 43, 889-897.

Offer, N.W., D.S. Percival, E.R. Deaville & I.C. Piotrowsk (1996). The potential for advisory silage analysis to be carried out entirely by near infra-red reflectance spectroscopy (NIRS). In: Jones, D.I.H., Jones R., Dewhurst R., Merry R. & Haigh P.M. (eds) *Proceedings of the XIth International Silage Conference*. pp. 68-69. Aberystwyth: Institute of Grassland and Environmental Research.

Park, R.S., R.E. Agnew & R.J. Barnes (1999). The Development of Near Infrared Reflectance Spectroscopy (NIRS) Calibrations for Undried Grass Silage and their Transfer to another Instrument using Multiple and Single Sample Standardization. *Journal of Near Infrared Spectroscopy*, 7, 117-131.

Park, R.S., R.E. Agnew, F.J. Gordon & R.J. Barnes (1998b). The development and transfer of undried grass silage calibrations between near infrared reflectance spectroscopy instruments. *Animal Feed Science and Technology*, 78, 325-340.

Park, R.S., R.E. Agnew, F.J. Gordon & R.W.J. Steen (1998a). The use of near infrared reflectance spectroscopy (NIRS) on undried samples of grass silage to predict chemical composition and digestibility parameters. *Animal Feed Science and Technology*, 72, 155-167.

Park, R.S., F.J. Gordon, R.E. Agnew, R.J. Barnes & R.W.J. Steen (1997a). The use of Near Infrared Reflectance Spectroscopy on dried samples to predict biological parameters of grass silage. *Animal Feed Science and Technology*, 68, 235-246.

Park, R.S., F.J. Gordon & R.W.J. Steen (1996). Organic matter digestibility and intake potential of grass silage predicted by Near Infrared Spectroscopy. Irish Grassland and Animal Production Association Meeting, Dublin, pp 141-142.

Pippard, C.J., M.G. Porter, R.W.J. Steen, F.J. Gordon, C.S. Mayne, E.F. Unsworth, & D.J. Kilpatrick (1996). A method for obtaining and storing uniform silage for feeding experiments. *Animal Feed Science and Technology*, 57, 87-95.

Reeves, J.B. III & T.H. Blosser (1991). Near infrared spectroscopy analysis of undried silages as influenced by sample grind, presentation method, and spectral region. *Journal of Dairy Science*, 74, 882-895.

Shenk, J.S. & M.O. Westerhaus (1995). Comparison of standardisation techniques. In: Davies A.M.C & Williams P. (eds) *Near Infra-red Spectroscopy: The Future Waves*. p. 112 Chichester: NIR Publications.

Shenk, J.S., K.H. Norris, R.F. Barnes & G.W. Fissel (1977). Forage and Feedstuff Analysis with an Infrared Reflectance Spectro-computer System. *Proceedings of the XIIIth International Grassland Congress.*

Shenk, J.S., M.O. Westerhaus & W.C. Templeton (1985). Calibration transfer between near infrared reflectance spectrophotometers. *Crop Science* 25, pp. 159-161.

Steen, R.W.J., F.J. Gordon, L.E.R. Dawson, R.S. Park, C.S. Mayne, R.E. Agnew, D.J. Kilpatrick & M.G. Porter (1998). Factors affecting the intake of grass silage by cattle and prediction of silage intake. *Animal Science* 66, pp. 115-127.

Ward, R.G., J.D. Wallace, N.S. Urquhart & J.S. Shenk (1982). Estimates of intake and quality of grazed range forage by near infrared reflectance spectroscopy. *Journal of Animal Science*, 54, 399-402.

Advances in silage quality in the 21st Century

D.R. Davies, M.K. Theodorou, A.H. Kingston-Smith and R.J. Merry
Institute for Grassland and Environmental Research, Plant, Animal and Microbial Science Department, Plas Gogerddan, Aberystwyth, SY23 3EB, UK, Email: david.davies@bbsrc.ac.uk

Key points

1. Sustainable production of silage
2. Enhanced use of forage N in the rumen
3. Breeding for enhanced herbage quality
4. Improved inoculants and chemical additives
5. Safety in the food chain

Keywords: forage breeding, ruminant, sustainability, environment, pathogens

Introduction

It was recently observed that in the future 'grazed and conserved forages are likely to be the bedrock of nearly all sustainable ruminant livestock systems' (Macrae & Theodorou, 2003). This presents considerable challenges in terms of enhancing the productive efficiency from these feedstuffs.

Globally, a major proportion of the forage and whole-crop cereals conserved for winter feeding are ensiled, with 200 million tonnes of silage dry matter (DM) produced annually (Wilkinson & Toivonen, 2003). Traditionally, silage quality has been judged in terms of its fermentation characteristics and a number of chemical measurements that reflect its energy and protein value (McDonald *et al.*, 1991). Over the past decade, under the influence of CAP reform and the move away from direct subsidies for agricultural produce in Europe, the characteristics used to describe the 'ideal' silage have broadened to embrace the concept of sustainability. Genome mapping and scientific advances in precision breeding make it more realistic to comply with these increasingly influential policy drivers enabling plant breeders to focus on producing forages with particular characteristics designed for the sustainable production of farm animals. Key areas for improving the sustainability of silage production in the 21st century will focus on a re-evaluation of how forage quality is defined, including methodologies for enhancing conversion efficiency both in the silo and rumen, in addition to farm animal welfare and issues relating to health and safety in the food chain. In this paper we will comment on how opportunities in plant breeding and advances in ensiling technology will address the future drivers for sustainable agriculture. It is not our intention to produce an exhaustive review of breeding for forage traits or to comment on the totality of factors that influence silage quality as this has been covered elsewhere (Buxton *et al.*, 2003; Charmley, 2001; Howarth & Goplen, 1983; Kingston-Smith & Thomas, 2003). Thus the main focus will be on the need to improve nutrient use efficiency (thereby reducing environmental impact) and increase safety in the food chain.

Environmental impact

During ensilage, soluble sugars in herbage are utilised by the microbial population to produce predominantly lactic acid, the main preservative agent, and plant proteins are extensively degraded to amino acids and ammonia (McDonald *et al.*, 1991). To an extent, products of silage fermentation become nutrients for a subsequent fermentation in the ruminant digestive tract where they are used to support microbial growth and the synthesis of microbial protein, which is the major source of protein available to the animal. However, the efficiency of microbial protein synthesis from silage-based diets in the rumen is low (Givens & Rulquin, 2004) and thus the conversion of N in silage-based diets to N in meat and milk is also low. An example illustrating the latter in dairy cows fed silage-based diets was recently provided by Dewhurst *et al.* (2003a), where values for conversion of feed N into milk N ranged from 18 to 26% for three legume silages and one grass silage. Thus the majority of N is lost to the environment, either in faeces or urine. This loss of N is thought to be due, in part, to an imbalance in nitrogen and energy supply (Chamberlain & Choung, 1995; Dewhurst *et al.*, 2000) caused by rapid degradation of plant protein to ammonia in both the silo and rumen, but only a slow release of energy yielding substrates derived from cell wall carbohydrates in the rumen. One of the consequences of this imbalance of nutrient supply is a scarcity of ATP during the time of maximum release of ammonia, which limits microbial protein synthesis such that only a modest proportion of the available N in ammonia is captured. The remainder is absorbed into the blood stream, converted to urea in the liver and excreted in urine, thus potentially contributing to environmental pollution.

Figure 1 provides a schematic representation of the asynchrony principle in the rumen, whereby rapid availability of protein degradation products is not balanced with an appropriate supply of readily available energy for microbial growth; as a consequence the efficiency of N incorporation into microbial protein is low.

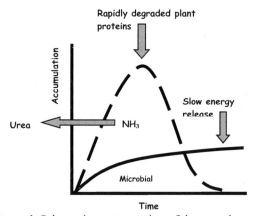

Figure 1 Schematic representation of the asynchrony principle in the rumen

The figure suggests two possible strategies for increasing N capture during microbial growth. One is to protect proteins in order to reduce the rate at which they become degraded to ammonia in the silo and rumen. The benefits of this approach are easy to demonstrate experimentally, but the translation of the principles into commercial practice has so far proved to be more difficult. The alternative strategy for improving microbial efficiency in animals given silage-based diets would be to increase the amount of readily available energy in silage for utilisation during the initial phase of fermentation in the rumen.

Progress in improving silage quality and nutrient use efficiency is influenced by a number of factors, but as implied from the above account, advances in plant breeding and additives for manipulating silage fermentation and ruminal processes present interesting opportunities. In the following two sections we address the issues relating to protein and carbohydrate supply and how they could be used to improve silage quality and gut function with the goal of reducing environmental impact, whilst maintaining or improving nutrition and production.

Silage quality

Protein

Much work has focussed on reducing the rate and extent of degradation of forage protein, both in the silo and in the rumen. A number of avenues have been explored, such as using different crops or crop mixtures or treatment of forages with additives to reduce their degradability (McDonald *et al.,* 1991). These approaches have had varying degrees of success but, as commented by Givens and Rulquin (2004), the manipulation of plant traits using modern breeding techniques offers another way forward, although this must be matched with the appropriate ensiling technology.

Tannins in forage plants are known to have a role in manipulating protein degradability in both the silo and the digestive tract of ruminants. Albrecht and Muck (1991) examined a number of legumes and demonstrated considerable variation in tannin concentrations in *Lotus* species with an inverse relationship between tannin concentration in the herbage and proteolysis in the silo. The potential for using tannins to protect herbage proteins and improve protein utilisation in ruminants is supported by the work of Barry & McNabb (1999), who concluded that for *Lotus corniculatus*, a tannin content of approximately 5% reduces the rate of degradation of proteins in the rumen and increases animal performance. Although *Lotus* species show varying degrees of success, it must be emphasised that these plants have specific soil-type requirements and are essentially best suited to sub-tropical conditions. Thus they have limited application because they are not suited for cultivation in more temperate, grassland regions. Whilst legumes such as *Lotus* have been selected and manipulated to produce differing concentrations of tannins (Miller & Ehlke, 1996) attempts to introduce tannins into other varieties (such as white clover) has proven difficult and the application of these approaches has been restricted to the use of crops which naturally contain tannins (Barry *et al.*, 2001).

The phenomenon of plant enzyme activity immediately after cutting and in the early stages of ensiling has long been recognised and much effort has gone into curtailing these activities by the use of additives and management practices (McDonald *et al.,* 1991). More recently plant-based approaches to restrict proteolytic activity have been examined. Kingston-Smith *et al.,* (2002) explored the variation in protease activity within herbage species and concluded that there is scope to utilise this variation to breed for plants with lower activity. Introgressed lines of *Festuca* and *Lolium* are also being used to identify sections of the ryegrass genome responsible for plant-mediated proteolysis. Large differences in proteolysis were found when these grasses were incubated *in vitro* under rumen-like conditions (A.H. Kingston-Smith, personal communication).

The observation that protein loss during the ensilage of red clover was low in comparison with lucerne led to speculation about a protein protection mechanism in red clover (Jones *et al.,* 1995a) and subsequently Jones *et al.,* (1995b) proposed that the plant enzyme, polyphenol oxidase (PPO), which is found in most plants, was involved. PPO converts phenols to quinones, which bind to proteins to produce stable protein-quinone complexes, altering their susceptibility to proteolysis and/or binding to endogenous proteases. Winters & Minchin (2002) have since isolated an isogenic red clover line with low PPO activity (cf. normal lines) and showed that, after ensiling, the soluble nitrogen and free amino acid contents were 25% and 20% greater respectively for the low PPO line, reflecting the difference in protein degradability. A study by Broderick *et al.* (2004) revealed an apparently wide and consistent distribution of values for the rumen degradability of red clover protein amongst cultivars. This suggests that new lines of red clover could be bred to manipulate protein degradability in the silo and rumen. Dewhurst *et al.* (2003b) reported that degradability of red clover N in the rumen of dairy cows was lower (65%) than for grass (70%), white clover (67%) and lucerne (72%) and speculated that this was in part due to inherent PPO activity. This may partly explain the improvement in rumen microbial growth efficiency observed when red clover was added to simulated rumen cultures compared to grass silage (Davies *et al.*, 1999). The finding by Lee *et al.* (2004) that PPO in red clover appears to inhibit lipolysis may confer another benefit, by reducing polyunsaturated fatty acid losses during field wilting of PPO-containing forages prior to ensiling.

Lucerne is an important silage crop in the drier climatic conditions of North America and parts of Western Europe, but has the disadvantage of very high protein degradability, linked to limited PPO activity (Jones *et al.*, 1995c). Lucerne has now been genetically modified to introduce the PPO gene from red clover (Sullivan *et al.*, 2004) but due to the lack of sufficient endogenous substrate, caffeic acid, which is needed to activate the enzyme (Hatfield *et al.*, 2003), other sources of caffeic acid, such as potato skins, are now being sought.

The stay-green mutation in grasses (Thomas & Smart, 1993) has been considered as a means of reducing protein breakdown in the silo and rumen. Although there was evidence of enhanced protein content in grass cut for hay, the *sid* mutation stay-green when incubated under either rumen-like conditions (Kingston-Smith *et al.*, 2003) or ensiled in the laboratory (D.R. Davies *et al.*, unpublished observations), showed no improvement in protein protection compared to the wild-type.

The plant breeding approaches outlined above present opportunities for manipulating forage protein content and degradability in the silo and rumen, but marked changes in nitrogenous fractions of forages occur during ensilage (McDonald *et al.*, 1991). Consequently, additive solutions have been devised for controlling the fermentation and hence proteolysis. One key factor that influences proteolysis in the silo is the speed of pH decline (McKersie, 1985). Application of additives such as formic and sulphuric acids etc to the crop before ensilage has been shown to reduce protein degradation in the silo (McDonald *et al.*, 1991), but health and safety and environmental considerations have made these additives less attractive. On the other hand, the use of silage inoculants containing homo-fermentative lactic acid bacteria to increase lactic acid production and enhance the rate and extent of pH decline (Kung *et al.*, 2003; McDonald, *et al.*, 1991; Weinberg & Muck, 1996) can also lead to a reduction in protein breakdown (Merry *et al.*, 1997). The introduction of freshly cultured inoculants (Merry *et al.*, 1995a; Nesbakken & Broch-Due, 1991) improved preservation rates in the silo in comparison with freeze dried preparations by enhancing the activity of the inoculant organisms during the initial stages of ensilage (Merry *et al.*, 1995a). This was accompanied by significantly lower acetic acid and ammonia-N concentrations and higher WSC concentrations in the mature silages (Merry *et al.*, 1995a). Other research with grasses and legumes confirmed the ability of freshly cultured silage inoculants to improve silage quality and, in particular, true protein content (Cussen *et al.*, 1995; Davies *et al.*, 1998).

Furthermore, Winters *et al.* (2001) showed that free amino acids accumulating in inoculated grass silages were 55% lower than in untreated silages and amongst the amino acids there were significantly greater levels of total (free plus bound) lysine and arginine in the inoculated compared to the untreated silage. When the inoculated silage was fed to steers without supplementary concentrates, intake and liveweight gain increased significantly by 0.81 kg/d and 0.23 kg/d respectively above that for the untreated silage. A partial explanation for this may lie in the results obtained in an *in vitro* rumen simulation study, where Davies *et al.* (1999) observed that inoculated red clover silage led to lower ammonia concentrations and higher efficiency of rumen microbial synthesis compared with untreated silage, implying that N use efficiency was improved. On the other hand, Charmley (1995) reviewed the literature and whilst he found potential for improved animal performance from animals fed inoculated compared to untreated silages, he found no evidence that inoculants reduced proteolysis.

Carbohydrate

The interaction between plants and micro-organisms is a central issue when forage carbohydrates are fermented to produce energy for rumen microbial growth, and a balanced supply of different forms of carbohydrate can be critical to ensure efficient capture of nitrogen (Dewhurst *et al.*, 2000). Equally, water soluble carbohydrates are essential substrates for lactic acid bacteria during ensilage, to enhance preservation and minimise protein degradation.

Water soluble carbohydrates

There is considerable genetic variation in the type and amount of water soluble carbohydrate (WSC) in commonly ensiled forages (McDonald *et al.*, 1991). This was recognised by Humphreys (1989) who selected ryegrass lines with the genetic potential to contain high leaf contents of WSC or used tetraploid lines to produce ryegrass cultivars with heritable, enhanced WSC content. Research at the *Institute for Grassland and Environmental Research* (IGER) has demonstrated the benefits of high WSC ryegrasses on *in vitro* rumen function parameters, forage intake, animal production and overall efficiency of N use in grazing ruminants (Miller *et al.*, 2001; Lee *et al.*, 2002). These findings were indirectly supported by the work of Fisher *et al.* (1999) who found that tall fescue grass hays cut at sun-down had higher WSC content, higher *in vitro* digestibility and lower NDF content than those cut at sun-up and increased preference and intake was associated with these parameters. The principles underlying the benefits of extra WSC on ruminal N use efficiency with grazed grasses apply equally, or perhaps more to silage, where readily available sugar is often in short supply (Rooke *et al.,* 1987; Chamberlain & Choung, 1995; Givens & Rulquin, 2004). While the target for producing good quality silage has generally been to maximise the conversion of WSC into lactic acid to prevent adverse microbial activity and restrict protein degradation (Davies *et al.*, 1998; Weinberg & Muck, 1996), little consideration has been given to retaining WSC in grass silage with the specific intention of increasing microbial growth and N use efficiency in the rumen. The availability of high-WSC grasses has thus stimulated a re-assessment of the 'ideal' composition for grass silages and strategies for manipulating silage fermentation. Indeed, Carpintero *et al.* (1979) have clearly demonstrated that restricting grass silage fermentation by addition of high levels of formic acid increases the amount of residual WSC in silage. More recently, Davies *et al.* (2002) confirmed this observation and showed that reducing the pH rapidly with homo-fermentative silage inoculants could have a similar effect, with 50% of the WSC remaining in silages made from the grass cultivar with elevated WSC content, compared to 12% with the control cultivar. The increase of >20% in rumen microbial N use efficiency when high sugar grass silages were fed under both *in vitro* and *in vivo* conditions (Merry *et al.*, 2002; 2003) confirmed the value of grasses of sufficiently high sugar content for increasing nutrient use efficiency in the rumen and also the potential to reduce N pollution.

The effectiveness of silage inoculants on fermentation and silage quality under difficult ensiling conditions (low WSC and/or DM forages) has been questioned (Davies *et al.*, 1998) and acid additives were recommended under these conditions. However, data from other workers (Nesbakken & Broch-Due, 1991) presented a contrary view. The importance of fructan, the main component of the grass WSC fraction during ensilage was later highlighted by Muller & Lier (1994) and subsequently the benefit of applying fructan utilising lactic acid bacteria to grass in order to enhance silage quality was demonstrated (Merry *et al.*, 1995b; Winters *et al.*, 1998). Furthermore, when high DM grass silage treated with a fructan degrading inoculant was fed to sheep, N-retention increased by greater than 70% in comparison with untreated silage (Theobald *et al.*, unpublished).

Cell wall carbohydrates

An alternative approach for improving energy supply is to make the recalcitrant fibrous fraction of forages more available for fermentation in the silo and rumen. Plant cell biologists are currently using transgenic approaches, as well as conventional mutagenesis (which is more acceptable in terms of public perception) to modify the rate and extent of digestion of cell walls in the plant (Chen *et al.*, 2003). These techniques involve the down-regulation of genes involved in lignification of plant tissues, bringing about a reduction in secondary thickening. In one such study (Chen *et al.*, 2003) down-regulation of the cinnamyl alcohol dehydrogenase enzyme brought about an increase in rumen *in vitro* DM digestibility of up to 9.5% in the transgenic compared to the parental lines. Further investigation of such transgenic material is needed to establish its ensiling potential and effect on cell wall digestion in the rumen.

Although their efficacy has often been questioned, cellulolytic, hemicelluloytic and amylolytic enzymes have been widely used as silage additives to increase the availability of forage cell wall and storage sugars in the silo and rumen (Muck, 1993). Several factors are likely to influence their activity. These include sub-optimal levels of enzyme, enzyme inactivation or degradation by resident proteases and inappropriate physiological conditions for expression of their activity (e.g. pH optima, moisture content). An alternative route to enzyme addition is expression of these activities by lactic acid bacteria used in silage inoculants. *Lactobacillus plantarum* is the most commonly used bacterium in silage inoculants but does not have cell wall carbohydrate degrading abilities. It has been targeted for the incorporation of cellulolytic and amylolytic genes from other bacteria such as *Clostridium spp.* (Bates *et al.*, 1989). Sharp *et al.* (1992) investigated the growth in silage of two genetically modified strains of *L. plantarum* in comparison with the unmodified parent strain. Although the recombinant strains out-competed the natural microflora, they were not as effective as the parent strain in improving silage preservation. The results suggested that the transgenic bacterium could out-compete the epiphytic microflora in the silo, but unfortunately no data were presented to indicate whether ruminal digestion of the resultant silages was altered. The use of transgenic inoculants to upgrade low digestibility forages could be of importance in the future where environmental schemes require forages to be harvested at more mature growth stages. Nevertheless, considerable work remains to be done in this area but it is unlikely to attract significant funding in those parts of the world where the public will not accept the use of transgenic micro-organisms in agriculture.

Aerobic stability

One of the consequences of improving silage preservation is the negative impact that its enhanced nutritional characteristics may have on aerobic stability. The subject area has been reviewed on a number of occasions (Woolford, 1990; McDonald et al., 1991) and it is accepted that well preserved, high quality silages, particularly those inoculated with homo-fermentative lactic acid bacteria, are more prone to spoilage than untreated silages (Weinberg et al., 1993). Aerobic spoilage of silage can affect both the efficiency of nutrient utilisation (through respiration of energetic fractions) and its hygienic quality (through production of mycotoxins by spoilage microflora).

Although there are no substitutes for good clamp management, additives have been used in an attempt to limit aerobic spoilage in the silo. A number of control strategies have been investigated (e.g. direct addition of formic, propionic and caprylic acids; McDonald et al., 1991), but most recent developments have focused on the use of (i) propionic acid bacteria (ii) heterofermentative lactic acid bacteria and (iii) a combination of homo-fermentative lactic acid bacteria with salts such as sorbate, sulphite or benzoate. None of these approaches could be described as being entirely successful. Propionic acid bacteria were considered to have considerable potential in preventing spoilage, but their use has failed to provide consistent results (Merry & Davies, 1999). The main problem appears to be the sourcing of suitable isolates, although recent molecular technological advances may provide a wider range of more appropriate strains (Romanov et al., 2004). In the latter two approaches, using either heterofermentative lactic acid bacteria (particularly *L. buchneri*) (Driehuis et al., 2001) or the combination of traditional homofermentative inoculants with chemical salts (Rammer et al., 1999), a potential for reduced spoilage has been shown. However, the mode of action of these additives could have negative impacts on silage quality by reducing the speed of fermentation and subsequently the true protein content.

Certain plant characteristics could make a positive contribution towards limiting aerobic spoilage in the silo. Lucerne silage has been shown to be more stable than maize silage (Muck & O'Kiely, 1992; O'Kiely & Muck, 1992) and our observations indicate that legume silages (red clover, lucerne and white clover) are more stable in comparison to grass silage (R.J. Dewhurst, personal communication). This suggests that legumes may contain a natural compound that inhibits spoilage micro-organisms. However, Muck and O'Kiely (1992) compared the aerobic stability of fresh and ensiled lucerne and concluded that the factor causing stability was produced during ensilage, as the fresh crop was not stable. A greater understanding of the factors involved may present opportunities for breeding or genetic manipulation of these and other forages to enhance aerobic stability.

The stay-green trait in grasses is a natural mutation which prevents or delays chlorophyll from progressing through the normal catabolic pathways upon senescence and cell death (Thomas & Smart, 1993) and although not considered in the literature, this behaviour may have application in reducing aerobic spoilage in the silo. A strong inverse relationship has been observed between different maize varieties that possess increasing resistance to the onset of senescence and accumulate different concentrations of the mycotoxin, zearalenone (Oldenburg, 1999). These observations suggest that the retarded senescence afforded by the stay green maize cultivars provided a less accessible supply of nutrients and consequently supported a smaller epiphytic bacterial and fungal flora. As fungal populations in the silo are the main instigator of aerobic spoilage, reducing the population size of these organisms on the crop is likely to reduce or remove the problem. Further research is needed to establish a relationship between the stay-green trait in maize and reduced aerobic spoilage in maize silage and whether the relationship is also evident in stay green grass silages.

Safety in the food chain

Lindgren *et al.* (2003) proposed that food safety should cover the entire food chain from primary production to the consumer and that the chain includes fodder production. Food borne pathogens such as enterohaemorrhagic *E. coli* (0157), *Listeria*, *Salmonella* and *Campylobacter* have received considerable attention over the past decade and Hazard Analysis and Critical Control Point (HACCP) principles have been developed and applied to assist in the control of pathogens from the point of presentation of livestock at the abattoir and beyond. However, much less effort has been focused on the control of pathogens on farm and there is considerable anecdotal evidence that suggests a significant number of the *E. coli* 0157 infections may be due to direct farm animal contact and not to food poisoning *per se* (I. Ogden, personal communication). A key target therefore is to develop effective strategies for inhibiting the growth of these potential human pathogens in both the silo and in the herbivore gut.

Although research is limited to just a few studies, some work has looked at the ability of potential digestive tract pathogens to survive in silage. By adding *E. coli* 0157:H7 to cut grass prior to ensilage, Byrne *et al.* (2002) were unable to detect viable cells in low pH forage after 19 days of ensiling, indicating that properly preserved grass silage is unlikely to be a vector for the transmission of this pathogen among cattle. However, *E. coli* 0157 has been reported as being able to survive in poorly fermented silage (Fenlon & Wilson, 2000).

Relative to work on pathogen proliferation during ensilage, more studies have been conducted on the effect of silage diets on faecal shedding of pathogens, particularly verotoxigenic *E. coli*. Most have focused on comparing high forage with high grain diets and whilst initial research suggested that forage-based diets provided considerable potential for reducing faecal shedding of *E. coli* 0157, more recently this has been questioned (Bach *et al.*, 2002). Despite equivocal data on the influence of diet on pathogen shedding, there is evidence to suggest that certain plant compounds are able to inhibit the growth of enteric pathogens. For example diets containing red clover have been shown to reduce the *E. coli* load in cattle (Garber *et al.*, 1995). More focused studies have found a role for plant compounds, such as esculetin and coumarin, in reducing the survival of verotoxigenic *E.coli* in rumen digesta (Duncan *et al.*, 1998). These compounds exist naturally in plants and in particular legumes such as red clover. Further screening of natural plant compounds to examine their ability to inhibit pathogens may provide breeding targets for crops with elevated levels of known active compounds. This approach could lead to feeding strategies that work at two levels, in the silo and the rumen, and offer the potential to reduce the risks of food borne pathogens entering the food chain by controlling their proliferation in animals during primary production and prior to slaughter.

Numerous positive effects on the health of animals and humans have been reported for lactic acid bacteria when they are used as probiotics. These include increased resistance to infectious disease, reduced blood pressure, stimulation of phagocytosis by blood leucocytes, and an improved balance of the intestinal microflora (Tannock, 1999). In addition to plant-based strategies for reducing pathogen load in farm animals, it is important that we understand the fate of silage lactic acid bacteria in the ruminant gut in order to exploit the potential of using silage inoculants as probiotics (ruminant gut modifiers). The main reason for inoculating herbage with lactic acid bacteria has been to increase the rate and extent of production of lactic acid during the initial stages of ensilage, thereby improving silage quality and animal performance (Kung *et al.*, 2003; Weinberg & Muck, 1996). Nevertheless, there are reports of increased intake and positive production responses in cattle fed inoculated silages that are not explained in terms of the current indicators of silage quality. In these circumstances, probiotic effects have been suggested as an explanation for the animal responses (Weinberg *et al.*, 2004). At the time of feeding, inoculated silage generally contains a minimum of 1×10^4 viable lactic acid bacteria per g of fresh silage. If the average dairy cow consumes 50 kg fresh weight of silage, a conservative estimate of the daily intake of lactic acid bacteria is 5×10^8. Although this value is low compared to the $10^9 - 10^{10}$ viable cells frequently quoted for 1 ml of rumen fluid, it is similar to that of 'direct-fed microbials' (which consisted of a mixture of *Enterococcus faecium, Lactobacillus plantarum* and *Saccharomyces cerevisiae*) used in a recent dairy cow study (Nocek *et al.*, 2002), where positive effects were seen in terms of stabilising rumen pH in animals fed potentially acidogenic diets. Weinberg *et al.* (2004) recently established that silage inoculant lactic acid bacteria could survive under *in vitro* conditions in rumen fluid. However, further research is required to define the exact role of inoculant bacteria in the live animal.

The possible use of dual purpose silage inoculants which exhibit preservative as well as probiotic characteristics offers considerable scope for tackling issues relating to silage quality and pathogens in the food chain. Such inoculants would also have the significant practical advantage of constant replenishment during feeding, at potentially no additional cost to the farmer. Overall there is a need for research to assess the potential for selected silage inoculant bacteria to survive the rumen environment, colonise the hindgut and initiate beneficial probiotic effects.

Conclusions

In his paper on silage quality, Charmley (2001) concluded that modern ensiling technology has increased the feeding value of silages close to that of the original un-ensiled forage and went on to suggest that the future may hold opportunities to produce silages with superior feeding value. While we support his contention, in our opinion it must be applied within the context of a sustainability agenda. We therefore suggest that in addition to technological developments to improve fermentation rate and digestibility, novel plant breeding criteria and the targeted use of inoculant bacteria to provide benefit not only in the silo but also the rumen will play an important part in helping to achieve these goals. Whilst there are many challenges ahead in advancing silage quality in the 21st century and developing livestock agriculture according to the sustainability concept advocated by scientific policy makers, there are a number of areas relating to farm animal production from grassland-based systems where consideration of the crop and the way we manipulate both its composition and the natural and added microflora, before and during ensilage, would seem to provide opportunities. Firstly, exploring and exploiting protein protection mechanisms and the mechanisms controlling carbohydrate storage in plants would not only enhance N-use efficiency in ruminants but increase the productive efficiency of ruminant enterprises and reduce the environmental footprint of livestock agriculture. Secondly, a deeper understanding of the role of inoculant bacteria and natural plant products in the silo and rumen would result in biological mechanisms of pathogen control and increased safety in the food chain. In our view, the key to success in these endeavours are three-fold and relate to (a) elaborating new criteria to breed grassland plants that are pre-disposed to behave in a particular way, or to cause a particular behaviour(s), both during ensilage and during their degradation and passage through the ruminant digestive tract, (b) obtaining a better understanding of digestive tract function within the context of the whole animal, and (c) taking a broader view of the role of 'beneficial' bacteria and plant products in silo and rumen processes.

Acknowledgements

The authors acknowledge the financial support of the Biotechnology and Biological Sciences Research Council, the Department for Environment, Food and Rural Affairs and the European Union.

References

Albrecht, K.A. & R.E. Muck (1991). Proteolysis in ensiled forage legumes that vary in tannin concentration. *Crop Science*, 31, 464-469.

Bach, S.J., T.A. McAllister, D.M Veira, V.P.J. Gannon & R.A. Holley (2002). Transmission and control of *Escherichia coli* O157:H7 – A review. *Canadian Journal of Animal Science*, 82, 475-490.

Barry, T.N. & W.C. McNabb (1999). The implication of condensed tannins on the nutritive value of temperate forages fed to ruminants. *British Journal of Nutrition*, 81, 263-272.

Barry, T.N., D.M. McNeill & W.C. McNabb (2001). Plant secondary compounds; their impact on forage nutritive value and upon animal production. In: J.A. Gomide, W.R.S. Mattos & S.C. da Silva (eds.) *Proceedings of the XIX International Grasslands Congress. Sao Paulo, Brazil*, 445-452.

Bates, E.M., H.J. Gilbert, G.P. Hazlewood, J. Huckle, J.I. Laurie & S.P. Mann (1989). Expression of a *Clostridium thermocellum* endoglucanase gene in *Lactobacillus plantarum*. *Applied and Environmental Microbiology*, 55, 2095-2097.

Broderick, G.A., K.A. Albrecht, V.N. Owens & R.R. Smith (2004). Genetic variation in red clover for rumen protein degradability. *Animal Feed Science and Technology*, 113, 157-167.

Buxton, D.R., R.E. Muck & J.H. Harrison (2003). Agronomy Series No. 42. *Silage Science and Technology*. Madison, Wisconsin, USA.

Byrne, C.M., P. O'Kiely, D.J. Bolton, J.J. Sheridan, D.A. McDowell & S.I. Blair (2002). Fate of *Escherichia coli* O157:H7 during silage fermentation. *Journal of Food Protection*, 65, 1854-1860.

Carpintero, C.M., A.R. Henderson & P. McDonald (1979). The effect of some pre-treatments on proteolysis during the ensiling of herbage. *Grass and Forage Science*, 34, 311-315.

Chamberlain, D.G. & J-J. Choung (1995). Recent Advances in Animal Nutrition. In: P.C. Garnsworthy & D.J.A. Cole (eds.) Nottingham University Press. Nottingham, UK, 3-27.

Charmley, E. (1995). Making the most of silage proteins. *Feed Mix*, 3, 28-31.

Charmley, E. (2001). Towards improved silage quality - A review. *Canadian Journal of Animal Science*, 81, 157-168.

Chen, L., C-K. Auh, P. Dowling, J. Bell, F. Chen, A. Hopkins, R.A. Dixon & Z-Y. Wang (2003). Improved forage digestibility of tall fescue (*Festuca arundinacea*) by transgenic down-regulation of cinnamyl alcohol dehydrogenase. *Plant Biotechnology Journal*, 1, 437-449.

Cussen, R.F., R.J. Merry, A.P. Williams & J.K.S. Tweed (1995). The effect of additives on the ensilage of forage of differing perennial ryegrass and white clover content. *Grass and Forage Science*, 50, 249-258.

Davies, D.R., D.K. Leemans & R.J. Merry (2002) Improving silage quality by ensiling perennial ryegrasses high in water soluble carbohydrate content either with or without different additives In: L.M. Gechie & C. Thomas (eds.) *Proceedings International Silage Conference XIII. Scottish Agricultural College (SAC), Auchincruive, Ayr*, 386-387.

Davies, D.R., R.J. Merry, A.P. Williams, E.L. Bakewell, D.K. Leemans & J.K.S. Tweed (1998). Proteolysis during ensilage of forages varying in soluble sugar content. *Journal of Dairy Science*, 81, 444-453.

Davies, D.R., A.L. Winters, D.K. Leemans, M.S. Dhanoa & R.J. Merry (1999). The effect of inoculant treatment of alternative crop forages on silage quality and *in vitro* rumen function. In: T. Pauly (eds.) *Proceedings of the XIIth International Silage Conference. Swedish University of Agricultural Sciences, Uppsala, Sweden*, 131-132.

Dewhurst, R.J., D.R. Davies & R.J. Merry (2000). Microbial protein supply from the rumen. *Animal Feed Science and Technology*, 85, 1-21.

Dewhurst, R.J., W.J. Fisher, J.K.S. Tweed, & R.J. Wilkins, (2003a). Comparison of grass and legume silages for milk production. 1. Production responses with different levels of concentrate. *Journal of Dairy Science*, 86, 2598-2611.

Dewhurst, R.J., N.D. Scollan, J.M. Moorby, R.T. Evans, R.J. Merry & R.J. Wilkins (2003b). Comparison of grass and legume silages for milk production. 2. *In vivo* and *in sacco* evaluations of rumen function. *Journal of Dairy Science*, 86, 2612-2621.

Driehuis F., S.J.W.H. Oude Elferink, & P.G. Van Wikselaar (2001). Fermentation characteristics and aerobic stability of grass silage inoculated with *Lactobacillus buchneri*, with or without homofermentative lactic acid bacteria. *Grass and Forage Science*, 56, 330-343.

Duncan, S.H., H.J. Flint & C.S. Stewart (1998). Inhibitory activity of gut bacteria against *Escherichia coli* 0157 mediated by dietary plant metabolites. *FEMS Microbiology Letters*, 164, 283-288.

Fenlon, D.R. & J. Wilson (2000). Growth of *Escherichia coli* 0157 in poorly fermented laboratory silage: a possible environmental dimension in the epidemiology of *E.coli* 0157. *Letters in Applied Microbiology*, 30, 118-121.

Fisher, D.S., H.F. Mayland & J.C. Burns (1999). Variation in the ruminant's preference for tall fescue hays cut at either sundown or sunup. *Journal of Animal Science*, 77, 762-768.

Garber, L.P., S.J. Wells, D.D. Hancock, M.P. Doyle, J. Tuttle, J.A. Shere & T. Zhao (1995). Risk factors for faecal shedding of *Escherichia coli* 0157:H7 in dairy calves. *Journal of the American Veterinary Medicine Association*, 207, 46-49.

Givens, D.I. & H. Rulquin (2004). Utilisation by ruminants of nitrogen compounds in silage-based diets. *Animal Feed Science and Technology*, 114, 1-18.

Hatfield, R.D., R.E. Muck, M.L. Sullivan & D.A Samac (2003). Nutrient boost for alfalfa silage. Agricultural Research, USDA Beltsville, USA, 51, 20-21.

Howarth R.E. & B.P. Goplen (1983). Improvement of forage quality through production management and plant breeding. *Canadian Journal of Plant Science*, 63, 895-902.

Humphreys, M.W. (1989). Water soluble carbohydrates in perennial ryegrass breeding. II. Cultivar and hybrid progeny performance in cut plots. *Grass and Forage Science*, 44, 237-244.

Jones, B.A., R.D. Hatfield & R.E. Muck (1995a). Characterisation of proteolytic activity in alfalfa and red clover extracts. *Crop Science*, 35, 537-541.

Jones, B.A., R.D. Hatfield & R.E. Muck (1995b). Red clover extracts inhibit legume proteolysis. *Journal of the Science of Food and Agriculture*, 67, 329-333.

Jones, B.A., R.D. Hatfield & R.E. Muck (1995c). Screening legume forages for soluble phenols, polyphenol oxidase and extract browning. *Journal of the Science of Food and Agriculture*, 67, 109-112.

Kingston-Smith, A.H. & H.M. Thomas (2003). Strategies of plant breeding for improved rumen function. *Annals of Applied Biology*, 142, 13-24.

Kingston-Smith, A.H., A.L. Bollard, H.M. Thomas, M.K. Theodorou (2002). The potential to decrease nitrogen losses from pasture by manipulating plant metabolism. In: J.L. Durand, J.C. Émile, C. Huyghe & G. Lemaire (eds.) Grassland in Europe, Volume 7. *Proceedings of the European Grassland Federation Conference. Poitiers, France: Imprimerie P. Oudin, La Rochelle, France*, 136-137.

Kingston-Smith, A.H., A.L. Bollard, I.P. Armstead, B.J. Thomas & M.K. Theodorou (2003). Proteolysis and cell death in clover leaves is induced by grazing. *Protoplasma*, 220, 119-129.

Kung, L., M.R. Stokes & C.J. Lin (2003). Silage Additives. In: D.R. Buxton, R.E. Muck & J.H. Harrison (eds.) Agronomy Series No. 42. *Silage Science and Technology*. Madison, Wisconsin, USA., 305-360.

Lee, M.R.F., L.J. Harris, J.M. Moorby, M.O. Humphreys, M.K. Theodorou, J.C. Macrae & N.D. Scollan (2002). Rumen metabolism and nitrogen flow to the small intestine in steers offered forage diets bred for elevated levels of water-soluble carbohydrates. *Animal Science*, 74, 587-596.

Lee, M.R.F., A.L. Winters, N.D. Scollan, R.J. Dewhurst, M.K. Theodorou & F.R. Minchin (2004). Plant-mediated lipolysis and proteolysis in red clover with different polyphenol oxidase activities. *Journal of the Science of Food and Agriculture*, 84, 1639-1645.

Lindgren, S.E., E. Oldenburg & G. Pahlow (2003). Influence of microbes and their metabolites on food and feed quality. In: J.L. Durand, J.C. Emile, C. Huyghe & G. Lemaire (eds.) Multi-Functional Grasslands. *Proceedings of the 19th European Grassland Federation, La-Rochelle, France*, 503 – 511.

Macrae, J. & M.K. Theodorou, (2003). Potentials for enhancing the animal and human nutrition perspectives of grazing systems. *Aspects of Applied Biology*, 70, 91-100.

McDonald, P., A.R. Henderson & S.J.E. Heron (1991). The Biochemistry of Silage. 2nd Edition Chalcombe publications, Marlow Bucks, UK, 340.

McKersie, D.B. (1985). Effect of proteolysis in ensiled legume forage. *Agronomy Journal*, 77, 81-86.

Merry, R.J. & D.R. Davies (1999). Propionibacteria and their role in the biological control of aerobic spoilage in silage. *Lait*, 79, 149-164.

Merry, R.J., M.S. Dhanoa & M.K. Theodorou (1995a). Use of freshly cultured lactic acid bacteria as silage inoculants. *Grass and Forage Science*, 50, 112-123.

Merry, R.J., M.R.F. Lee, D.R. Davies, J.M. Moorby, R.J. Dewhurst & N.D. Scollan (2003). Nitrogen and energy use efficiency in the rumen of cattle fed high sugar grass and/or red clover silage. *Aspects of Applied Biology*, 70, 87-92.

Merry, R.J., D.K. Leemans & D.R. Davies (2002). Improving the efficiency of silage-N utilisation in the rumen through the use of perennial ryegrasses high in water-soluble carbohydrate content. In: L.M. Gechie & C. Thomas (eds.). *Proceedings International Silage Conference XIII. Scottish Agricultural College (SAC), Auchincruive, Ayr*, 374-375.

Merry, R.J., K.F. Lowes & A.L. Winters (1997). Current and future approaches to biocontrol in silage. In: V. Jambor, L, Klapil, P. Chromec & P. Prochazka (eds.). *Proceedings of the 8th International Symposium on Forage Conservation, Brno. Research Institute of Animal Nutrition Ltd., Pohorelice, Czech Republic*, 17-27.

Merry, R.J., A.L. Winters, P.I. Thomas, M. Muller & T. Muller (1995b). Degradation of fructans by epiphytic and inoculated lactic acid bacteria and by plant enzymes during ensilage of normal and sterile hybrid ryegrass. *Journal of Applied Bacteriology*, 79, 583-591.

Miller, L.A., J.M. Moorby, D.R. Davies, M.O. Humphreys, N.D. Scollan, J.C. Macrae & M.K. Theodorou (2001). Increased concentration of water-soluble carbohydrate in perennial ryegrass (*Lolium perenne L.*). Milk production from late-lactation dairy cows. *Grass and Forage Science*, 56, 4, 383-394.

Miller, P.R. & N.J. Ehlke (1996). Condensed tannins in birds foot trefoil; genetic relationships with forage yield and quality in NC-83 germplasm. *Euphytica*, 92, 383-391.

Muck (1993). The role of silage additives in making high quality silage. In: NRAES-67 Silage Production - from seed to animal. *Proceedings from the National Silage Production Conference. New York*, 106-114.

Muck, R.E. & P. O'Kiely (1992). Aerobic deterioration of Lucerne (*Medicago sativa*) and Maize (*Zea mais*) silages - Effects of fermentation products. *Journal of Science and Food Agriculture*, 59, 145-149.

Muller, M. & D. Lier (1994). Fermentation of fructans by epiphytic lactic acid bacteria. *Journal of Applied Bacteriology*, 76, 406-411.

Nesbakken, T. & M. Broch-Due (1991). Effects of a commercial inoculant of lactic acid bacteria on the composition of silages made from grasses of low dry matter content. *Journal of the Science of Food and Agriculture*, 54, 177-190.

Nocek, J.E., W.P. Kautz, J.A.Z. Leedle & J.G. Allman (2002). Ruminal supplementation of direct-fed microbials on diurnal pH variation and *in situ* digestion in cattle. *Journal of Dairy Science*, 85, 429-433.

O'Kiely, P, & R.E. Muck (1992). Aerobic deterioration of Lucerne (*Medicago sativa*) and Maize (*Zea mais*) silages - Effects of Yeasts. *Journal of Science and Food Agriculture*, 59, 139-144.

Oldenburg, E. (1999). Fungal secondary metabolites in forages: Occurrence, biological effects and prevention. *Landbauforschung Volkenrode*, 206, 91-109.

Rammer, C., P. Lingvall & I. Thylin (1999). Combinations of biological and chemical silage additives. In: T Pauly (eds.). *Proceedings of the 12th Silage Conference. Uppsala, Sweden*, 327-328.

Romanov, M.N., R.V. Bato, Yokoyama & S.R. Rust (2004). PCR detection and 16S rRNA sequence-based phylogeny of a novel *Propionibacterium acidipropionici* applicable for enhanced fermentation of high moisture corn. *Journal of Applied Microbiology*, 97, 38-47.

Rooke, J.A., N.H. Lee & D.G. Armstrong (1987). The effects of intraruminal infusions of urea, casein, glucose syrup and a mixture of casein and glucose syrup on nitrogen digestion in the rumen of cattle receiving grass-silage based diets. *British Journal of Nutrition*, 57, 89-94.

Sharp, R., A.G. O'Donnell, H.G. Gilbert & G.P. Hazlewood (1992). Growth and survival of genetically manipulated *Lactobacillus plantarum* in silage. *Applied and Environmental Microbiology*, 58, 2517-2522.

Sullivan, M.L., R.D. Hatfield, S.L. Thoma & D.A. Samae (2004). Cloning and characterization of red cover polyphenol oxidase cDNA's and expression of active protein in Escherichia coli and transgenic alfalfa. *Plant Physiology*, 136, 3234-3244.

Tannock, G.W. (1999). Introduction. In: G.W. Tannock (eds.) Probiotics: A Critical Review. Horizon Scientific Press, Wymondham, UK, 1-4.

Thomas, H. & C.M. Smart (1993). Crops that stay green. *Annals of Applied Biology*, 123, 193-219.

Weinberg, Z.G. & R.E. Muck (1996). New trends and opportunities in the development and use of inoculants for silage. *FEMS Microbiology Reviews*, 19, 53-68.

Weinberg, Z.G., G. Asbell, Y. Hen & Azriela (1993). The effect of applying lactic acid bacteria at ensiling on the aerobic stability of silage. *Journal of Applied Bacteriology*, 75, 512-518.

Weinberg, Z.G., R.E. Muck, P.J. Weimer, Y. Chen & M. Gamburg (2004). Lactic acid bacteria used in inoculants for silage as probiotics in ruminants. *Applied Biochemistry and Biotechnology*, 118, 1-9.

Wilkinson, J.M. & M.I. Toivonen (2003). World Silage. Chalcombe Publications, Lincoln, UK, 204.

Winters, A.L. & F.R. Minchin (2002). The effect of PPO on the protein content of ensiled red clover. In: L.M. Gechie & C. Thomas (eds.). *Proceedings International Silage Conference XIIIth Scottish Agricultural College (SAC), Auchincruive, Ayr*, 84-85.

Winters, A.L., R. Fychan & R. Jones (2001). Effect of formic acid and a bacterial inoculant on the amino acid composition of grass silage and on animal performance. *Grass and Forage Science*, 56, 181-192.

Winters, A.L., R.J. Merry, M. Muller, D.R. Davies, G. Pahlow & T. Muller (1998). Degradation of fructans by epiphytic and inoculant lactic acid bacteria during ensilage of grass. *Journal of Applied Microbiology*, 84, 304-312.

Woolford, M.K. (1990). The detrimental effects of air on silage. *Journal of Applied Bacteriology*, 68,101-116.

Section 1

A. Effect of conserved feeds on milk production

The effect of grass silage chop length on dairy cow performance

The Norwegian University of Life Sciences, P.O. Box 5003, NO-1432 Ås, Norway, Email: ashild.randby@umb.no

Keywords: grass silage, chopping length, dairy cows, intake, naked oats

Introduction Good compaction of herbage is essential for production of well fermented silage. Self loading wagons, which often have rather poor chopping capacities, may be a challenge for the silage quality in bunker silos. The objective of this study was to evaluate the effect of silage chop length on dairy cow performance.

Materials and methods The herbage from a timothy-dominated re-growth sward (D-value 650 and 120 g/kg DM CP) was wilted for 5-12 h to 240 g/kg DM and treated with 4 l/t of GrasAAT (Norsk Hydro; 645 g/kg formic acid). The crop was harvested with either a precision chop harvester (PC, Taarup 602B) with 32 knives and a theoretical chopping length (TCL) of 19 mm or a self loading wagon (SLW, Krone Titan 4/25 L) with 25 knives and a TCL of 56 mm. The median chopping lengths (MCL) of the harvested crop was measured to be 41 mm for PC and 107 mm for SLW. In each of two bunkers of 6 m x 24 m, 88 t were ensiled during a total of 22 h on August 7-8 in 2001. The last loads (12 t) for each silo were harvested at 157 g/kg DM due to rain. The herbage was continuously compacted between the emptying of each load into the silos, using two 6.4-t tractors. The total compaction time was 6.9 and 8.4 min/t for the PC and SLW silo, respectively. Each of the two silages was fed *ad libitum* to 18 mid lactation Norwegian Red dairy cows in a continuous design experiment. Two concentrates (176 g/kg DM CP), produced from the same ingredients, but containing 246 g/kg of either common oats or naked oats (var. *Bikini*), were also evaluated.

Results and discussion Both silages were well fermented with no butyric acid. The concentrations of DM, water-soluble carbohydrates, lactic acid, acetic acid and ethanol, NH_3 and pH, were 244 g/kg, 37 g/kg DM, 70 g/kg DM, 28 g/kg DM, 9.0 g/kg DM, 62 g/kg N and 3.85 for PC silage and 249 g/kg, 43 g/kg DM, 78 g/kg DM, 28 g/kg DM, 6.6 g/kg DM, 65 g/kg N and 3.88 for SLW silage respectively. Prior to feeding, the silages were further chopped using Serigstad RBK 1202 roundbale chopper. MCL was measured to be 22 mm for PC silage and 67 mm for SLW silage when fed to the cows. Precision chopped silage increased feed intake ($P<0.001$) and milk yield ($P<0.01$) relative to SLW. The lower fibre concentration and higher fat and energy concentration in naked vs. common oats did not influence silage intake or milk yield, but increased the BW gain and the body condition score (BCS) of the dairy cows.

Table 1 Effect of grass silage chop length on feed intake, milk yield and composition, and changes in body weight and condition score

	Treatment					PC vs SLW silage[§]		Naked vs common oats[§]	
Silage	PC	PC	SLW	SLW	s.e.m.	Effect	Sig.	Effect	Sig.
Type of oats	comm.	naked	comm.	naked					
Feed intake (kg DM/d)									
Silage	12.4	12.2	11.1	11.2	0.28	1.1	***	-0.1	NS
Concentrates	5.65	5.69	5.66	5.74					
Total	18.0	17.9	16.8	16.9	0.28	1.1	***	± 0	NS
DM (g/kg BW)	31.1	31.3	29.7	29.7	0.65	1.5	*	0.1	NS
NDF (g/kg BW)	12.9	12.6	12.0	11.7	0.30	0.9	**	-0.3	NS
Animal performance									
Milk (kg)	22.0	21.4	20.6	21.0	0.32	0.9	*	-0.2	NS
ECM (kg)	22.4	22.0	21.1	21.3	0.30	1.0	**	-0.1	NS
Fat (g/kg)	42.9	43.2	43.3	42.8	0.8	± 0	NS	-0.1	NS
Protein (g/kg)	32.4	32.9	32.6	32.3	0.3	0.2	NS	0.1	NS
Lactose (g/kg)	45.2	45.4	45.5	45.2	0.2	-0.1	NS	± 0	NS
Milk taste score[#]	4.02	3.52	4.24	4.20	0.13	-0.45	**	-0.27	NS
FFA (meq/l)	0.89	0.96	0.88	0.80	0.05	0.08	NS	± 0	NS
Urea (mM)	3.85	3.83	4.10	4.26	0.11	-0.34	**	0.06	NS
α-tocoph. (mg/l)	0.68	0.77	0.91	0.80	0.06	-0.13	*	-0.01	NS
BW change (g/d)	194	359	66	258	75	115	NS	179	*
BCS/100 d	-0.01	0.21	-0.04	0.04	0.06	0.09	NS	0.15	*

[#] Five-point scale where 1 = poor quality milk and 5 = high quality milk with no deviation from normal taste.
[§] No silage by concentrate interaction.

Conclusions Well fermented grass silage may be produced using a self loading wagon, if the silage mass is well compacted. Silage intake and animal production may be decreased compared with feeding of precision chopped silage due to the effect of chop length, if the silage is not thoroughly chopped (20-30 mm MCL) prior to feeding.

Silage production and utilisation

Whole crop silage from barley fed in combination with red clover silage to dairy cows

J. Bertilsson and M. Knicky
The Swedish University of Agricultural Sciences (SLU), Department of Animal Nutrition and Management, Kungsängen Research Centre, S-753 23 Uppsala, Sweden. Email: jan.bertilsson@huv.slu.se

Keywords: feeding, nitrogen partitioning, protein efficiency, stage of maturity, milk stage, dough stage

Introduction Grass silage is the basic feed in Swedish dairy cow rations. The nitrogen utilisation in this type of diet is, however, low. A combination of forage legume protein and whole crop silage carbohydrates might be a solution to this problem. From other countries in Northern Europe the experience from feeding barley whole crop silage in combination with legumes is that it is possible to maintain a reasonably high milk production and at the same time have a good protein utilisation (Kristensen, 1992).

Material and methods Whole crop silage (WCS) of barley was made at two stages of maturity; either at milk stage or at early dough stage. Red clover silage was from a second cut. These three silages were made in the form of round big bales covered with 6 layers of plastic. Kofasil Ultra[TM] was used as an additive. Barley silage from both cuts was mixed with clover silage either at 40/60 or 70/30 (DM basis), giving four experimental treatments. Precision chopped clover/grass silage of high quality (10.8 MJ ME; 18% CP) from a first cut stored in a tower silo was used as a control. All silages were fed *ad libitum* to dairy cows in mid to late lactation in combination with a fixed amount of 7.2 kg DM concentrate. The feeding was according to a balanced, incomplete changeover design with 15 cows, 3 blocks, 3 periods and 5 treatments. Total collection of faeces (5 days) and urine (3 days) was performed for five of the cows in each period. The cows were of the Swedish Red and White Breed and had an average live weight of 670 kg.

Results The later cut of barley led to an increase in DM content from 32.0 to 37.5%, an increase for starch from 13.3 to 16.6% (in DM) and a decrease in sugar from 15.8 to 10.2% (in DM). Contents of ash, protein and fibre showed relatively small changes between cutting dates. The inclusion of 40% WCS gave very similar production results as feeding a pure clover/grass silage, while 70% WCS in the mix gave lower milk production. Protein content in milk increased at the highest inclusion of WCS. N in milk and faeces increased, while N in urine decreased drastically as WCS increased as a proportion of the silage.

Table 1 Production results and nitrogen efficiency. LS-means per cow and day

	n[1]	Grass silage	WCS1-40[2]	WCS2-40	WCS1-70	WCS2-70	s.e.	P<	LSD[3]
Silage intake (kg DM)	9	13.2	13.0	14.9	12.7	14.6	0.5	0.0007	1.1
Milk (kg)	9	23.4	22.4	23.3	20.6	21.9	1.1	0.04	1.9
Protein content (%)	9	3.47	3.49	3.55	3.69	3.58	0.11	0.02	0.13
% of N in feeds									
N in milk	3	21.3	22.8	24.5	26.1	26.1	1.6	0.08	4.0
N in faeces	3	28.8	43.5	43.4	46.6	51.2	2.5	0.004	6.7
N in urine	3	46.8	40.4	39.6	34.3	29.2	2.9	0.05	10.6

[1]n = no of observations behind a LS-mean; [2]WCS1-40 = whole crop silage, cut 1, 40% of DM in mix; [3]LSD = least square difference

Conclusions Combinations of whole crop silage from barley and red clover were consumed at the same level as clover/grass silage. Milk production tended to be lower at high proportions of WCS while protein content in milk and protein efficiency increased. N in urine decreased at the same time. This gives prerequisites for a lower nitrogen loss to the environment.

References
Kristensen, V.F. (1992). The production and feeding of whole-crop cereals and legumes in Denmark. Chapter 12 in Whole-crop cereals (eds. Stark and Wilkinson), Chalcombe publications, 21-37.

Responses to grass or red clover silages cut at two stages of growth in dairy cows

A. Vanhatalo[1], K. Kuoppala[2], S. Ahvenjärvi[2] and M. Rinne[2]
[1]University of Helsinki, Department of Animal Science, P.O. Box 28, FI-00014 University of Helsinki, Finland, Email: aila.vanhatalo@helsinki.fi, [2]MTT Agrifood Research Finland, FI-31600 Jokioinen, Finland

Keywords: legume, maturity, D-value, plasma metabolite

Introduction Red clover has an important role in organic farming, and also potential to reduce dependence on N fertilisers in conventional farming. This experiment compared dairy cow responses to grass and red clover silages cut at two stages of growth.

Materials and methods Four silages were made from primary growth: two grass silages from timothy (*Phleum pratense*) and meadow fescue (*Festuca pratensis*) grass (G) and two red clover (*Trifolium pratense*) silages (R). G silages were harvested on 17 June at early (G_E) and on 26 June at late (G_L) growth stage and R silages on 2 July at early (R_E) and on 16 July at late (R_L) growth stage. The preparation and ensiling of G and R silages are further described in the companion paper (Kuoppala *et al.*, 2005). These four pure silages and a mix of G_L and R_E ($G_L R_E$, 1:1 on dry matter (DM) basis) were fed *ad libitum* with 9 kg/d concentrates to five rumen cannulated cows in a 5x5 Latin square experiment. During the collection period silage DM intake was restricted to 95% of the adaptation period intake. The *in vivo* D-values (digestible organic matter in feed DM) of the silages were measured with sheep. Feed intake and milk yield of cows were recorded daily. Blood from the tail vein was sampled three times during the feeding cycle on the last day of the experimental period.

Results The D-values and crude protein concentrations (g/kg DM) of the experimental silages were 714, 673, 678, 610 and 134, 111, 212 and 181 for G_E, G_L, R_E and R_L, respectively. The advancing growth stage tended to increase rather than decrease DM intake of R silages while opposite was true for G silages (Table 1). However, the highest DM intake was found with $G_L R_E$ mix. Despite differences in silage DM intakes no significant differences in milk yields of cows between the treatments were observed. Milk fat and protein contents were lower but lactose content higher with R than with G silages. Plasma non-esterified fatty acids (NEFA), acetate, essential amino acids (EAA) and urea concentrations were higher for R than for G silages.

Table 1 Feed intake, milk production and plasma metabolites

	G_E	G_L	R_E	R_L	Mix $G_L R_E$	SEM[#]	Statistical significance C_1	C_2	C_3	C_4
Intake										
Silage DM (kg/d)	13.2	12.0	11.3	12.1	14.0	0.49	o		o	**
Total DM (kg/d)	21.2	20.1	18.8	20.2	21.5	0.59	o		o	*
Production										
Milk (kg/d)	27.1	25.6	27.7	27.4	27.8	0.94				
ECM (kg/d)	26.8	25.4	25.5	26.2	26.6	0.79				
Fat (g/kg)	40.9	41.2	37.4	39.4	38.6	0.76	**			
Protein (g/kg)	32.6	32.6	30.4	30.8	31.5	0.38	***			
Lactose (g/kg)	46.1	46.6	47.0	46.7	46.9	0.21	*			
Urea (mM)	2.38	2.32	6.15	5.94	4.15	0.186	***			
Milk N/Feed N	0.285	0.315	0.223	0.237	0.248	0.0077	***	*		o
Plasma metabolites (mM)										
Glucose	3.67	3.61	3.57	3.68	3.63	0.104				
NEFA	0.105	0.107	0.131	0.157	0.117	0.0121	*			
Acetate	1.34	1.03	1.30	1.48	1.20	0.08	*		*	
EAA	0.78	0.79	1.27	1.07	1.02	0.072	***			
Urea	3.02	2.84	7.10	6.93	5.10	0.301	***			

[#]SEM for diet $G_L R_E$ should be multiplied by 1.19. Contrasts: C_1=G *vs.* R; C_2=E *vs.* L; C_3=C_1 x C_2; C_4=G_L, R_E *vs.* Mix. Significance: *** = $P<0.001$, ** = $P<0.01$, * = $P<0.05$, o = $P<0.10$

Conclusions Advancing maturity decreased DM intake with grass silage, whereas the opposite was true for red clover silage. In spite of tendency for lower DM intakes, R silages supported as high ECM yields as G silages. This was associated with higher plasma concentrations of NEFA, acetate and EAA with R rather than G silages. Lower feed N efficiencies with R silages were reflected in high plasma and milk urea concentrations.

References
Kuoppala, K., S. Ahvenjärvi, M. Rinne & A. Vanhatalo (2005). NDF digestion in dairy cows fed grass or red clover silages cut at two stages of growth. *Proceedings of the XIV International Silage Conference*. Paper 1132.

The effect of chop length and additive on silage intake and milk production in cows

V. Toivonen and T. Heikkilä
MTT Agrifood Research Finland, Animal Nutrition, FIN-31600 Jokioinen, Finland, Email: vesa.toivonen@mtt.fi

Keywords: silage, chop length, additive, intake, dairy cows

Introduction Effects of reduced silage chop length on silage intake and milk production by dairy cows have been variable. Chopping of grass at harvest generally improves silage fermentation quality and consequently intake. The objective of this study was to assess the effect of chopping flail harvested silage ensiled with acid or biological additive prior to feeding on intake, milk yield and milk composition in dairy cows.

Materials and methods Eight Ayrshire cows were used in two 4 x 4 Latin squares, one with cows in their first lactation and one with cows in their $2^{nd} – 4^{th}$ lactation. Silages were made directly cut with a flail-harvester from the 2nd cut of meadow fescue-cocksfoot-timothy grass using either lactic acid bacteria + cellulase enzyme (Biol) or formic acid-based (Acid = 800 g/kg formic acid + 20 g/kg orthophosphoric acid, 5 l/t) additive. Silages were made in bunker silos and stored about 3 months before the feeding experiment started. Silages were fed *ad libitum* unchopped or chopped with a precision chopper prior to feeding. Experimental treatments were: unchopped flail-harvested Biol silage (Flail/Biol), precision chopped Biol silage (Precis/Biol), unchopped flail-harvested Acid silage (Flail/Acid) and precision chopped Acid silage (Precis/Acid). The mean chop length of Flail-silages was 13.5 cm and that of Precis-silages 6 cm. A concentrate mixture of barley-oats-rapeseed meal-minerals (40.5-40.5-15-4 g/kg) was given 8 kg/d. Digestibility of feeds was determined with four wether sheep and used to calculate metabolisable energy (ME). Data were analysed statistically using the SAS GLM procedure. Square, animal, period and treatment effects and square*period and square*treatment interactions were used in the model, and the sums of squares for the treatment effects were further divided into contrasts.

Results Chemical composition of silages is presented in Table 1. Fermentation quality of both silages was good. Biol silage had lower pH and sugar content and higher fermentation acid and ammonia contents than Acid silage. Silage chop length had no significant effect on silage intake while the intake of Acid silages was higher than that of Biol silages. Chop length had only minor and insignificant effect on milk yield but additive had a more prominent effect Acid silage treatments produced a higher yield than Biol silage treatments, apparently because of differences in silage intake between additive treatments. Chop length had no clear-cut effect on fat, protein or lactose content of milk. The same was true for additive, except a higher fat content of milk in Acid than Biol silage treatments. Milk energy output (MJ/d) per metabolisable energy intake (MJ/d) and efficiency of utilisation of dietary crude protein for milk protein production was better with precision chopped silage than with flail harvested silage. Biol silage tended to increase feed energy conversion to milk energy compared with Acid silage

Table 1 Chemical composition and fermentation quality of silages (dry matter g/kg, others g/kg DM)

Additive	Dry matter	Ash	Crude protein	Neutral detergent fibre	Acid detergent fibre	pH	Sugar	Lactic acid	Acetic acid	Prop. acid	Butyr acid	NH₄-N g/kg N
Biol silage	225	101	148	511	290	4.09	38	91	20	1.0	1.5	69
Acid silage	227	99	144	535	313	4.12	78	39	11	0.0	0.1	36

Table 2 The effect of chop length and silage additive on silage intake, milk yield and milk composition in cows

Chopper/ Additive	Intake				Yield			Milk composition			Utilisation	
	Silage	Conc.	Milk	ECM	Fat	Prot.	Lact.	Fat	Prot.	Lact.	Milk E/ Feed E	Milk N/ Feed N
	----------------kg/d----------------				-------------g/d-----------			---------g/kg----------				
Flail / Biol	10.2	7.1	20.5	21.4	896	658	992	44.1	32.3	48.4	0.341	0.243
Precis / Biol	10.0	7.0	20.5	21.6	913	661	999	44.8	32.4	48.8	0.360	0.247
Flail / Acid	11.5	7.0	21.0	22.4	953	680	1024	45.5	32.4	48.7	0.332	0.239
Precis /Acid	11.4	7.0	21.3	23.0	984	698	1038	46.3	32.8	48.6	0.348	0.247
SEM	0.29	0.03	0.31	0.31	15.6	9.6	16.4	0.67	0.30	0.19	0.005	0.003
Flail vs Precis	NS		NS	NS	NS	NS	NS	NS	NS	NS	**	*
Biol vs Acid	***		*	**	**	*	*	*	NS	NS	P=0.08	NS

Significance: NS = non significant, * $P<0.05$, ** $P<0.01$, *** $P<0.001$

Conclusions Chopping of silage prior to feeding, when fermentation quality was similar had no effect on intake or milk yield and milk composition but utilisation of feed energy and protein were better with precision-chopped silage. Restriction of silage fermentation with formic acid compared with stimulation with lactic acid bacteria + cellulase enzyme additive increased silage intake and milk production.

Effect of supplementing grass silage with incremental levels of water soluble carbohydrate on *in vitro* rumen microbial growth and N use efficiency

D.R. Davies, D.K. Leemans and R.J. Merry

Institute of Grassland and Environmental Research, Plas Gogerddan, Aberystwyth, SY23 3EB, UK, Email: david.davies@bbsrc.ac.uk

Keywords: silage, water soluble carbohydrate, rumen, *in vitro*, N use efficiency

Introduction The efficiency of utilisation of N for milk production in dairy cows is often less than 25% and a shortfall in readily available carbohydrate to provide substrate for microbial growth is often cited as a potential problem. Grasses bred for their high water soluble carbohydrate (WSC) content have potential to address this issue but there is limited information on the level of sugar required to optimise rumen microbial growth efficiency. The objective of this experiment was to examine the effect of different concentrations of WSC on the efficiency of use of grass silage N under *in vitro* conditions.

Materials and methods Silage was prepared from a third cut of perennial ryegrass (cv Fennema) after treatment with a silage inoculant (Powerstart; Genus plc, Nantwich; 10^6 cfu/g FM). An 8-vessel simulated rumen culture system (Rusitec) (Czerkawski & Breckenridge, 1977) was used and the experiment repeated on two occasions with fresh rumen fluid, giving 4 replicates per treatment. There were 4 treatments, basal silage alone (15 g DM/d) or mixed with 56, 97 or 168 g/d of a sucrose and inulin (fructan) mixture, ratio 60:40, prior to feeding. The silages ± sugar were fed to cultures every 24 h. Artificial saliva plus ($^{15}NH_4)_2SO_4$ as a microbial marker was infused continuously into the culture vessels at a rate of 0.7 volumes/d. Samples of washed bacteria, vessel contents and effluent were taken 7 d after the addition of ($^{15}NH_4)_2SO_4$. Flows of organic matter and ammonia-N concentration in the effluents were estimated and daily microbial N flows and efficiency were calculated from values for ^{15}N enrichment of the effluent and harvested microbial fractions. Statistical analysis utilised Genstat (2003) two-way analysis of variance and least significant differences.

Results The silage composition, prior to WSC supplementation was: DM, 270 g/kg; pH, 3.59; total N, 29.1 g/kg DM; ammonia-N, 75.6 g/kg TN; lactic acid, 121.7 g/kg DM and WSC, 14.9 g/kg DM. Thus, actual feed WSC concentrations were the equivalent of 15 (basal silage) and 71, 112 and 183 g of WSC/kg silage DM after sugar supplementation. Values for rumen parameters are shown in Table 1. Organic matter digestion and microbial N production increased up to the highest WSC level of 183 g/kg DM. This was reflected by the values for efficiency of microbial protein synthesis in terms of energy supply, where a small but non significant increase from 25 to 26 g microbial N (MN)/kg organic matter apparently digested (OMAD) was generally observed across all levels of supplementation, compared to the basal silage. A marked response was observed in N use efficiency (g MN/g feed input) for all levels of WSC supplementation, rising from 57% with the basal silage to 74% at 183 g WSC/kg DM.

Table 1 Effect of silage WSC concentration on rumen fermentation parameters in Rusitec

	WSC Concentrations (g/kg DM)				
	15	71	112	183	sed
OMAD (g/d)	9.39[a]	10.42[b]	10.75[b]	11.94[c]	0.167
Microbial N (MN; g/d)	0.250[a]	0.295[bc]	0.271[ab]	0.326[c]	0.0158
EMPS (g MN/kg OMAD)	26.96	28.65	25.84	27.37	1.735
Ammonia-N (mmol/l)	8.1[a]	6.79[b]	6.17[b]	3.72[c]	0.393
N use Efficiency (g MN/g feed N input)	0.57[a]	0.68[bc]	0.62[ab]	0.74[c]	0.037

OMAD, organic matter apparently digested; EMPS, efficiency of microbial protein synthesis. Values in rows with different superscripts differ significantly $P<0.001$.

Conclusions Microbial protein synthesis and N use efficiency increased up to the highest level of 18% WSC, which supports the view that low residual WSC in grass silage can contribute to the poor efficiency with which silage-N is used by ruminants. It also provides useful information for breeders developing new grass cultivars.

Acknowledgements Funding was provided by DEFRA and the EU framework V (QLK5-CT-2001-0498) programme.

References

Czerkawski, J. & G. Breckenridge (1977). Design and development of a long-term rumen simulation technique (Rusitec). *British Journal of Nutrition*, 38, 371–385.

Genstat (2003). Genstat 7. Lawes Agricultural Trust: Clarendon Press.

Effects of access time to feed and sodium bicarbonate in cows given different silages

T. Heikkilä and V. Toivonen

MTT Agrifood Research Finland, Animal Nutrition, FIN-31600 Jokioinen, Finland, Email: terttu.heikkila@mtt.fi

Keywords: silage, access time, sodium bicarbonate, additive, dairy cow

Introduction Animal performance is closely related to silage intake, which might be affected by access time to feed, silage fermentation quality or using neutralising agents. The aim of this study was to assess the effects of access time to silage or addition of sodium bicarbonate on silage intake, milk yield and milk composition in diets based on restrictively fermented acid-treated or extensively fermented enzyme-treated grass silage.

Materials and methods Sixteen Ayrshire cows were used in four (4 x 4) Latin square designs with four 3-wk periods of which the results of the last week were used. The treatments in squares 1 and 2 were 8 or 20 h free access either to acid treated (Acid) or enzyme treated (Enz) grass silage. Addition of $NaHCO_3$ 0 or 100 g/d and 0 or 200 g/d in concentrate (12.5 or 25 g/kg) with Acid or Enz silages were studied in separate squares 3 and 4. Results are presented over squares because treatment effects did not differ significantly between squares. The first cut timothy dominated (67%) grass was direct cut with a flail-harvester using either formic acid-based (800 g/kg HCOOH + 20 g/kg H_3PO_4 4.8 l/t) or enzyme additive (cellulase + hemicellulase) and ensiled into bunker silos. A concentrate mixture (barley-oats-rapeseed meal-minerals: 430-430-100-40 g/kg) was given 8 kg/d and hay 1 kg/d. Digestibility of silages was determined with four wethers.

Results Acid and Enz silages contained dry matter (DM) 224 and 219 g/kg, crude protein 168 and 159 and neutral detergent fibre 529 and 484 g/kg DM, respectively. Both silages were well preserved: pH 4.03 vs 3.95, sugar 54 vs 19, lactic acid 37 vs 103, acetic acid 14 vs 27, butyric acid 0.1 vs 0 g/kg DM and NH_4-N 38 vs 68 g/kg N, respectively. Organic matter digestibility of Enz silage tended to be lower ($P<0.1$) than that of Acid silage (0.757 vs 0.740) mainly due to lower fibre digestibility (cellulose 0.814 vs 0.777, $P<0.05$). Longer access time to silage (20 vs 8 h) tended to increase silage intake with Acid (4%) and Enz (8%) silages but the effect on milk (2 or 3%), fat, protein and lactose yields was statistically insignificant. Sodium bicarbonate had no significant effect on silage intake or milk production parameters either with Acid or Enz silage. Instead, silage fermentation quality affected significantly silage intake and, consequently, milk, energy corrected milk (ECM), fat and protein yield and fat and protein content, being higher in cows given Acid than Enz silage. Milk coagulation was affected so that curd firmness time was shorter (K_{20} 15.6 vs 19.5 min) and curd firmness better (A_{10}, 27.5 vs 24.0 mm) with Acid than Enz silage. No differences between treatments were found in milk energy output (MJ/d) per metabolisable energy (ME) intake (MJ/d), but efficiency of utilisation of dietary crude protein for milk protein production was better with Enz than Acid silage.

Table 1 Effect of access time or $NaHCO_3$ on silage intake and milk production in cows fed Acid or Enz silage

	Intake			Yield				Milk composition			Utilisation	
Access time /	Silage	Conc.	Milk	ECM	Fat	Prot	Lact	Fat	Prot	Lact	Milk E/	Milk N/
Additive	---------------kg/d---------------				------------g/d---------			----------g/kg--------			Feed E	Feed N
8 h / Acid	10.6	6.8	22.8	26.4	1176	772	1128	51.6	33.8	49.4	0.393	0.274
20 h / Acid	11.0	6.8	23.2	26.4	1158	780	1158	49.8	33.6	49.9	0.387	0.272
8 h / Enz	8.6	6.9	21.5	23.6	1005	710	1087	46.7	33.1	50.6	0.398	0.295
20 h / Enz	9.3	6.9	22.2	24.3	1032	722	1133	46.5	32.7	51.0	0.392	0.286
SEM	0.25	0.06	0.44	0.42	24.3	14.0	22.6	1.16	0.23	0.20	0.007	0.005
8 h vs 20 h	P=0.06									*		
Acid vs. Enz	***		*	***	***	***		**	**	***		**
$NaHCO_3$/Additive		Significance	* $P<0.05$.	** $P<0.01$.	*** $P<0.001$							
– / Acid	11.9	7.0	25.2	28.1	1216	837	1256	48.6	33.4	49.9	0.390	0.271
+ / Acid	12.0	6.8	24.7	27.6	1198	819	1239	48.6	33.1	50.2	0.388	0.267
– / Enz	10.0	6.9	23.5	25.1	1053	762	1183	45.1	32.6	50.3	0.390	0.285
+ / Enz	9.8	6.9	23.5	24.9	1046	753	1184	44.9	32.3	50.5	0.395	0.289
SEM	0.14	0.05	0.43	0.44	20.0	11.9	23.5	0.53	0.20	0.19	0.007	0.004
$NaHCO_3$ – vs +												
Acid vs Enz	***		**	***	***	***	*	***	**	P=0.07		***

Conclusions Access time to silage of 20 h compared with 8 h only tended to increase silage intake. Instead, silage fermentation quality affected significantly animal performance. Restricted silage fermentation resulted in higher silage intake and milk production than extensive fermentation. Sodium bicarbonate had no benefit for intake or milk production parameters.

Dairy cow performance associated with two contrasting silage feeding systems

C.P. Ferris, D.C. Patterson, R.C. Binnie and J.P. Frost
The Agricultural Research Institute of Northern Ireland, Hillsborough, Co. Down BT26 6DR, UK,
Email: conrad.ferris@dardni.gov.uk

Keywords: dairy cows, silage feeding systems

Introduction As a result of increasing labour costs, the lack of skilled labour, and the desire of many farmers to reduce their working hours, there is considerable interest in using simple feeding systems for dairy cows. A study was conducted to compare two silage feeding systems that differed in complexity.

Materials and methods This study involved eighty-six winter calving (mid September – late February) Holstein-Friesian dairy cows. Twenty-four of these animals were primiparous, while the remainder were in their second lactation. Animals were of high genetic merit, and had a predicted transmitting ability for fat + protein yield (PTA_{2000}) of 37 kg. Animals were allocated to two winter feeding systems, CD and EF (43 animals per system), within 48 h of calving, and remained on these systems until 9 April, a mean of 146 d. The rations offered comprised grass silage, maize silage (introduced into the ration at proportionally 0.3 of forage DM from 13 November onwards) and concentrates. The level of concentrate supplementation was increased incrementally from calving until d 20 post-calving (4.0, increasing to 10.4 kg/d with primiparous animals; and 6.0, increasing to 13.0 kg/d with second lactation animals). Of the daily concentrate allowance, 1.0 kg/d was offered through the milking parlour at the time of milking, 0.5 kg at each milking, the remainder being offered as detailed below. With treatment CD, the forage and concentrate components were offered in the form of a 'complete diet', with this ration prepared using a 'mixer wagon'. This mixed ration was prepared daily, and offered via a series of feed boxes, access to which was controlled via a Calan gate feeding system. An average of three animals shared each Calan gate. With system EF, animals were offered the forage component of the ration twice weekly, in quantities sufficient for the following three- or four-day period. Silage blocks were placed along a feed passage, perpendicular to a series of feed barriers, with maize silage and grass silage blocks 'inter-mixed' along the barriers. The feed barriers used were mounted on wheels, while a hinge mechanism allowed the barriers to extend 112 cm beyond their 'resting' position. Thus cows were able to push the barriers out whilst eating their way through the blocks of silage placed along the barriers. The feed barriers were subdivided into individual 'dovetail' feed spaces, with an average of three animals sharing each feed space. With system EF, the concentrate component of the ration, was offered via electronic out-of-parlour feed stations.

Results While the grass silage offered had a relatively poor feed value (dry matter and crude protein concentrations of 216 g/kg and 109 g/kg DM respectively), the maize silage offered had a high feed value (dry matter and starch concentrations of 282 g/kg and 241 g/kg DM respectively). As a consequence of the feeding systems used, individual animal intakes were not measured. Nevertheless total DM intakes, based on group intake data, were similar for both treatments. With treatment EF, it was noticeable that animals initially selected maize silage in preference to grass silage, with grass silage consumed once maize silage became inaccessible. Silage feeding system had no significant effect on milk yield or milk composition and body tissue reserves were similar with both treatments at the end of the study ($P>0.05$). These findings are in close agreement with those of an earlier study (Ferris *et al.*, 2002).

Table 1 Animal performance with two winter feeding systems

	CD	EF	s.e.m.	Significance
Total dry matter intake (kg/d)	18.7	18.5		
Total milk output during winter period (kg)	4170	4264	106.5	NS
Daily milk yield (kg)	30.0	30.6	0.75	NS
Milk fat (g/kg)	41.8	40.2	0.75	NS
Milk protein (g/kg)	33.9	33.9	0.38	NS
Live weight at end of study (kg)	561	556	5.5	NS

Conclusions Animal performance was unaffected by the two different feeding systems compared in this study. The results suggest that simple silage feeding systems can be adopted without adverse effects on animal performance.

Acknowledgements Funded by DARDNI, AgriSearch, John Thompson and Sons Ltd and Devenish Nutrition.

Reference:
Ferris, C.P., R.C. Binnie, J.P. Frost & D.C. Patterson (2002) A comparison of two silage feeding systems, involving different labour inputs, for dairy cows. *Proceedings of the XIIIth International Silage Conference,* Auchincruive, Scotland, pp. 382–383.

Pea-barley bi-crop silage in milk production

M. Tuori, P. Pursiainen, A.-R. Leinonen and V. Karp
Helsinki University, Department of Animal Science, FIN-00014 Helsinki, Finland,
Email: mikko.tuori@helsinki.fi

Keywords: pea, barley, bi-crop silage, milk production, microbial protein supply

Introduction Whole crop silage (WCS) from barley or wheat has many advantages as roughage feed. The possibility to use the same harvest machinery as in harvesting grass reduces investment costs. The farms which are specialised in grass production may have shortage of open field area for manure spreading, in which case WCS can be the answer. However, digestibility and protein content of WCS is usually lower than in grass silage, which is limiting the feed intake and performance of the dairy cows. Cultivation of grains with grain legumes increases digestibility and protein content of the stand (Lunnan, 1988). Feeding of bi-crop pea-wheat silages has increased forage intake and milk yield compared to grass silage (Salawu *et al.*, 2002; Adesogan *et al.*, 2004). In this experiment pea-barley bi-crop silage was studied since in Finland barley harvested for WCS is more digestible than wheat.

Materials and methods Barley (*Hordeum vulgare* var. Mette) and pea (*Pisum sativum* var. Perttu) were intercropped on a 2.4 ha field in Helsinki, Finland (60E N 13', 25E 0' E). The seed rates used were 55 and 250 germinating seeds/m^2, respectively. The bi-crops were harvested at 10 wk after sowing. The growth stages of the peas and wheat was early to full pod and the early dough stage, respectively. The control timothy-fescue (TF) sward was harvested at booting. Both silages were wilted, ensiled to round bales using formic acid-based additive 5 l/t. Silages were fed to 8 multiparous dairy cows (four cows fitted with a rumen cannula) in two 4x4 latin squares. Treatments differed in the proportion of pea-barley silage (PB) to grass silage: 0, 33, 67 and 100% of the total amount of silage dry matter (DM). Silages were fed *ad libitum* and concentrate (192 g CP/kg DM) was given at fixed amount of 12 or 14.5 kg/d in the different squares.

Results The proportion of pea was 74% in the PB silage DM, which in laboratory scale has shown to improve ensiling compared to pure pea sward (Pursiainen *et al.*, 2002). D-value was about the same for PB and TF silage (645 and 658 d/kg DM). Dry matter content was lower in PB silage compared to TF silage (255 and 559 g/kg) and crude protein content was higher (170 and 131 g/kg DM). PB silage was more abundantly fermented than TF silage: amount of total fermentation acids was 120 vs 12 g/kg DM, respectively. Ammonia-N content was higher (108 vs 31) and pH lower (3.96 vs 5.16) in PB than TF silages. Butyric acid content was low in both silages. Thus silage intake index (Huhtanen *et al.*, 2002) was lower for PB silage (85 and 101). However increasing PB silage in the diet increased silage intake up to the 33% diet, but after that it decreased (9.2, 9.7, 9.0, 7.1, SEM 0.72 kg DM/d for 0, 33, 67 and 100). Milk production was the highest with the 100% diet (28.7, 28.5, 29.5, 30.3, SEM 2.27 kg/d, lin. $P<0.05$). The increase of liveweight gain decreased with increasing amount of PB silage (0.58, 0.49, 0.46, 0.10, SEM 0.24 kg/d). However, the protein content in milk decreased with PB (38.5, 37.8, 37.0, 37.1, SEM 1.03 g/kg, lin. $P<0.05$). The effect of PB silage on protein yield tended to be quadratic (1099, 1063, 1083, 1115, SEM 75.0 g/d, quadr. $P<0.10$). Microbial protein synthesis estimated from urine secreted purine derivatives per kg digested carbohydrates was the highest in the 100% diet (lin. effect $P<0.05$). NDF pool in the rumen, measured in the evacuation technique, decreased with increasing PB silage ($P<0.01$). However, INDF pool in the rumen was the highest on 100% diet.

Conclusions Pea-barley bi-crop silage increased milk yield, decreased silage intake and liveweight gain of the cows compared to grass silage diet. Decreased intake of pea-barley silage was likely due to lower dry matter content and more abundant fermentation products in the pea-barley silage.

References

Adesogan, A.T., M.B. Salawu, S.P. Williams, W.J. Fisher & R.J. Dewhurst (2004). Reducing concentrate supplementation in dairy cow diets while maintaining milk production with pea-wheat intercrops. *Journal of Dairy Science,* 87, 3398-3406.
Huhtanen, P., H. Khalili, J.I. Nousiainen, M. Rinne, S. Jaakkola, T. Heikkilä & J. Nousiainen (2002). Prediction of the relative intake potential of grass silage by dairy cows. *Livestock Production Science,* 73, 111-130.
Lunnan, T. (1988). Blandningar av bygg og ulike belgvekstar til grønfôr. *Norsk Landbruksforskning,* 2, 219-232.
Pursiainen, P., M. Tuori, J. Nousiainen, H. Miettinen & U. Kämäräinen (2002). Field bean, field pea and common vetch ensiled with whole crop wheat using formic acid or inoculant. In: Gechie, L.M. & Thomas, C. (eds.): *The XIIIth International Silage Conference*, September 11-13, 2002, Auchincruive, Scotland. p. 120-121.
Salawu, M.B., A.T. Adesogan & R.J. Dewhurst (2002). Forage intake, meal patterns, and milk production of lactating dairy cows fed grass silage or pea-wheat bi-crop silages. *Journal of Dairy Science,* 85, 3035-3044.

Conjugated linoleic acid content of milk from cows fed different diets

E. Staszak and J. Mikołajczak

Department of Animal Nutrition, University of Technology and Agriculture, ul. Mazowiecka 28, 85-084 Bydgoszcz, Poland, ewas@atr.bydgoszcz.pl

Keywords: maize silage, grass hay, dairy cows, conjugated linoleic acid, milk

Introduction Conjugated linoleic acid (CLA) is a mixture of positional and geometric isomers of linoleic acid (c-9, c-12 $C_{18:2}$). Conjugated linoleic acid occurs naturally in foods, however the main dietary sources are dairy products and other foods derived from ruminants. Continuous interest in CLA is attributed to its potential health benefits such as anticarcinogenic, antiatherogenic, antidiabetic and antiadipogenic effects (Dhiman *et al.*, 1999, Staszak *et al.*, 2001). Typical consumption of CLA by humans is far lower than the dose that has been shown to be effective in reducing tumours in animal models (Dhiman *et al.*, 1999), so it is very important to increase the CLA content of ruminants edible products. The CLA concentration can be positively influenced by animal diet. Grazing cows on pasture, feeding fresh cut pasture, addition of fish oil etc demonstrate positive effects on CLA content in milk (Bessa *et al.*, 2000, Staszak *et al.*, 2001). The objective of this research was to determine the CLA content of milk from cows fed diets containing different proportions of conserved forages.

Material and methods The experiment was conducted from July 2003 to July 2004. Twenty primiparous Polish black and white cows were used. Cows which were randomly assigned to one of two groups consumed diets containing different proportions of conserved forages. Group S was fed diet based on maize silage (more than 50% of daily dry matter intake) and group H was offered diet based on grass hay (more than 50% of dry matter intake). Chemical composition and nutritive value of feeds were determined. Samples of milk from individual cows were collected monthly and analysed for chemical composition, including the CLA content (analysed every two months). Milk yield was recorded.

Results The average CLA content in milk analysed during experiment is shown in Table 1. The CLA content varied significantly between individual cows, which suggest animal related factors. Milk yield and composition was similar for both groups. Diets containing grass hay resulted in significantly ($P<0.01$) larger CLA concentration in July, September, November 2003 and in July 2004. Slightly higher CLA content was observed in cows fed a maize silage-based diet in January, March (significant at $P<0.01$) and May 2004. In summer months both diets contained 20% grass cut fresh, which resulted in an increase in CLA levels in milk. Green pasture may contain up to 3% fatty acids on a dry matter basis, of which about 90% will be unsaturated C_{18} acids (Bessa *et al.*, 2000). Higher levels of CLA in cows fed grass hay-based diets may be explained by the fact that lipid composition of the preserved forage remains relatively unchanged from that prior to preservation unless there is gross deterioration (during normal drying and storing the $C_{18:3}$ content decreases and $C_{16:0}$ increases in forage) (Dhiman *et al.*, 1999).

Table 1 The CLA concentration in milk

		The CLA content in milk (mg/kg of milk)						
		July 2003	September 2003	November 2003	January 2004	March 2004	May 2004	July 2004
Group S	Mean	191.60[a]	223.13[b]	235.47[c]	193.09	228.11[d]	169.51	275.07[e]
Pooled SD = 96.19	SD	61.74	120.48	80.93	48.81	112.63	82.46	134.29
Group H	Mean	310.99[a]	290.78[b]	295.35[c]	187.71	170.41[d]	145.12	429.76[e]
Pooled SD = 141.60	SD	207.87	189.89	122.16	38.16	59.92	114.34	167.48

[a, b, c, d, e] means in the same column with the same superscripts differ significantly at $P<0.01$

Conclusion Both crop type (maize silage vs grass hay) and conservation method altered the CLA content in bovine milk.

References

Bessa R.J.B., J. Santos-Silva, J.M.R. Ribeiro & A.V. Portugal (2000). Reticulo-rumen biohydrogenation and the enrichment of ruminant edible products with linoleic acid conjugated isomers. *Livestock Production Science* 63, 201-211.

Dhiman T.R., G.R. Anand, L.D. Satter & M.W. Pariza (1999). Conjugated linoleic acid content of milk from cows fed different diets. *Journal of Dairy Science* 82, 2146-2156.

Staszak E., J. Forejtová & P. Český (2001). Conjugated linoleic acid (CLA) a new ruminants meat and milk modifying factor. *Collection of Scientific Papers, Faculty of Agriculture in České Budějovice, Series for Animal Sciences* 18 (1), 29-32.

Feeding with badly preserved silages and occurrence of subclinical ketosis in dairy cows

F. Vicente[1], B. de la Roza[1], A. Argamentería[1], M.L. Rodríguez[2] and M. Peláez[2]
[1]Servicio Regional de Investigación y Desarrollo Agroalimentario (SERIDA). PO Box 13; 33300 Villaviciosa (Asturias), Spain, Email: fvicente@serida.org, [2]Sociedad Asturiana de Servicios Agropecuarios, S.L. Polígono Bravo-Sierra de Granda s/n; 33199 Granda Siero (Asturias), Spain.

Keywords: ketosis, dairy cows, silage fermentation

Introduction Ketosis in dairy cows is due to high levels of circulating ketone bodies in blood (Duffiel, 2000). In early lactation, the capacity of voluntary dry matter intake does not allow dairy cows to cover the total energy requirements for maintenance and production, and then the body reserves are mobilised. However, the amounts of fatty acids that can be metabolised in the liver are limited, later they are converted to ketone bodies (Tveit *et al.*, 1992). During forage ensiling, acetic and lactic fermentations only are desirable, but frequently butyric and alcoholic fermentations appear. When the animals ingest these silages, the butyric acid is metabolised to ketone bodies (Chalupa, 1974). The ketosis problems could be due to both causes simultaneously. The objective of this paper was to establish the incidence of subclinical ketosis in dairy herds of Asturias (Spain) and its relationship with the nutritive and fermentative characteristics of silages used in the ration.

Materials and methods Twenty representative dairy herds were recorded over a 12-month period. Monthly, the feeding was recorded and samples of total ration and its components were taken. Simultaneously, the urine of all cows that were between the previous month of parturition and three months later was sampled to determine the incidence of subclinical ketosis. The proximate analyses of feed samples were determined by techniques proposed by Wende (AOAC, 1984) and Van Soest *et al.* (1991). The volatile fatty acid concentration was determined by HPLC. The urine ketone bodies level was determined immediately with quantitative test strips.

Results A total of 2831 urine samples were taken from 1112 dairy cows. Of the total, 79.94% did not show a detectable excretion of ketone bodies by urine, whereas 11.18% presented slight subclinical ketosis and 5.14% and 3.74% presented moderate and high levels of excretion (Figure 1). The metabolisable energy content of the diet in the four levels of ketone bodies excretion was similar (10.73 ±0.141 MJ ME/kg DM). However, the concentration of butyric acid in the diet was significantly higher in the animals with subclinical ketosis ($P<0.001$, Figure 2).

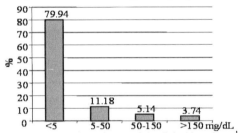

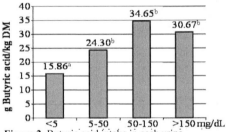

Figure 1 Proportions of cases of sbuclinical ketosis according to content in ketone bodies in urine

Figure 2 Butyric acid intake in each urine ketone bodies excretion

Conclusions The analysis of frequencies in which the positive cases appear indicates that the probability of subclinical ketosis in cows is higher when they are feeding a ration elaborated with silages with high butyric acid content. A low energy concentration of the ration (<10.5 MJ ME/kg DM), does not imply a higher probability of ketosis incidence, because, in our study, the distribution of positive cases are not related with that. However, the incidence in our cases can be further reduced by the use of maize silage well preserved as an energy supplement.

Acknowledgement This work is supported by Spanish PROFIT project FIT-06000-2004-1

References
AOAC, (1984). "Official methods of analysis" Ed: Association of official agricultural chemist. 14th ed.
Chalupa, W. (1974). In: D.C. Church (ed.) Fisiología digestiva y nutrición de los rumiantes. Vol. 3, Nutrición práctica. Acribia, Zaragoza, 320-351.
Duffiel, T. (2000). In: T.H. Herdt (ed.) Metabolic disorders of ruminants. W.B. Saunders, Philadelphia, 231-253.
Tveit, B., F. Lingaas, M. Svendsen & Ø.V. Sjaastad (1992). Etiology of acetonemia in Norwegian cattle. 1. Effect of ketogenic silage, season, energy level, and genetic factors. *Journal of Dairy Science*, 75, 2421-2432.
Van Soest, P.J., J.B. Robertson, & B.A. Lewis (1991). Methods of dietary fibre, neutral detergent fibre, and nonstarch polysaccharides in relation to animal nutrition. *Journal of Dairy Science*, 74, 3583-3597.

Modelling contamination of raw milk with butyric acid bacteria spores

M.M.M. Vissers[1], F. Driehuis[1], P. de Jong[1], M.C. te Giffel[1] and J.M.G. Lankveld[2]
[1]NIZO food research, PO Box 20, 6710 BA Ede, The Netherlands, Email: Marc.Vissers@nizo.nl, [2]Wageningen University, Chair of Dairy Science, PO Box 8129, 6700 EV Wageningen, The Netherlands.

Keywords: butyric acid bacteria, spores, raw milk, risk assessment, farm management

Introduction Raw milk contains low concentrations of bacterial endospores, originating from the farm environment (e.g. soil, feeds, faeces). Spores of *Clostridium tyrobutyricum*, also called butyric acid bacterium (BAB), are of great interest to the dairy industry. They survive milk pasteurisation and cause off-flavours and texture defects in various cheese types. The contamination pathway of BAB spores is well known. Their primary origin is soil. In silage the number of spores will increase if conditions permit BAB growth. The spores are excreted in the cows faeces and are transferred to milk by contaminated teat surfaces. Many factors are involved in the contamination of milk with BAB spores. In this study, the contamination pathway was described using a combination of predictive models. The objective of the study was to quantitatively assess the importance of the different steps of the contamination pathway and to identify the most effective control points.

Materials and methods The contamination of milk was described as a process of sequential unit-operations of carriers of BAB spores, starting with the sources (soil, feeds) and ending with bulk tank milk. For the unit-operations (storage, mixing, concentration, removal) basic mathematical equations were used to describe transmission between carriers. Microbial growth during storage in silage was described as a function of time, pH, a_w and temperature using the gamma-concept (Zwietering *et al.*, 1996). Input variables included controllable variables, which are influenced by the farmers management (e.g. silage quality, silage proportion in ration and teat cleaning strategy and efficiency), and uncontrollable variables (e.g. spore level in soil and amount of dirt on teats before cleaning). Variable values were based on literature data, experimental data available at NIZO and expert estimates.

Results Monte Carlo simulations were performed to quantitatively assess the importance of the different BAB spore sources and transmission steps for contamination of milk. The results showed that the concentration of BAB spores in silage is significantly more important than other factors, including cattle-house and milking hygiene. Obviously, the importance of the latter factors increase when silage with a low spore concentration is fed. The part of the model from faeces to milk was validated with experimental data (Stadhouders and Jorgensen, 1990; Witlox, 1983). Good agreement was observed.

Simulations were conducted with fixed BAB spore concentrations in silage and other controllable variables set at either poor, average or optimal management. The results show that with optimal management the critical level of 1 BAB spore per ml of milk is exceeded when silage contains more than 10^6 BAB spores per g (Figure 1). With average management this level is exceeded when silage contains more than 10^4 spores per g.

Conclusions The developed model simulates the entire contamination pathway of BAB spores and was applied to predict the effectiveness of control measures. Factors related to cattle-house and milking hygiene proved less important than silage quality. The model can assist farmers in taking effective management decisions with respect to raw milk quality. It may be integrated in on-farm dairy quality management systems.

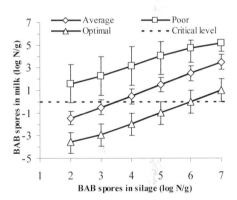

Figure 1 Model simulations with poor, average and optimal farm management

References
Stadhouders, J. & K. Jorgensen (1990). Prevention of the contamination of raw milk by a hygienic milk production. *Bulletin of the International Dairy Federation*, 251, 32-36.
Witlox, P. (1983). De besmetting van melk met sporen van boterzuurbakteriën. Thesis, Wageningen University.
Zwietering, M.H., J.C. de Wit, & S. Notermans (1996). Application of predictive microbiology to estimate the number of *Bacillus cereus* in pasteurised milk at the point of consumption. *International Journal of Food Microbiology*, 30, 55-70.

Use of a dairy whole farm nutrient balance education tool to teach the importance of forages in the context of nutrient management concepts at the whole-farm level

J.H. Harrison and T.D. Nennich
Washington State University Puyallup Research and Extension Center, 7612 Pioneer Way, Puyallup, WA, 98371, Email:jhharrison@wsu.edu

Keywords: forages, nutrient management, education tool

Introduction Prior to 2003 dairy farms in Washington state were required by law to have a certified nutrient management plan that was nitrogen-based. In early 2003, the national Environmental Protection Agency released new Concentrated Animal Feeding Operation guidelines to require that nutrient management plans consider phosphorus as well. To better prepare producers and their advisers for this change, a spreadsheet-based education tool was developed in Microsoft Excel to demonstrate whole farm concepts related to nutrient balance.

Materials and methods The goals in developing the tool were: 1) to use a simple interface viewed on a single page (Figure 1), 2) to use input information readily available on most dairy farms, and 3) to use terminology and calculations consistent with a program developed by the Natural Conservation Resource Service that is used in writing nutrient management plans. The inputs required to determine a farm balance are herd milk production; number of milking cows, dry cows, and heifers; dry matter intake of lactating cows; nitrogen and phosphorus content of lactating cow rations; fertiliser import; land in forage crops; yield, nitrogen and phosphorus content of forage crops; and estimated availability of nitrogen and phosphorus in manure. Output of the analysis includes the manure nitrogen and phosphorus available to crops and the whole farm balance of nitrogen and phosphorus.

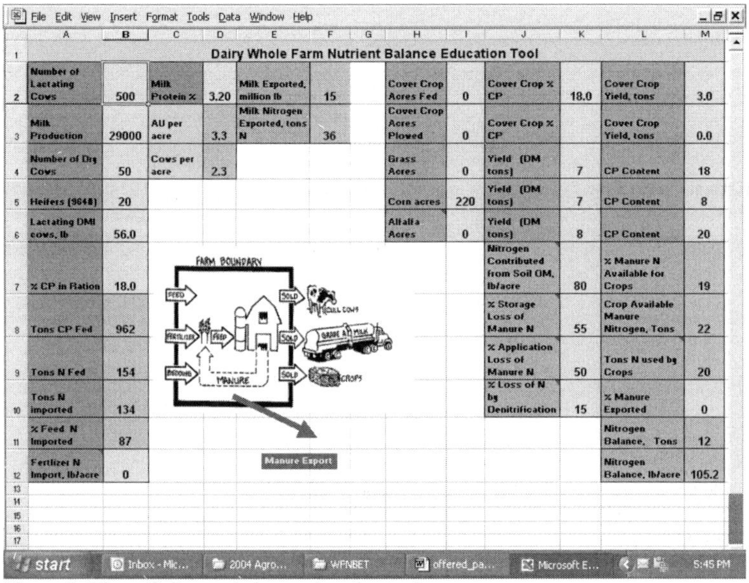

Figure 1 View of nitrogen balance education tool

Results This educational tool has been used to demonstrate successfully the effects of management changes such as reduction in diet phosphorus, level of milk production, custom raising of heifers, forage yield and type, and the use of winter cover crops. An advantage of the tool is that real farm data can quickly be entered at a workshop for a farm represented by a participant and thus create teachable moments.

Conclusions Forages are a critical component of whole farm nutrient balance of nitrogen and phosphorus. The Microsoft Excel-based tool provides the opportunity to quickly and effectively demonstrate the valuable role that forages play in whole farm nutrient balance.

Feeding mixed grass-clover silages with elevated sugar contents to dairy cows

J. Bertilsson
The Swedish University of Agricultural Sciences (SLU), Department of Animal Nutrition and Management, Kungsängen Research Centre, S-753 23 Uppsala, Sweden, Email: jan.bertilsson@huv.slu.se

Keywords: sugar, water-soluble carbohydrates, grass, clover, dairy cows

Introduction Grasses with high sugar content (WSC) have been claimed to increase feed intake and milk production and at the same time give more efficient N utilisation and thus reduce pollution (e.g. Miller *et al.*, 2001). In an ongoing EU-supported project ("Sweetgrass"), we have grown the new varieties of perennial ryegrasses and fed them to dairy cows. Results from the first year's experiment when pure ryegrass silages made from standard or high-sugar varieties were fed, showed small differences in feed intake, milk production and N partitioning. In order to test the principle, it was therefore decided to increase the difference in sugar content in the following experiment by mixing sucrose into the silage before feeding.

Material and methods High-sugar perennial ryegrass (*Lolium perenne)* (cv.Aberdart), standard perennial ryegrass (cv. Fennema) and red clover (*Trifolium pratense*) (cv. Vivi) were grown in pure swards. The crops were cut with a mower conditioner and wilted for up to 24 h and then ensiled in bales. The bales were covered with six layers of plastic and Kofasil Ultra[TM] was used as an additive. The differences in WSC content between varieties of perennial ryegrass within cuts were small. Grass silages from the second and third cuts were mixed with red clover (75/25 or 50/50 on a dry matter (DM) basis). At each mixing occasion samples were drilled out of the bales and DM content was rapidly determined using a microwave-oven. Grass silages were combined in order to give an even WSC content in the combined silage. Finally an addition of 10% sucrose (DM basis) was mixed into two of the silage mixes, giving in total four treatments. The silages were fed *ad libitum*, while concentrate was fed at a fixed amount of 6.5 kg DM/cow/d.

Results The analyses of silages fed are presented in Table 1 and the animal performance in Table 2. The intended difference in WSC content was achieved. When comparing sugar levels in silage, only the N efficiency differed.

Table 1 Composition of the silage mixes as fed (g/kg DM)

Clover (% DM)	25	50	25	50
Sugar	no addition	no addition	addition	addition
DM	320	325	332	334
CP	175	169	162	158
WSC	121	107	207	197
NDF	405	403	376	364

Table 2 Feed intake, milk production and N partitioning for cows fed silage mixtures with and without addition of sugar (LS-means)

Clover (% DM)	25	50	25	50	Se	(significance *P<*)		
Sugar	no addition	no addition	addition	addition		sugar	clover	Sugar* clover
Feed intake (kg DM)								
Silage	13.9	14.4	14.5	14.3	0.7	0.44	0.63	0.27
Silage plus conc.	19.9	20.9	20.5	21.4	0.8	0.29	0.07	0.17
Milk production								
Milk (kg)	24.1	25.4	24.7	25.9	1.1	0.39	0.08	0.17
Protein (g/kg)	33.8	33.6	34.0	34.7	0.6	0.16	0.61	0.70
Fat (g/kg)	48.0	43.6	46.7	45.0	1.4	0.99	0.005	0.004
N in milk/N in feeds	0.232	0.230	0.247	0.260	0.01	0.02	0.52	0.89

Conclusions When fed to dairy cows in combination with moderate to high levels of grain-based concentrate, the effects on milk production of added sugar to silage were minor, while the effect on N efficiency (N in milk/N in feeds), although numerically small, was statistically significant.

Reference

Miller, L.A., J.M. Moorby, D.R. Davies, M.O. Humphreys, N.D. Scollan & J.C. Macrae (2001). Increased concentration of water-soluble carbohydrate in perennial ryegrass (*Lolium perenne* L.). Milk production from late-lactation dairy cows. *Grass and Forage Science*, 56, 383-394.

Section 1

B. Effect of conserved feeds on meat production

An evaluation of grain processing and storage method, and feed level on the performance and meat quality of beef cattle offered two contrasting grass silages

T.W.J. Keady[1,2], F.O. Lively[1] and D.J. Kilpatrick[2]
[1]Agricultural Research Institute of Northern Ireland, Hillsborough, Co Down BT26 6DR, U.K. Email tim.keady@dardni.gov.uk, 2 Department of Agriculture and Rural Development for Northern Ireland, Newforge Lane, Belfast BT9 5PX

Keywords wheat, crimped, urea, grass silage, beef production, meat quality

Introduction Traditionally cereals have been dried or treated with propionic acid and processed prior to feeding to finishing beef cattle. Recently new techniques have been developed for storing and feeding grain to beef cattle. The objective of the current study was to evaluate the effects of grain storage and processing method, and grain feed level on performance and meat quality of beef cattle offered two contrasting feed value grass silages.

Materials and methods The study involved a total of 132 continental cross beef cattle, which were allocated to 12 treatments in a continuous design, randomised block experiment. High and low feed value silages were supplemented with either 3.5 or 5.9 kg concentrate dry matter/head/d. The concentrate consisted of 850 g/kg dry matter (DM) wheat and 150 g/kg DM citrus pulp. Wheat was harvested and ensiled either crimped and treated with 4.5 l/t fresh weight of a proprietary acid-based additive, ensiled whole mixed with 20 kg urea and 30 l of water/t fresh weight or harvested conventionally and treated with propionic acid. Cattle were slaughtered and meat quality assessments were undertaken after 7 days ageing and are detailed by Keady *et al.* (2005). Data were analysed as 3 (grain storage/processing methods) x 2 (grain feed levels) x 2 (grass silage feed values) experiment.

Results The effects of grain processing/storage method, grain feed level and grass silage feed value on animal performance and meat quality are presented in Table 1. Urea treatment increased silage and total DM intake, and cooking loss, and tended to decrease (P=0.09) carcass gain. Increasing silage feed value increased (P<0.05 or greater) feed intake, final live weight, carcass weight, liveweight gain, carcass gain, fat classification and kill out proportion. Increasing grain feed level increased (P<0.05 or greater) total DM intake, final live weight, carcass weight, liveweight gain, carcass gain and conformation. Grain processing method, silage feed value or grain feed level did not alter (P>0.05) ultimate pH, sarcomere length, lean L*, a* and b*, or fat L*, a* and b*.

Table 1 Effect of grain storage and processing method, and feed level and silage feed value on feed intake and animal performance

	Processing method (PM)				Silage feed value (SIL)		Grain feed level (GFL)			Significance[1]		
	Convent-ional	Urea	Crimped	Sem	Low	High	Low	High	Sem	PM	SIL	GFL
SDMI (kg/d)[2]	4.16[a]	4.74[b]	4.35[a]	0.129	3.84[a]	4.99[b]	5.22[b]	3.61[a]	0.092	**	***	***
TDMI (kg/d)[3]	8.85[a]	9.43[b]	9.04[a]	0.129	8.70[a]	9.51[b]	8.72[a]	9.49[b]	0.092	**	***	***
Final LW (kg)[4]	625	618	625	4.74	613[a]	633[b]	616[a]	630[b]	3.4	NS	***	**
LWG (kg/d)[5]	1.04	0.98	1.04	0.036	0.93[a]	1.11[b]	0.96[a]	1.08[b]	0.026	NS	***	**
Carcass wt (kg)	338	333	341	2.66	330[a]	346[b]	334[a]	341[b]	1.9	NS	***	*
Car. gain (kg/d)	0.60	0.55	0.61	0.020	0.52[a]	0.66[b]	0.56[a]	0.61[b]	0.014	0.09	***	*
Kill out (%)	54.2	54.0	54.5	0.358	53.8[a]	54.7[b]	54.4	54.1	0.026	NS	*	NS
Conformation[6]	3.07	3.07	2.93	0.635	2.98	3.07	2.94[a]	3.11[b]	0.045	NS	NS	*
Fat class[7]	3.47	3.38	3.33	0.115	3.21[a]	3.59[b]	3.29	3.51	0.082	NS	**	0.07
Cooking loss	26.8[a]	28.0[b]	27.5[ab]	0.333	27.61	27.3	27.31	27.56	0.272	*	NS	NS
WBSF (kg/cm²)	2.59	2.59	2.63	0.058	2.61	2.60	2.53	2.68	0.048	NS	NS	*

[1]There was a PM x GFL interaction (P<0.05) for SDMI and TDMI; SIL x GFL interaction (P<0.05) for final LW and LWG. There were no PM x SIL or PM x SIL x GFL interactions. [2]SDMI = Silage dry matter intake; [3]TDMI = Total dry matter intake; [4]LW – Live weight; [5]LWG = Liveweight gain; [6]= EUROP scale: 5, 4, 3, 2, 1 respectively; [7]EU fat classification, where 5 = fat, 1 = lean, WBSF = Warner Bratzler shear force

Conclusions It is concluded that ensiling crimped grain did not alter meat quality or animal performance relative to conventionally processed and stored grain. However, ensiling urea-treated grain increased cooking loss and tended (P=0.09) to decrease carcass gain by 9%.

Reference

Keady, T.W.J., F.O. Lively, D.J. Kilpatrick & B.W. Moss (2005). Effect of replacing grass silage with either maize or whole crop wheat silages on the performance and meat quality of beef cattle offered two levels of concentrates. *Journal of Animal Science* (Submitted for publication).

Nutritive value for finishing beef steers of wheat grain conserved by different techniques

P. Stacey[1,2], P. O'Kiely[1], A.P. Moloney[1] and F.P. O'Mara[2]
[1]Teagasc, Grange Research Centre, Dunsany, Co. Meath, Ireland, Email: pokiely@grange.teagasc.ie, [2]Faculty of Agri-Food and the Environment, University College Dublin, Belfield, Dublin 4, Ireland.

Keywords: wheat, moist grain, conservation, nutritive value, cattle

Introduction Wheat grain harvested at dry matter (DM) concentrations above 860 g/kg is slow to deteriorate during long-term storage. However, high moisture grain (HMG) ranging from below 600 to 750 g DM/kg is conserved on some farms in the form of anaerobic storage of acid-treated, rolled wheat (AR) and urea-treated whole-wheat (UN) (Stacey et al., 2003). This experiment quantified the nutritive value for beef cattle of standard wheat grain (propionic acid-treated and rolled:PR) compared to AR and UN at different levels of intake.

Materials and methods The experiment was a 3 (forms of wheat: AR, UN, PR) x 3 (levels of wheat offered: low (L), medium (M), high (H)) factorial arrangement of treatments, and with a control group of animals on grass silage only (GS). Friesian steers (n=120) were allocated to 10 treatments in a randomised complete block design. For 144 days, all animals were offered grass silage ad libitum as the sole diet (GS) or supplemented with either PR, UN or AR at 3 kg/head (L), 6 kg/head (M) or ad libitum (H). Total faecal collections were made on all animals over a 24 h duration between days 102 and 109, and assessed for DM and starch concentration. Carcass weight (hot carcass x 0.98) was recorded after slaughter and carcass weight gain was estimated as the difference between final carcass weight and 0.48 of initial live weight. Samples of M. longissimus dorsi were taken 24 h post-mortem from between ribs 5 to 7 and stored at 3°C for a further 24 h. Colour measurements (lightness (l), redness (a) and yellowness (b) of the muscle and subcutaneous fat) were made using a Minolta ChromaMeter CR 100. Animal data were analysed as a factorial arrangement of nine treatments (3 wheat forms x 3 wheat levels) and as 10 treatments within a randomised complete block design using Genstat 5.0.

Results The mean (s.d) DM, pH, crude protein (CP) and organic matter digestibility (OMD) values for GS at feedout were 226 (9.7) g/kg, 3.9 (0.11), 152 (4.6) g/kg DM and 679 (14.1) g/kg, respectively. The mean (s.d.) DM (g/kg), pH, CP (g/kg DM, starch (g/kg DM) and OMD (g/kg) values at feed out for AR were 693 (10.1), 4.3 (0.15), 116 (2.4), 671 (18.5) and 925 (7.4). The corresponding values for UN were 738 (9.1), 9.3 (0.07), 145 (3.9), 664 (39.0) and 934 (9.7) and for PR were 827 (8.1), 4.8 (0.26), 111 (4.8), 655 (23.4) and 933 (9.4), respectively. GS had the highest (P<0.001) silage DM intake (SDMI) but the lowest (P<0.001) daily live weight (DLG) and daily carcass weight (DCG) gains (Table 1). Increasing levels of wheat consumption progressively reduced SDMI and increased DLG and DCG. SDMI was equally lower (P<0.001) with AR and PR compared to UN whereas DLG and DCG were equally higher (P<0.05) with AR and PR compared to UN. For steers offered wheat ad libitum, wheat DM intake was lower (P<0.001) with AR than UN or PR, while DLG and DCG were lower (P<0.001) with UN than AR or PR. UN had the highest (P<0.001) amount of starch in the faeces indicating considerable loss of undigested grains. Muscle redness ('a value') was not influenced by method of wheat management but was higher at M compared to L level of supplementation. Fat yellowness ('b value') was higher (P<0.01) with UN than AR, while M>L>H.

Table 1 Performance, DM intakes, faecal results and meat data from 144 day feeding trial

Diet	GS	AR			UN			PR			10 treatments		9 treatments[1]	
Wheat level (W_L)	0	L	M	H	L	M	H	L	M	H	s.e.	Sig.	s.e.	Sig.
SDMI[3] (kg/ d)	7.4	5.4	3.7	1.3	5.9	4.6	1.5	5.8	3.9	1.2	0.15	***	0.15	NS
WDMI[4] (kg/d)	0	2.5	4.9	7.8	2.4	4.8	8.3	2.4	4.9	8.2	0.10	***	0.11	*
DLG[5] (g)	100	719	887	983	612	724	843	622	870	1043	65.5	***	64.1	NS
KO[6] (g/kg)	484	503	502	516	495	502	501	497	511	520	4.4	***	4.1	NS
DCG[7] (g)	64	421	517	629	351	433	491	362	545	676	35.6	***	35.1	NS
Faecal DM (g/kg)	143	158	160	184	155	162	204	147	152	175	6.0	***	5.9	NS
Starch[8] (g/kg DM)	8	9	15	31	51	99	118	9	14	20	10.2	**	10.4	**
Muscle 'a' value	13.0	13.1	13.4	14.3	13.1	14.0	13.5	12.9	14.2	13.3	0.38	NS	0.38	NS
Fat 'b' value	13.7	12.6	13.2	11.4	13.6	14.5	12.4	13.2	14.3	11.5	0.39	NS	0.39	NS

[1]W_F x W_L; [2]Wheat form; [3]silage DM intake; [4]wheat DM intake; [5]daily liveweight gain; [6]killout; [7]daily carcass gain; [8]in faeces

Conclusions AR replaced PR in finishing beef rations without compromising performance or meat colour (qualitative conservation losses for both forms of wheat were restricted). The severe faecal losses of undigested grains with UN resulted in inferior growth rates compared to AR or PR. The relative magnitude of the decrease in performance appeared greater as the level of wheat ingestion increased.

Effect of feeding red clover, lucerne and kale silage on the voluntary intake and liveweight gain of growing lambs

R. Fychan, C.L. Marley, M.D. Fraser and R. Jones
Institute of Grassland and Environmental Research, Plas Gogerddan, Aberystwyth SY23 3EB UK, Email: rhun.fychan@bbsrc.ac.uk

Keywords: legume, brassica, ensiling, lamb production, growth

Introduction Despite a need for alternative forages to provide home-grown sources of protein (Wilkins & Jones, 2000), there have been few studies comparing the effects of such forages on lamb production when fed as silage. In this experiment the effects of offering ensiled red clover (*Trifolium pratense*), lucerne (*Medicago sativa*) and kale (*Brassica oleracea*) on voluntary intake and liveweight gain in growing lambs were compared.

Material and methods Silages were produced from 0.5 ha plots of red clover (cv. Merviot) and lucerne (cv. Vertus) (*Rhizobium meliloti* inoculated seed), sown on 2 September 2002 at a rate of 14.5 kg/ha and 18.5 kg/ha respectively, and kale (Kaleage, a blend of Pinfold and Keeper) sown on 22 April 2003 at a rate of 7.5 kg/ha. The red clover and lucerne plots were harvested on 29 May 2003 and 13 July 2003 (first and second cut silage, respectively) and the kale was harvested on 12 August 2003. After wilting for 24 h, all forages were chopped, treated with Sil-all 4x4™ (Alltech, Stamford, UK; applied at a rate of 10^6 colony forming units/g fresh matter) and ensiled in large round bales. Sixty Suffolk-cross lambs aged 8 months were restrictively allocated to each forage treatment according to gender and live weight. After a 5-wk covariate period on grass silage, animals were group housed for 14 d and offered *ad libitum* access to their treatment silage. The lambs were then split into four groups of five lambs for each forage treatment. Dry matter (DM) intake and live weight were recorded every 7 d over an 8-wk period. Lambs on red clover and lucerne silage were fed first cut silage during weeks 1-4 and second cut silage during weeks 5-8. Silage and DM intake data were analysed by analysis of variance and liveweight gain data were analysed as a complete block with group pens as the blocking structure.

Results DM content differed among all silages, but kale had a notably lower DM content than lucerne or red clover (Table 1). All forages were well preserved, although the higher pH and lower lactate concentrations of red clover and lucerne silage indicate a more restricted fermentation during ensiling compared to kale. Lambs offered red clover and lucerne silage had a higher DM intake and a tendency for a higher liveweight gain than lambs offered kale silage (Table 2).

Table 1 Chemical composition (all values g/kg DM, except DM content (g/kg FM) and ammonia-N (NH_3-N) (g/kg TN)) of different silage treatments as fed to growing lambs

	Kale	Red Clover	Lucerne	s.e.d.	F effect
DM	187[a]	410[b]	508[c]	26.9	***
pH	3.97[a]	4.18[b]	4.63[c]	0.047	***
NH_3-N	53	49	46	6.76	NS
Crude protein	161[a]	188[b]	222[c]	7.21	***
ME	11.2[a]	10.6[b]	10.4[c]	0.07	***
Lactate	42[a]	25[b]	21[b]	3.06	***
Acetate	19[a]	6[b]	8[b]	1.4	***
Propionate	0.8[a]	0.3[b]	0.3[b]	0.12	***
Butyrate	0.16[a]	1.29[b]	0.25[a]	0.401	*

F effect; forage effect; NS, not significant; * = $P<0.05$; *** = $P<0.001$

Table 2 Lamb dry matter intake (DMI) (g/d) and liveweight gain (LWG) (g/d)

	Kale	Red Clover	Lucerne	s.e.d.	F effect
DMI (g/d)	738[a]	1012[b]	1053[b]	33.6	***
LWG (g/d)	103	132	137	14.3	†

F effect; forage effect; † = $P<0.10$; *** = $P<0.001$

Conclusions Lambs fed red clover and lucerne silage had a higher DM intake and a tendency for a higher liveweight gain than lambs fed kale silage. These effects are probably due to the lower DM content and crude protein concentration of the kale silage compared to the legume silages.

References
Wilkins, R & R Jones. (2000). Alternative home-grown protein sources for ruminants in the United Kingdom. *Animal Feed Science and Technology*, 85, 23-32.

The effects of alfalfa silage harvesting systems on dry matter intake of Friesland dairy ewes in late pregnancy

H.F. Elizalde

Instituto de Investigaciones Agropecuarias, Centro Regional de Investigación Tamel Aike, Casilla 296, Coyhaique, Chile E- mail: helizald@tamelaike.inia.cl

Keywords: silage, alfalfa, harvesting, dairy ewes, chop length

Introduction With the recent introduction of alfalfa in Chilean Patagonia (Aisén), its utilisation as silage has to be reviewed relative to animal performance. The effect of silage chop length on the voluntary intake has been evaluated in different species, with sheep being more sensitive to chop length than cattle (Dulphy *et al.*, 1984). The objective of this experiment was to evaluate the effects of different alfalfa silage chop lengths on dry matter (DM) intake and eating behaviour of Friesland dairy ewes in late pregnancy.

Materials and methods Twenty-four synchronised Friesland ewes, weighing 75.3 kg (s.d. 10.1 kg) were used to evaluate the following wilted alfalfa silage treatments: T1) single chop, flail; T2) double chop, flails plus rotary chopper-blower fitted with six knives and; T3) double chop, flails plus rotary chopper-blower fitted with twelve knives. Animals were allocated to a randomised block design, with 8 ewes per treatment, and housed in individual pens 1.5x1.5 m, arranged in three rows and separated by metal wire partitions. A daily sample for each silage was taken for chemical composition and silage chop length measurements (Elizalde, 1993). Silage DM intake and eating rate (Forbes, 1972) was recorded daily; live weight (LW) and condition score recorded once per week

Results Animals offered T1 silage had lower (*P*<0.05) eating rate during the first meal than those offered T2 and T3 silages. Overall silage DM intake was significantly higher (*P*<0.05) for T3 compared to T1, being intermediate with T2 (Table 1). Higher condition score (*P*<0.05) was observed when feeding the twelve knives silage compared to six knives or single chop silages. However, no differences (*P*>0.05) were observed in liveweight gain across treatments. With the exception of DM content, WSC and ammonia N, there were no differences (*P*>0.05) in chemical composition between silages (Table 2).

Table 1 Treatment effects on silage DM intake, feeding behaviour, condition score and LW gain

	T1	T2	T3	s.e.
Silage dry matter intake (g/kg $W^{0.75}$)	43.1[a]	47.7[b]	56.2[c]	1.0
Eating rate during the first meal (g DM/min)	8.3[a]	15.2[b]	18.1[b]	1.6
Condition score	2.19[a]	2.28[a]	2.47[b]	0.02
Liveweight gain (kg/day)	0.30[a]	0.28[a]	0.21[a]	0.02

Table 2 Chemical composition and mean particle lengths of silages as removed from the silos (g/kg DM, unless otherwise stated)

	T1	T2	T3
Dry matter	557.0[a]	640.0[b]	700.0[c]
pH	5.8[a]	5.8[a]	5.9[a]
Crude protein	153[a]	158[a]	156[a]
Ammonia N (g/kg TN)	175.0[b]	199.0[b]	118.0[a]
WSC	7.0[a]	7.9[b]	7.1[a]
Ash	103[a]	102[a]	105[a]
ADF	319.0[a]	308.0[a]	325.0[a]
Mean particle length (mm)	250	70	20

Conclusions The results of the present study indicate that DM intake was affected with different alfalfa silage harvesting systems. Higher DM intake and eating rate were observed as the mean particle length decreased. A better condition score was observed when animals were offered silage harvested with 12 knives.

References

Dulphy, J.P., B. Michalet–Doreau & C. Demarquilly (1984). Étude comparé des quantités ingérées et du comportement alimentaire et mérycisme d'ovins et de bovins recevant des ensilages d'herbe réalisés selon différentes techniques. *Annales de Zootechnie*, 33(3), 291-320.

Elizalde, H.F. (1993). *Studies on the effects of chemical and physical characteristics of grass silage and degree of competition per feeding space on the eating behaviour of lactating dairy cows.* Ph.D. Thesis. The Queen's University, Belfast, U.K. 272 p.

Forbes, J.M., J.S. Wright & A. Bannister (1972). A note on rate of eating in sheep. *Animal Production*, 15, 211-214.

Replacement of maize/soybean meal concentrate by high moisture maize grain plus wholeseed soybean silage for cattle

C.C. Jobim, A.F. Branco, V.F. Gai and U. Cecato
Universidade Estadual de Maringá (www.uem.br)- Maringá-PR, Brasil. ccjobim@uem.br

Keywords: high moisture maize, wholeseed soybean, total digestibility, partial digestibility

Introduction Ensiling high moisture maize grain with wholeseed soybean can increase quality of silage, mainly in relation to protein and energy (Jobim *et al.*, 2002) working as concentrate. This fact contributes to reduced use of concentrate and costs for milk and beef production, and costs related to grain storage on the farms. The objective of this study was to evaluate the nutritive value of high moisture maize grain plus wholeseed soybean silage through partial and total digestibility in cattle.

Materials and methods Three ruminal and duodenal cannulated steers (Nelore x Red Angus) (305 kg live weight) were fed to 1.5% of live weight. Treatments consisted of maize silage (60%) plus a concentrate (40%) as follows: CGSBM (maize grain + soybean meal); HMS33 (high moisture maize grain plus wholeseed soybean silage, 3:1); HMS66 (high moisture maize grain plus wholeseed soybean silage, 1:3). At ensiling wholeseed soybean and high moisture grain were ensiled at a ratio of 1:7. Experimental period lasted 14 days, with 10 days for adaptation and 4 days for digesta (200 mL) and faeces (50 g) sampling. Samples were collected every 4 hours, for 4 days, advancing 2 hours per day, totalling 12 digesta and 12 faecal samples per animal per period. Samples were analysed for dry matter (DM), crude protein (CP) and starch (S).

Results Ruminal DM digestibility was not influenced ($P>0.05$) by inclusion of high moisture grain silage in the diet. HMS 66 increased ($P<0.05$) intestinal DM digestibility relative to CGSBM and HMS 33. HMS 66 had higher ($P<0.05$) total DM digestibility than HMS 33. Ruminal CP digestibility was not influenced ($P>0.05$) by inclusion of high moisture grain silage in the diet, but HMS 66 increased digestibility in relation to CGSBM. Intestinal CP digestibility was not influenced ($P>0.05$) by inclusion of high moisture silage in the diet. HMS 66 increased ($P<0.05$) total CP digestibility relative to CGSBM and HMS 33. There was no difference ($P>0.05$) among treatments in relation to ruminal, intestinal and total starch digestibility.

Table 1 Ruminal dry matter digestibility (RDMD), intestinal DMD (IDMD), total DMD (TDMD), ruminal crude protein digestibility (RCPD), intestinal CPD (ICPD), total CPD (TCPD), ruminal starch digestibility (RSD), intestinal SD (IST) and total SD (TSD)

	CGSBM	HMS33	HMS66	SE	VC(%)
RDMD (%)	61.1	59.6	60.3	3.17	9.1
IDMD (%)	25.3[b]	25.3[b]	36.1[a]	4.04	24.2
TDMD (%)	71.1[ab]	69.9[b]	74.6[a]	9.95	2.4
RCPD (%)	34.4	37.1	41.3	8.41	38.5
ICPD (%)	40.1	37.2	45.6	6.43	27.1
TCPD (%)	61.5[b]	61.3[b]	68.2[a]	1.21	3.2
RSD (%)	75.3	79.4	78.9	3.11	6.9
ISD (%)	80.1	75.6	80.1	3.72	5.3
TSD (%)	95.3	95.2	96.1	3.84	1.7

a and b, within a line, are different ($P<0.05$) by Tukey test.

Conclusions Replacement of maize grain plus soybean meal by high moisture maize grain silage with 14% of wholeseed soybean in cattle concentrate increased crude protein digestibility without effects on dry matter and starch digestibility.

References
Jobim, C.C., G. Barrin & W. Reis (2002). Composição química da silagem de grãos úmidos de milho com adição de grãos de soja. In: 39 Reunião Anual da Soc. Bras. de Zootecnia. Anais. *Recife*, 1, 25-28.

Effect of additive treatment on meat quality

V. Vrotniakiene and J. Jatkauskas
Lithuanian Institute of Animal Science, R. Žebenkos 12, LT-5125 Baisogala, Radviliškis distr., Lithuania. Email: lgi_pts@siauliai.omnitel.net

Keywords: inoculant, chemical additive, meat, fatty acids

Introduction Major components of meat quality are physico-chemical properties (including visual appearance and tenderness) and dietetic properties (i.e. fat content and fatty acid composition) (Razminowicz *et al.*, 2004). Physico-chemical and technological properties of meat are influenced by feeding system, feeds quality and various feeds additives (Brzoska *et al.*, 1999). The aim of the present study was compare the influence of untreated, inoculated and chemically-treated legume-grass silage on carcass composition and physico-chemical properties of meat when fed to fattening bulls.

Materials and methods Roundbale silage of a second cut red clover-dominated sward was produced. Unwilted herbage was baled at 180 g DM/kg fresh matter. Every third bale was left untreated (C), treated with inoculant Feedtech[TM] (10^5 cfu/g fresh herbage) (I) or treated with a formic acid-based silage additive AIV-2000 (6 l/t fresh herbage) (A). Fifteen Lithuanian Black-and-White bulls on average 312 (±13) kg initial weight were used in factorial designed production experiment with 3 silages and 3 blocks in 126-d experimental period. Silages were offered *ad libitum* in two daily feeds on an individual bull basis. All animals received some quality of concentrate feed (2.24 kg/d) offered 2 times per day. At the end of the trial, (n=3 from each group) bulls were slaughtered for control data. The morphological composition of carcass was calculated by weighing bones, tendons and meat separately, and by dividing these weights by the chilled carcass weight.

Results The I and A bulls tended to have higher carcass yield compared with C group. The meat:bone ratio in I and A groups was numerically higher than that in the C group (Table 1). The chemical composition of ground meat and *M. longissimus dorsi* showed no significant differences between the groups. In I and A groups, the pH values of the *M. longissimus dorsi* was 0.41 (*P*<0.001) and 0.31 (*P*<0.001) unit lower, water binding capacity 0.05 and 0.13% higher, cooking losses 0.73 and 0.1% lower and protein value index 0.22 (*P*<0.025) and 0.15 unit higher in comparison with the C group (Figure 1).

Table 1 Control slaughter data

	Control	Inoculant	Acid
LW gain (kg/d)	1.12 ±0.07	1.214±0.09	1.206±0.04
Final weight (kg)	488.3±16	461.7±15.9	460.0±18.0
Carcass weight (kg)	249.2±11.4	238.9±9.3	235.4±9.1
Killing out (%)	51.00±0.6	51.74±0.5	51.20±1.0
AFY (%)	2.00±0.1	2.05±0.1	2.09±0.1
Muscles and fat (%)	78.83±0.3	80.27±0.9	79.87±0.5
Bones (%)	18.68±0.3	17.78±0.9	18.14±0.6
Tendons (%)	1.95±0.1	1.96±0.1	2.00±0.0
M S	4.22±0.5	4.51±0.4	4.40±0.3

LW- liveweight, AFY-abdominal fat yield, MS- muscling score

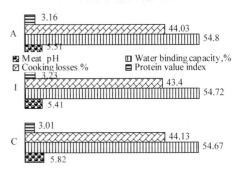

Figure 1 Physico-chemical indicators of *Musculus longissimus dorsi*

Conclusions Inoculant and chemical additive did not affect the chemical composition of ground meat and *M. longissimus dorsi*, however the nutritive value and cooking loss, pH, water binding capacity of these muscles tended to be higher.

References

Brzoska, F., W. Brejta & R. Gasior (1999). Fattening efficiency, carcass parameters and meat composition in bulls. Polish Annals of Animal Science, 26, 1, 141–154.

Razminowicz, R.H., M. Kreuzer, K. Lerch & M.R.L. Scheeder (2004). Quality of beef from grass-based production systems compared with beef from intensive production systems. In: A. Lüscher et al. (eds.) Proceedings of the 20th General Meeting of European Grassland Federation, Lucern, Switzerland, 9, 1151-1153.

Blood meal as a source of histidine for cattle fed grass silage and barley

R. Berthiaume[1] and C. Lafrenière[2]

[1]Agriculture & AgriFood Canada, Lennoxville, QC, Canada, J1M 1Z3. Email: berthiaumer@agr.gc.ca,
[2]Agriculture & AgriFood Canada, Kapuskasing, ON, Canada, J9X 2K3. Email: carole.lafreniere@uqat.qc.ca

Keywords: blood meal, beef cattle, nitrogen, partitioning

Introduction Previous research has shown that cattle fed grass silage are responsive to protected amino acids (Veira *et al.*, 1991). Methionine and lysine were suggested as the most limiting amino acids for grass silage diets. Recently, Korhonen *et al.* (2000) have shown that histidine is the first limiting amino acids for dairy cows fed grass silage and barley-based concentrates. However, histidine is not available in a rumen protected form and needs to be provided through dietary ingredients. Blood meal is rich in histidine. This trial was designed to determine the effect of increasing histidine supply through blood meal on N metabolism of cattle fed a grass silage and barley-based diet.

Materials and methods Thirty-two crossbred steers, mean initial live weight 362 (sem 7.1) kg, were blocked by live weight and allocated to one of four dietary treatments in a randomised complete block design. This experiment lasted until the steers reached a minimum of 8 mm backfat. Diets were fed once daily through Calan gates. Animals were weighed after 24 h of fasting on day 0 and two days before slaughter. Dietary treatments consisted of *ad libitum* grass silage:barley mixture (0.50:0.50 DM basis) with four isonitrogenous combinations of blood meal and urea. Diet digestibility and N balance were determined in a 4-period latin square design experiment with 4 animals. Data were analysed using the GLM procedure of SAS. Polynomial contrasts were used to determine if the effect of blood meal was linear, quadratic or cubic.

Results The nitrogen (N), neutral detergent fibre (NDF) and starch contents (g/kg DM) of grass silage and barley were 23.8, 532 and <1 and 19.8, 226 and 339, respectively. Silage NH_3-N was 0.127 of total N, indicating that the silage was not well preserved. Intake, growth and N partitioning results are presented in Table 1.

Table 1 Effect of isonitrogenous combinations of blood meal and urea on steer performance, apparent digestibility and nitrogen partitioning

| Item | Blood meal: Urea combinations (g/d) | | | | Sem | P | |
	0:85	100:55	200:30	300:0		Linear	Quad.
Liveweight gain (kg/d)	1.25	1.23	1.28	1.26	0.05	0.66	0.96
Carcass gain (kg/d)	0.74	0.75	0.72	0.74	0.03	0.78	0.94
DMI (kg/100 kg LW)	2.15	2.05	2.13	2.15	0.06	0.72	0.27
OM digestibility (g/kg)	727	719	714	732	1.76	0.60	0.05
N intake (g/d)	188	186	183	180	12.57	0.20	0.83
Faecal N (g/d)	50	52	52	48	4.23	0.59	0.21
Urine N (g/d)	110	101	98	102	7.40	0.05	0.04
N retained (g/d)	28	33	33	28	8.08	0.78	0.25
Urinary PD (mmol/d)	111	118	102	107	14.94	0.94	0.86
Plasma Urea-N (mM)	11.3	10.4	11.2	10.0	0.59	0.21	0.79

There were no significant effects of blood meal on dry matter intake and animal performance. However, the addition of blood meal to the diet had a quadratic effect on OM digestibility, and on the amount of N excreted in urine. The addition of blood meal had no effect on microbial protein synthesis, estimated from urinary excretion of purine derivatives, and did not affect circulating concentrations of urea, suggesting that blood meal reduced N excretion by improving the supply and/or AA profile of dietary rumen undegradable protein (RUP). Other factors, such as energy are likely to have negated any other improvement in animal performance.

Conclusions The addition of blood meal to a grass silage and barley diet had a quadratic effect on OM digestibility and on the amount of N excreted in the urine. In both cases the lowest levels were reached with the addition of 200 g/d of blood meal.

References

Korhonen, M., A. Vanhatalo, T. Varvikko & P. Huhtanen (2000). Responses to graded postruminal doses of histidine in dairy cows fed grass silage diets. *Journal of Dairy Science*, 83, 2596-2608.
Veira, D.M., J.R. Seoane & J.G. Proulx (1991). Utilisation of grass silage by growing cattle: Effect of a supplement containing ruminally protected amino acids. *Journal of Animal Science*, 69, 4703-4709.

An evaluation of the inclusion of alternative forages with grass silage-based diets on carcass composition and meat quality of beef cattle offered two contrasting grass silages

T.W.J. Keady[1,2,3], F.O. Lively[1], D.J. Kilpatrick[2,3] and B.W. Moss[2,3]
[1]Agricultural Research Institute of Northern Ireland, Hillsborough, Co. Down BT26 6DR, U.K, Email: tim.keady@dardni.gov.uk, 2Department of Agriculture and Rural Development for Northern Ireland, Newforge Lane, Belfast BT9 5PX, [3]The Queen's University of Belfast, Newforge Lane, Belfast BT9 5PX

Keywords: maize, whole crop wheat, beef cattle, meat quality

Introduction Recent studies have shown that the inclusion of some alternative forages with grass silage-based diets can increase animal performance of beef cattle. The aim of the present study was to evaluate the effects of including either maize or whole crop wheat (WCW) silages with grass silage-based diets on meat quality of beef cattle offered two levels of concentrate.

Materials and methods Grass silage was offered either as the sole forage or in addition to either maize or WCW silages at a ratio of 40:60 alternative forage:grass silage and supplemented with either 3 or 5 kg concentrate/head/d. The six treatments were offered to 66 continental cross beef cattle (mean initial live weight 523 (sd 37.2 kg) in a continuous design, randomised block experiment. The forages were offered *ad libitum* following mixing in a diet wagon once per day, whilst the concentrate was offered in two equal feeds daily. Carcasses were hung tenderstretch and were chilled under standard commercial conditions. The methods used for meat quality assessment are described by Keady *et al.* (2005).

Results Animal performance data from this study have been presented by Keady *et al.* (2005). The main effects of alternative forage and concentrate feed level on meat quality and carcass composition are presented in Table 1. Inclusion of maize silage increased ($P<0.05$) carcass weight and daily carcass gain. Inclusion of either maize or WCW did not alter ($P>0.05$) fat colour, lean colour, pH, sarcomere length, cooking loss or Warner Bratzler shear force (WBSF). Increasing concentrate feed level increased lean a*, b* and Chroma, and sarcomere length. Otherwise concentrate feed level did not alter ($P>0.05$) fat colour, pH, cooking loss or WBSF.

Table 1 Effects of forage type and concentrate feed level on fat and lean colour and meat quality

| | Forage (F) | | | Sem | Concentrate (kg/d) (C) | | | Significance[+] | |
	Grass	Grass + maize	Grass + WCW		3	5	Sem	F	C
Animal performance									
Carcass weight (kg)	326[a]	334[b]	325[a]	3.0	326	331	2.4	*	NS
Carcass gain (g/d)	514[a]	602[b]	496[a]	31.4	515	560	25.6	*	NS
Fat colour									
L*	71.4	71.9	72.6	2.07	73.9	70.0	1.69	NS	NS
a*	5.7	7.1	5.5	1.01	5.7	6.5	0.83	NS	NS
b*	16.6	17.9	17.9	0.89	17.3	17.6	0.72	NS	NS
Chroma	17.9	19.5	18.8	1.11	18.5	18.9	0.90	NS	NS
Hue	71.7	69.6	73.7	2.47	73.0	70.4	2.01	NS	NS
Lean colour									
L*	42.2	40.7	41.4	0.87	41.0	41.9	0.71	NS	NS
a*	20.7	21.4	20.9	0.51	20.0	22.0	0.42	NS	**
b*	15.9	16.2	15.9	0.37	15.4	16.6	0.31	NS	*
Chroma	26.1	26.8	26.3	0.59	25.2	27.6	0.48	NS	**
Hue	37.6	37.2	37.4	0.52	37.6	37.2	0.43	NS	NS
PH	5.57	5.56	5.55	0.011	5.56	5.56	0.009	NS	NS
Sarcomere length (μm)	2.28	2.29	2.34	0.055	2.23	2.38	0.050	NS	*
Cooking loss (%)	26.2	26.5	26.1	0.97	26.1	26.4	0.79	NS	NS
WBSF (kg/cm²)	1.99	2.07	2.00	0.172	2.06	1.98	0.141	NS	NS

There were no significant ($P>0.05$) forage type x concentrate interactions.

Conclusions It is concluded that whilst the inclusion of forage maize with grass silage-based diets increased animal performance, inclusion of either maize or WCW did not alter fat or lean colour or meat quality. However increasing concentrate feed level altered lean colour and increased sarcomere length.

References
Keady, T.W.J., F.O. Lively, D.J. Kilpatrick & B.W. Moss (2005). Effects of replacing grass silage with either maize or whole crop wheat silages on the performance and meat quality of beef cattle offered two levels of concentrates. *Journal of Animal Science* (Submitted for publication).

Section 2

Alternative forages

Effects of feeding legume silage with differing tannin levels on lactating dairy cattle

U.C. Hymes Fecht, G.A. Broderick and R.E. Muck
United States Department of Agriculture, Agricultural Research Service, U.S. Dairy Forage Research Center, 1925 Linden Drive West Madison, Wisconsin 53706-1108 U.S.A. Email: uchymesfecht@wisc.edu

Keywords: condensed tannins, birdsfoot trefoil, N utilisation, milk production

Introduction Condensed tannins (CT) bind to plant proteins in the rumen, reducing protein degradation to ammonia and increasing milk production and milk protein (e.g. Waghorn, 1987). Previous research showed that the reduced soluble non-protein nitrogen (NPN) content of red clover (*Trifolium pratense*) silage (RCS) was related to its greater N efficiency relative to lucerne (*Medicago sativa*) silage (LS) (Broderick *et al.*, 2001). Commercial cultivars of birdsfoot trefoil (*Lotus corniculatus*; BFT) contain modest levels of CT which reduce NPN formation in silage (Albrecht & Muck, 1991). The objective was to compare silages made from BFT with RCS and LS for milk production and N efficiency in lactating dairy cows.

Materials and methods Twenty-five lactating Holstein cows (5 fitted with ruminal cannulae) were randomly assigned to incomplete 5x5 Latin squares to assess effects on milk production and N utilisation. Diets contained (DM basis) 50% of LS, RCS or one 3 BFT lines that contained low (LTBFT), normal (NTBFT) or high (HTBFT) concentrations of CT. The HTBFT and LTBFT lines were developed from the NTBFT by selecting for high and low CT by Dr Nancy Ehlke (University of Minnesota, USA). The remainder of the ration consisted of maize silage, high moisture maize and soybean meal 48% CP.

Results Characteristics of the silages and animal performance on the silages are shown in Table 1. There were differences in CP among silages: LS and LTBFT were highest, NTBFT and HTBFT intermediate, and RCS lowest ($P<0.01$). The levels of NDF were higher in RCS and LS than in the BFT silages ($P<0.01$). There were no differences in DM intake or in milk composition due to silage source ($P>0.01$). However, yield of milk and FCM was higher on NTBFT and HTBFT than LTBFT, which was higher than that on LS or RCS ($P<0.01$). Fat yield was 0.19 kg/d higher on NTBFT than on LS, with the other 3 diets being intermediate ($P<0.01$). Protein yield on all 3 BFT diets, regardless of CT level, was higher than on LS and RCS, despite the fact that the BFT diets contained about 1% less CP ($P<0.01$). Milk urea nitrogen (MUN) was lower on NTBFT and HTBFT than on LTBFT, LS and RCS ($P<0.01$). Differences in milk yield may have been confounded by the BFT diets being lower in fibre (27% NDF) than the LS and RCS diets (29% NDF) ($P<0.01$). However, these results suggest CT concentration was directly related to improved utilisation of CP in BFT silages.

Table 1 Silage characteristics, intake, milk production and milk constituents

Item	LS	RCS	LTBFT	NTBFT	HTBFT	SE	P Value
Silage NDF (% DM)	35.3[b]	42.7[a]	32.4[c]	32.4[c]	32.2[c]	0.4	<0.01
Silage CP (% DM %)	22.0[a]	18.1[c]	21.7[a]	20.4[b]	20.1[b]	0.3	<0.01
DMI (kg/d)	24.4	25.6	25.2	23.3	24.5	0.7	0.41
Milk (kg/d)	30.2[c]	31.1[c]	32.9[b]	34.6[a]	34.3[a]	0.5	<0.01
3.5% FCM (kg/d)	31.4[d]	32.6[cd]	33.8[bc]	36.3[a]	35.3[ab]	0.6	<0.01
Fat (kg/d)	1.13[c]	1.17[bc]	1.20[bc]	1.32[a]	1.24[ab]	0.03	<0.01
Protein (kg/d)	0.94[b]	0.96[b]	1.04[a]	1.09[a]	1.07[a]	0.02	<0.01
MUN (mg/dl)	10.8[a]	11.0[a]	10.8[a]	9.3[b]	9.2[b]	0.3	<0.01

[a,b,c] Means within a row with unlike superscripts differ ($P<0.05$)

Conclusions The results indicate that the condensed tannins in BFT improved N utilisation in the dairy cow with no ill effects on milk production.

References

Albrecht, K.A. & R.E. Muck (1991). Proteolysis in ensiled forage legumes that vary in tannin concentration. *Crop Science*, 31,464-469.

Broderick, G.A., R.P. Walgenbach & S. Maignan (2001). Production of lactating dairy cows fed alfalfa or red clover silage at equal dry matter or crude protein contents in the diet. *Journal of Dairy Science*, 84, 1728-1737.

Waghorn, G.C., M.J. Ulyatt, A. John, & M.T. Fisher (1987). The effect of condensed tannins on the site of digestion of amino acids and other nutrients in sheep fed on *Lotus corniculatus* L. *British Journal of Nutrition*, 57, 115-126.

NDF digestion in dairy cows fed grass or red clover silages cut at two stages of growth

K. Kuoppala[1], S. Ahvenjärvi[1], M. Rinne[1] and A. Vanhatalo[2]
[1]MTT Agrifood Research Finland, Animal Production Research, FI-31600 Jokioinen, Finland, Email: kaisa.kuoppala@mtt.fi, [2]University of Helsinki, Department of Animal Science, P.O. Box 28, FI-00014 University of Helsinki, Finland

Keywords: legume silage, cell wall structure, maturity, rate of digestion

Introduction Increasing demand for organic dairy products has encouraged research on red clover, as it is an important plant species in organic farming systems. The objective of this experiment was to investigate the effects of plant species and growth stage on NDF digestion in dairy cows.

Materials and methods Four silages were made from primary growth: two grass silages (G) from mixed timothy (*Phleum pratense*) meadow fescue (*Festuca pratensis*) swards and two red clover (*Trifolium pratense*) silages (R), in 2003 in Jokioinen, Finland (61°N). G silages were harvested on 17 June at early (G_E) and on 26 June at late (G_L) growth stage and R silages on 2 July at early (R_E) and on 16 July at late (R_L) growth stage. The sward was cut with a mower conditioner, wilted for approximately 4 h and harvested with a precision chop harvester. Silages were preserved with a formic-acid based additive (5 l/t for G and 6 l/t for R silages) in bunker silos or clamps. These four pure silages and a mixture of G_L and R_E ($G_L R_E$, 1:1 on DM basis) were fed with 9 kg/d concentrates to five rumen cannulated dairy cows in a 5x5 Latin square design. Indigestible NDF (INDF) content of the silages was measured with a 12-day rumen incubation in nylon bags. Rate of digestion (k_d) was determined by rumen evacuation method. Total tract digestibility was determined by 4-day total collection of faeces.

Results The content of NDF was 500, 570, 375 and 463 g/kg DM and the content of INDF was 57, 84, 70 and 138 g/kg DM for G_E, G_L, R_E and R_L, respectively. The proportion of INDF in NDF was higher in R than in G. Dry matter (DM) intake was lowest with R_E and highest with $G_L R_E$ (Table 1). The intake of NDF and potentially digestible NDF (DNDF) was lower and intake of INDF was higher in R diets. The advancing stage of growth increased intake of NDF and INDF with both forages, but more markedly with R. Rate of digestion was significantly faster in R than in G. Postponed harvest decreased k_d of G but increased that of R. Total tract digestibility of NDF was similar for both plant species, but postponed harvest decreased it significantly. Digestibility of DNDF was higher in R silages.

Table 1 Intake of feeds, NDF, DNDF and INDF, digestion rate and total tract digestibility

					Mix		Statistical significance			
	G_E	G_L	R_E	R_L	$G_L R_E$	SEM#	C_1	C_2	C_3	C_4
Intake										
Silage DM (kg/d)	13.2	12.0	11.3	12.1	14.0	0.49	o		o	**
Total DM (kg/d)	21.2	20.1	18.8	20.2	21.5	0.59	o		o	*
NDF (kg/d)	8.2	8.4	5.8	7.2	8.1	0.23	***	**	*	
DNDF (kg/d)	6.9	6.9	4.5	5.0	6.6	0.20	***			**
INDF (kg/d)	1.28	1.54	1.31	2.23	1.59	0.063	***	***	***	o
Rate of digestion in rumen, k_d										
DNDF (1/h)	0.034	0.030	0.037	0.043	0.035	0.0017	***		*	
Digestibility										
NDF (g/kg)	621	580	602	547	609	14.4		**		
DNDF (g/kg)	705	680	740	728	726	15.3	*			

#SEM for diet $G_L R_E$ should be multiplied by 1.19. INDF=indigestible NDF; DNDF=potentially digestible NDF (NDF-INDF). Contrasts: C_1=G vs. R; C_2= E vs. L; C_3=C_1 x C_2; C_4=G_L, R_E vs. Mix. Significance: *** $P<0.001$, ** $P<0.01$, * $P<0.05$, o $P<0.10$

Conclusions The content of NDF in red clover was lower than in grass and the composition of NDF was different. Red clover contained more indigestible and less potentially digestible NDF. However, the rate of digestion of DNDF was higher with red clover leading to higher total tract digestibility.

The effects of maize and whole crop wheat silages and quality of grass silage on the performance of lactating dairy cows

D.C. Patterson[1] and D.J. Kilpatrick[2]
[1]Agricultural Research Institute of Northern Ireland, Hillsborough, Co. Down BT26 6DR, U.K., Email: arini@dardni.gov.uk, [2]Biometrics Division, Department of Agriculture and Rural Development, Newforge Lane, Belfast BT9 5PX, UK

Keywords forage maize, fermented whole crop wheat, urea-treated whole crop wheat, milk production

Introduction Patterson *et al.* (2004) obtained positive milk production responses to the inclusion of maize silage in grass silage-based diets under Northern Ireland conditions. By contrast, while inclusion of fermented whole crop wheat increased total forage intake, it had no significant effect on milk production of dairy cows. More recently, a newer technique has been developed of harvesting the wheat crop at high DM content, with milling of the grain during harvesting and treatment with an urea/urease mixture at ensiling (alkalage treatment). The aim of the present study was to investigate the milk production potential of high DM whole crop wheat as a partial replacement for grass silage.

Material and methods The feeding study (2x5 factorial design) was based on two qualities (medium and high) of grass silage (GS) offered as the sole forage with either 7 or 10 kg concentrates/d, or as a 50:50 DM mixture of grass silage with forage maize silage (MS), fermented whole crop wheat silage (FW) or high DM urea/urease-treated milled whole crop wheat (UW). The forages were offered *ad libitum* and forage mixtures were supplemented with 7 kg concentrates/d. The medium and high quality grass silages had: DM of 185 and 234 g/kg respectively and ME of 10.7 and 12.5 MJ/kg DM respectively. The MS, FW and UW had DM; 305, 459 and 751 g/kg, and starch; 359, 350 and 420 g/kg DM. The 10 dietary treatments were offered to 40 lactating dairy cattle in a partially balanced, changeover design consisting of 2 periods each of 5 weeks, with the final 2 weeks of each period being used as the main recording period. The forages were mixed in a diet mixer and individual intakes were recorded. The concentrate was offered separately through out-of-parlour feeders. The results were subjected to statistical analysis using the REML technique in Genstat 5.

Results There were no significant forage type x quality of grass silage interactions, and main treatment effects are presented in Table 1. The high quality grass silage significantly increased both silage intake and performance. FW, UW and MS produced major proportionate increases in forage DM intake of 0.25, 0.29 and 0.39 respectively, but only MS produced a significant increase in milk yield. However, the ratio milk energy output/total forage intake was similar for all of the alternative forage mixtures, and was significantly lower than for the low concentrate grass silage treatment.

Table 1 The effects of forage treatment and concentrate level on performance

	Forage type							Grass silage			
	GS Low conc.	GS High conc.	FW	UW	MS	SED	Sig	Med. quality	High quality	SED	Sig
Forage DMI (kg/d)	9.80[a]	8.93[a]	12.25[b]	12.60[b]	13.58[c]	0.437	***	10.37	12.49	0.242	***
Total DMI (kg/d)	15.88[a]	17.45[b]	18.37[bc]	18.64[c]	19.61[d]	0.446	***	16.97	19.01	0.248	***
Milk yield (kg/d)	28.5[a]	30.5[b]	29.6[ab]	28.7[a]	30.9[b]	0.76	**	27.9	31.5	0.43	***
Fat (g/kg)	39.9	39.1	39.0	39.5	40.3	1.34	NS	40.7	38.5	0.73	**
Protein (g/kg)	30.6	32.0	32.0	32.0	32.2	0.64	NS	30.8	32.8	0.36	***
Fat+protein yield (g/d)	1989[d]	2129[b]	2053[ab]	2057[ab]	2236[b]	66.0	**	1961	2224	37.2	***
Milk energy output/total forage DMI (MJ/kg)	8.80[b]	11.01[c]	7.36[a]	7.20[a]	7.25[a]	0.534	***	8.08	7.84	0.337	NS

Conclusions The alternative forages produced significant increases in total forage intake but only maize silage increased milk yield. The concentrate sparing effects of FW, UW and MS relative to grass silage only were 1.3, 1.4 and 5.0 kg/d on a fat plus protein basis.

References
Patterson, D.C., D.J. Kilpatrick & T.W.J. Keady (2004). The effects of maize and whole crop silages on the performance of lactating dairy cows offered two levels of concentrates differing in protein concentration. *Proceedings of the British Society of Animal Science*, p. 4.

The feeding value of conserved whole-crop wheat and forage maize relative to grass silage and *ad-libitum* concentrates for beef cattle

K. Walsh[1,2], P. O'Kiely[1] and F. O'Mara[2]

1Teagasc, Grange Research Centre, Dunsany, Co. Meath, Ireland, Email: kwalsh@grange.teagasc.ie, [2]Faculty of Agri-Food and the Environment, University College Dublin, Belfield, Dublin 4, Ireland

Keywords: cattle, maize, whole-crop wheat, silage, *ad-libitum* concentrates

Introduction   Grass is the predominant forage ensiled in Ireland. However, the relatively modest yields achieved in a single harvest allied to variability in digestibility and ensilability (and thus in intake and animal performance response) and the likelihood of effluent production create disadvantages for grass silage compared to the potential of some alternative forage crops. Thus, alternative forages are worthy of consideration on many farms. The objectives of this study were to quantify the relative intake, digestibility and performance of beef cattle offered grass silage, forage maize silage and whole-crop wheat (fermented or urea-treated), rank these relative to cattle offered an *ad libitum* concentrate-based diet and compare the "alkalage" system of urea-treated processed whole-crop wheat with whole-crop wheat silage.

Materials and methods   Seventy continental cross-bred beef steers, mean initial live weight 424 (sd 33.0) kg, were blocked for live weight and breed and allocated to one of 5 dietary treatments in a randomised complete block design. Treatments were grass silage (GS), maize silage (cv. Benecia) (MS), fermented whole-crop wheat (cv. Soissons) (FWCW), alkalage whole-crop wheat (cv. Soissons) (ALK) and *ad libitum* concentrates (ALC). The four forages were precision-chop harvested. The ALK harvester was fitted with a grain processor and ensiled with 45 kg Home 'N' Dry (Volac International Ltd.)/t DM. Forages were offered *ad libitum* through individual Calan gates and supplemented with 3 kg concentrates/head/d. The ALC treatment was supplemented with 5 kg grass silage/head/d throughout the 160-d trial period. The mean DM (g/kg) (uncorrected for volatiles) of the GS, MS, FWCW and ALK were 161, 303, 391 and 705 respectively. *Ad libitum* concentrate composition was 830 g rolled barley, 100 g soya-bean meal, 50 g molasses and 20 g minerals and vitamins/kg (DM 838 g/kg) and the concentrate supplement was 650 g rolled barley, 280 g soya-bean meal, 50 g molasses and 20 g minerals and vitamins/kg (DM 839 g/kg). Live weight was recorded every 3 weeks and starting and finishing live weight calculated as the mean of two consecutive day's weighings. Blood samples were taken from all animals mid-way through the experiment. The data were analysed using analysis of variance taking account of diet and block.

Results   Total DM intake and carcass growth were lowest for GS ($P<0.001$) (Table 1). Relative to ALC, GS, FWCW and ALK had a poorer ($P<0.05$) FCE, lower live weight ($P<0.05$) and carcass ($P<0.01$) gain and a poorer ($P<0.05$) kill-out proportion. Despite ALK having the highest ($P<0.05$) forage DM intake, kill-out proportion and rate of carcass gain were lower ($P<0.05$) than MS. MS had a better FCE than the ALK ($P<0.001$) or the FWCW ($P<0.05$). Plasma urea concentration was lowest for MS and highest for ALK ($P<0.001$).

Table 1 Feed DM intake, growth, kill-out proportion, feed conversion efficiency (FCE) and plasma urea

	GS	MS	FWCW	ALK	ALC	s.e.m.	Sig.
Forage DM intake (kg/d)	4.54[c]	6.75[b]	7.07[b]	7.56[a]	0.95[d]	0.166	***
Total DM intake (kg/d)	7.07[c]	9.27[b]	9.59[ab]	10.06[a]	9.86[a]	0.194	***
Liveweight gain (g/d)	802[c]	1200[ab]	1149[b]	1132[b]	1302[a]	48.3	***
Carcass gain (g/d)	479[d]	776[ab]	723[bc]	686[c]	851[a]	30.9	***
Carcass weight (kg)	290[c]	335[ab]	329[b]	321[b]	348[a]	5.2	***
Kill out (g/kg)	523[d]	547[ab]	539[bc]	532[cd]	551[a]	4.2	***
FCE[1]	15.2[a]	12.1[c]	13.5[b]	14.8[ab]	11.9[c]	0.5	***
Plasma urea (mmol/l)	4.6[b]	2.7[c]	5.0[b]	6.8[a]	4.9[b]	0.21	***

[1](kg DM intake/kg carcass gain)
Within row, means with the same superscripts are not significantly different ($P>0.05$), ***$P<0.001$

Conclusions   Forage maize and whole crop wheat silages supported superior levels of growth by cattle compared to grass silage (*in vitro* DMD 698 g/kg). The FCE with maize silage and *ad libitum* concentrates were greater than for the other forages. There was no animal productivity advantage with alkalage compared to fermented whole-crop wheat.

Sustained aerobic stability of by-products silage stored as a total mixed ration

N. Nishino, H. Hattori and H. Wada
Dept of Animal Science, Okayama University, Okayama 700-8530, Japan. Email: j1oufeed@cc.okayama-u.ac.jp

Keywords: aerobic stability, by-products, silage, mixed ration

Introduction Ensiling a total mixed ration (TMR) has been practiced in Japan when high-moisture by-products are used as ruminant feed. Wet brewers grains (BG) are a common feed resource and approximately one million t are produced annually. Nishino *et al.* (2003; 2004) reported that, although silage would easily deteriorate in the presence of air when wet BG were ensiled alone, the spoilage could be avoided when stored as a TMR. Interestingly, the resistance to deterioration was consistently found whether high ($>10^6$ cfu/g) or no ($<10^2$ cfu/g) yeasts were detected at unloading. In this study, changes during ensilage and after exposure to air were examined in fermentation products and microbial composition of wet BG stored as a TMR.

Materials and methods Wet BG were mixed with lucerne hay, dried beet pulp, cracked maize, wheat bran and molasses at a ratio of 5:1:1:1:1:1 on a fresh weight basis. A 400 g mixture was ensiled in plastic pouches with and without inoculation of *Lactobacillus casei* (10^5 cfu/g) or *Lactobacillus buchneri* (10^5 cfu/g). *L. casei* and *L. buchneri* were used to impair and fortify the aerobic stability of silage, respectively. Silage was sampled at 14 d, because around this time high numbers of yeasts can be found after which numbers may decrease (Nishino *et al.*, 2004). A 200 g sub-sample of the silage from each treatment was put into a 0.5 l polyethylene bottle and subjected to aerobic deterioration for 14 d. Chemical and microbial compositions were determined at 1, 3, 5, 7 and 14 d. There were three replicates of each treatment.

Results Lactic acid dominated the fermentation in untreated TMR silage. Addition of *L. casei* increased the acid production, while that of *L. buchneri* enhanced acetic acid and produced small amounts of 1,2-propanediol (12 g/kg DM). Although more than 10^5 cfu/g of yeasts and acetic acid bacteria were detected at unloading, no heating was observed in untreated and *L. casei*-treated silage for 14 and 7 d respectively. Yeasts were not found ($<10^2$ cfu/g) in *L. buchneri*-treated silage, while the numbers of acetic acid bacteria were comparable to other silages. Chemical and microbial composition appeared stable in untreated TMR silage during the deterioration test for 14 d; the acids, alcohols and yeasts were almost unchanged while the acetic acid bacteria decreased as the air exposure was extended. In *L. casei*-treated silage, yeast numbers increased steadily after unloading while the pH increased rapidly after 7 d. Yeast numbers were below the detectable level in *L. buchneri*-treated silage; no apparent changes were found in the chemical composition for 14 d after exposure to air.

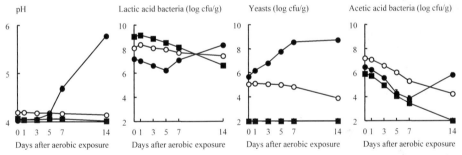

Figure 1 Changes during aerobic deterioration in pH and numbers of lactic acid bacteria, yeasts and acetic acid bacteria in TMR silage inoculated without (○) and with *L. casei* (●) or *L. buchneri* (■) at ensiling

Conclusions Ensiling as a total mixed ration can be a good option to preserve by-products with high stability in the presence of air. The stability is sustained even when high population of yeasts and acetic acid bacteria are counted.

References

Nishino, N., H. Harada & E. Sakaguchi (2003). Evaluation of fermentation and aerobic stability of wet brewers grains ensiled alone or in combination with various feeds as a total mixed ration. *Journal of the Science of Food and Agriculture*, 83, 557-563.

Nishino, N., H. Wada, M. Yoshida & H. Shiota (2004). Microbial counts, fermentation products and aerobic stability of whole crop corn and a total mixed ration ensiled with and without inoculation with *Lactobacillus casei* or *Lactobacillus buchneri*. *Journal of Dairy Science*, 87, 2563-2570.

Evaluation of narrow-row forage maize in field-scale studies

W.J. Cox, J.H. Cherney and D.J.R. Cherney
Cornell University, Department of Crop and Soil Science, Ithaca, NY, USA 14850, Email: wjc3@cornell.edu

Keywords: forage maize, nitrogen management, participatory research, forage quality

Introduction Some dairy producers in the north-eastern USA adopted narrow row (0.38 m) maize forage production in the mid-1990s because of its 5% dry matter (DM) yield advantage (Cox *et al.*, 1998). These dairy producers, however, continued to plant forage maize at high plant densities (125,000 plants/ha) under high N fertility (225 kg N/ha), despite research that indicated that forage maize had optimum DM yields and forage quality when planted at the recommended 100,000 plants/ha under 175 kg/ha of N fertility (Cox & Cherney, 2001). We evaluated forage maize at 0.38 and 0.76 m (conventional) row spacing under recommended vs. high plant densities and N fertility on a large dairy farm in New York. The objective of the study was to demonstrate to dairy producers that narrow-row forage maize does not require high plant densities and N fertility for optimum DM yield and forage quality.

Materials and methods We formed a farmer-researcher partnership to conduct field-scale studies (5-10 ha) on a large dairy farm with field-scale narrow-row equipment. We evaluated first, second, and third-year forage maize at recommended vs. high plant densities and N fertility for three years for a total of nine comparisons. The work crew on the farm performed all field operations, including applications of dairy manure, tillage, planting, spraying and harvesting. We sampled for soil NO_3-N and plant N concentrations at the 6th leaf stage (V6), silking and at harvest. We also measured neutral detergent fibre (NDF), NDF digestibility and *in vitro* true digestibility (IVTD) at harvest. Years were considered random and year in rotation and row spacing were fixed in a combined analysis of variance (ANOVA). A mixed model was used to analyse the data using PROC MIXED (SAS Inst., 1999). Mean separations were conducted using Fisher's Protected LSD ($P= 0.05$).

Results When averaged across years and rotations, narrow-row maize at high vs. recommended plant densities and N fertility had greater soil NO_3-N concentrations at planting (Table 1). All treatments, however, had similar soil NO_3-N and whole plant NO_3-N concentrations at the V6 stage, ear-leaf N concentrations at silking, plant N concentrations at harvest and DM yields at harvest (Table 1). Also, NDF, NDF digestibility, and IVTD did not differ significantly between narrow-row maize at high vs. recommended plant densities and N fertility (data not shown). Narrow-row maize at high vs. recommended N fertility, however, had more than twice the residual soil NO_3-N concentrations at harvest (Table 1). The doubling of residual soil NO_3-N concentrations and the non-significant 3.25% DM yield advantage of narrow-row maize at high N fertility demonstrated to dairy producers that narrow-row forage maize did not benefit from high vs. recommended plant densities and N fertility.

Table 1 Soil NO_3-N, plant N, and DM yields of forage maize when averaged across rotations and years

Row Spacing	Soil NO_3-N			Plant N			DM yield
	Planting	V6	Harvest	V6	Silking	Harvest	
	----------------mg/kg-----------------			---------------------g/kg---------------------			t/ha
0.76 m	21	54	11	42.1	26.0	10.5	17.6
0.38 m	27	49	10	41.7	25.8	10.6	18.3
0.38 m High	37	49	21	41.2	26.2	10.6	18.9
LSD 0.05	10	NS	9	NS	NS	NS	0.7

Conclusions Dairy producers in New York with more than 700 cows are classified as a Concentrated Animal Feeding Operation (CAFO) and must have a Nutrient Management Plan that follows Cornell University guidelines. Based on the results of this study, Cornell maintained guidelines of a 175 kg N/ha limit for forage maize production, regardless of row spacing. The results of this study helped dairy producers in New York with more than 700 cows understand why there is a 175 kg N/ha recommended limit for narrow-row maize production in New York.

References

Cox, W.J, & D.J.R. Cherney (2001). Row spacing, plant density, and nitrogen effects on corn forage. *Agronomy Journal*, 93, 597-602.
Cox, W.J., D.J.R. Cherney & J.J. Hanchar (1998). Row spacing, hybrid, and plant density effects on corn silage yield and quality. *Journal of Production Agriculture*, 11, 128-134.
SAS Institute (1999). SAS User's Guide. Statistics. SAS Institute, Cary, NC.

Ensiling safflower (*Carthamus tinctorius*) as an alternative winter forage crop in Israel

Z.G. Weinberg[1], S.Y. Landau[2], A. Bar-Tal[3], Y. Chen[1], M. Gamburg[1], S. Brener[2] and L. Devash[2]
[1]*Dept. of Food Science* [2]*Dept. of Natural Resources* [3]*Inst. Of Soil, Water and Environmental Science, The Volcani Center, Bet Dagan 50250, Israel. Email: zgw@volcani.agri.gov.il*

Keywords: safflower, silage

Introduction   Israel is a subtropical country in which the rainy season is in winter, with frequent droughts. Wheat is the major winter forage crop in Israel, along with legumes as rotation crops. Alternative forage crops are sought that would be suitable for semi-arid areas. Safflower (*Carthamus tinctorius*) is usually grown as a source for oil and pigments but spineless cultivars could be used as fodder. Leshem *et al.* (2001) reported DM yields up to 22 t/ha and high DM digestibility when used for heifers. When safflower silage substituted maize and wheat silage in the rations of lactating cows, milk yields and milk fat were similar in the two groups (Landau *et al.*, 2004). Safflower was preserved satisfactorily by ensiling in mini-silos (Weinberg *et al.*, 2002). However, on some farm scale trials, safflower silages spoiled upon aerobic exposure. The objective of the current experiments was to further study the ensiling characteristics of safflower.

Materials and methods   Safflower was grown in two locations in southern Israel. In one location (S) it was grown on winter rain (210 mm rainfall) and ensiled in a bunker silo. On the day of ensiling, two rows of 5 dacron bags that contained the chopped crop were buried in the centre along the bunker silo. When the unloading front reached the bags they were brought to the laboratory for analysis and subjected to an aerobic stability test lasting 5 days. In addition, the safflower from S was ensiled in fifteen 1.5 l anaerobic jars, three of which were sampled on days 2, 5, 8, 13 and 120 after ensiling. In the other place (L) experimental plots received fresh or sewage irrigation and various levels of nitrogen fertilisation. The safflower from L was wilted for 24 h and samples from each treatment were ensiled in nine 1.5 l anaerobic jars, three of which were sampled on days 2, 6 and 30 after ensiling. The final silages from all treatments were subjected to an aerobic stability test.

Results   The mean DM and water soluble carbohydrate contents of the fresh safflower in S and L was 350 and 250 g/kg, and 100 and 35 g/kg, respectively. The safflower from S ensiled well in the jars and the pH decreased rapidly and was around 3.9 already after 8 days of ensiling. The samples from L did not ensile well and the pH remained between 4.5 and 5.8 throughout the ensiling period. The samples from the bags were of good quality, their pH was around 4.0 and there were no visible yeasts or moulds. The aerobic stability of the samples from the jars varied from fair to unstable according to treatment; the samples from the bunker silo were quite stable (Table 1).

Discussion   The safflower has potential as a forage crop and can be ensiled successfully. It might well be that the safflower from L did not ensile well because of its low WSC content. The reason for the difference in WSC between the two locations is not as yet clear. Commercial safflower silages might be unstable upon aerobic exposure; therefore, the use of lactic acid producing bacteria such as *L. buchneri* should be considered.

Table 1 Results of the aerobic stability test of safflower silages after 5 days

Source	Sealed silage pH	Exposed silage pH	CO_2 (g/kg DM)	Yeasts	Moulds
L (jars)	4.8-5.8	4.6-6.9	10.8-35.0	5.1-9.1	4.5-8.0
S (jars)	4.0	4.0	1.6±0.2	3.4	3.7
S (bunker silo)	4.0±0.1	4.1±0.1	7.4±5.7	9.5	3.7

Yeast and mould numbers are given as $\log_{10}$ CFU/g DM.

References

Leshem, Y., I. Bruckental, S. Landau, G. Ashbell & Z.G. Weinberg (2001). Safflower: a promising forage crop for semi-arid regions. *Proceedings of the 19th International Grassland Congress*, February 11-21, Sao Pedro, Sao Paulo, Brazil. pp. 303-304.

Landau, S., S. Friedman, S. Brenner, I. Bruckental, Z.G. Weinberg, G. Ashbell, Y. Hen, L. Dvash, & Y. Leshem (2004). The value of safflower (*Carthamus tinctorius*) hay and silage grown under Mediterranean conditions as forage for dairy cattle. *Livestock Production Science*, 88, 263-271.

Weinberg, Z.G., G. Ashbell, Y. Hen, Y. Leshem, S. Landau & I. Bruckental (2002). A note on ensiling safflower forage. *Grass and Forage Science*, 57, 184-187.

Effect of variety and species on the chemical composition of *Lotus* when ensiled

C.L. Marley, R. Fychan and R. Jones
Institute of Grassland and Environmental Research, Plas Gogerddan, Aberystwyth SY23 3EB UK, Email: christina.marley@bbsrc.ac.uk

Keywords: *Lotus* varieties, *Lotus* species, birdsfoot trefoil, ensiling

Introduction Research has shown that there are positive benefits from using *Lotus* as a grazing forage for ruminants. These findings warrant studies into the suitability of different varieties for silage production. In this experiment, we investigated the chemical composition of 13 birdsfoot trefoil varieties and 1 greater birdsfoot trefoil when ensiled.

Materials and methods Replicate 2.4 x 6 m plots of pure stands of 13 varieties of *L. corniculatus* or 1 variety of *L. uliginosus* (previously *L. pedunculatus*) were sown at 12 kg/ha in a randomised block design. Plots were maintained by cutting to a height of 100 mm using a Haldrup 1500 plot harvester, on 2 and 3 occasions in the establishment and first harvest year, respectively. At cut 2 of the first harvest year, sub-samples of forage were taken and separated according to *Lotus* and unsown species. Approximately 3 kg of the *Lotus* only sample was then wilted *in situ* for 24 h, chopped (to approximately 20 mm) using a stationary modified precision chop forage harvester, before being treated with a *Lactobacillus plantarum* inoculant (Live System, Genus Ltd., UK), applied at 10^6 colony forming units per gram fresh matter. One kg of this forage was weighed, compacted and ensiled in laboratory scale silos (four replicates of each *Lotus*) of PVC drain-pipe 100 mm diameter and 450 mm long and sealed as described by Jones (1988). The silos were opened after 110 days and the contents of each tube weighed and thoroughly mixed before representative sub-samples were collected and analysed for chemical composition.

Results The results show differences in crude protein (CP), water-soluble carbohydrate (WSC), ammonia-N and lactate concentrations exist among varieties and species of *Lotus* (Table 1). Overall, the silages were well preserved, with pH and ammonia-N within acceptable ranges. All silages had lower than expected ammonia-N and lactate concentrations, indicating reduced proteolysis in *Lotus* silages compared to other forages.

Table 1 Effect of *lotus* species or variety on silage chemical composition (all values g/kg DM, except DM content (g/kg FM) and Ammonia-N (NH_3-N) (g/kg TN)) at second cut of the first harvest year

Cultivar	Country of origin	DM	CP	WSC	NDF	pH	NH_3-N	Lactate
Oberhaunstaedter	Germany	327	192[cf]	15[d]	416	4.43	27[bc]	18[ab]
Lotar	Czech Rep.	316	200[def]	17[cd]	423	4.28	28[bc]	19[a]
Emlyn	Hungary	315	187[f]	17[cd]	432	4.36	36[b]	18[ab]
Leo	Canada	306	219[ab]	19[bcd]	413	4.30	27[bc]	18[ab]
Upstart	Canada	314	226[a]	18[bcd]	400	4.36	27[bc]	19[a]
Steadfast	USA	308	217[abc]	21[abc]	397	4.37	33[b]	12[d]
Georgia-1	USA	336	190[ef]	20[abcd]	397	4.37	20[c]	12[d]
Dawn	USA	304	216[abc]	23[ab]	412	4.34	34[b]	14[bcd]
Norcen	USA	310	211[bcd]	25[a]	440	4.27	27[bc]	16[abcd]
AU-Dewey	USA	336	203[cde]	18[bcd]	406	4.39	29[bc]	13[cd]
Inia Draco	Uruguay	nd	nd	nd	nd	nd	nd	nd
San Gabriel	Uruguay	nd	nd	nd	nd	nd	nd	nd
Grasslands Goldie	New Zealand	342	204[cde]	18[bcd]	398	4.27	26[bc]	19[a]
Grasslands Maku[†]	New Zealand	316	225[ab]	7[e]	450	4.49	49[a]	17[abc]
SED		16.9	7.1	2.4	18.6	0.070	5.2	2.1
C		NS	***	***	NS	NS	**	**

[†]*L. uliginosus*; C, effect of species/cultivar; nd, not done due to insufficient sample available; NS, not significant; **, $P<0.01$; ***, $P<0.001$. Means in the same column with different superscripts are significantly different.

Conclusions Differences in chemical composition exist among *L. corniculatus* varieties when ensiled. Further studies are now needed to investigate protein degradation in *Lotus* species during the ensiling process.

References
Jones, D.I.H. (1988). The effect of cereal incorporation on the fermentation of spring and autumn silages in laboratory silos. *Grass and Forage Science*, 43, 167-172.

Effect of additives at harvest on the digestibility in lambs of whole crop barley or wheat silage

S. Muhonen, I. Olsson and P. Lingvall
Swedish University of Agricultural Sciences, Department of Animal Nutrition and Management, Kungsängen Research Centre , SE-753 23 Uppsala, Sweden, Email: Sara.Muhonen@huv.slu.se

Keywords: bacterial additives, digestibility, formic acid, whole crop silage

Introduction There are very few published articles about how silage additives affect digestibility of whole crop silage. In this experiment, male lambs were given whole crop barley or wheat silage harvested at dough stage with a number of different acid-based and bacterial additives.

Materials and methods Spring varieties of barley (37% dry matter) (DM) or wheat (45% DM) were harvested as whole crop silage at the middle of dough stage (24-27 July 2003). At harvest either no additive (control), acid-based additives (Promyr, Promyx and Kofasil Ultra) or bacterial additives (Lactisil wholecrop and Lactisil NB200), were used. Fifteen male lambs (Swedish finewool) with an average live weight (LW) of 45 kg were used in three subsequent digestibility experiments. The digestibility of each silage was determined on three animals. Each of the three experiments comprised at least one of the control silages. Feed allowances were adjusted according to the LW to meet the requirements for 0.1 kg daily weight gain. Soybean meal was added to cover the protein requirements and constituted 10% of total diet DM. The statistical analyses were performed using the mixed procedure of SAS (SAS, 1999).

Results The silages used for all treatments were well fermented with low contents of ammonia, butyric acid and ethanol. Control silages and silages treated with acid-based additives had higher pH and lower content of lactic acid than treatments with bacterial additives. All lambs ate the diets with good appetite, but in most cases the allowances were not completely consumed. There were no significant differences in DM, organic matter (OM) and gross energy (GE) digestibility between the two crops. Neutral detergent fibre digestibility was higher ($P<0.001$) for barley than wheat diets. Crude protein (CP) digestibility was significantly ($P<0.05$) higher for wheat than barley diets. There were no significant interactions between crops and additives. There were no differences in DM, OM, CP and GE digestibility (Table 1) as an effect of different additives.

Table 1 Dry matter (DM), organic matter (OM), crude protein (CP) and gross energy (GE) digestibility (%) of whole crop silage diets preserved with different additives (means over both crops). Least squares means (LSM), standard error of the mean (SEM) and significance level for pair-wise comparisons between additives using Bonferroni correction. NS – not significant ($P>0.05$)

		Additives						Significance level
		Control	Promyr	Promyx	Kofasil Ultra	Lactisil wholecrop	Lactisil NB200	
DM	LSM	60.9	62.8	62.5	62.6	60.8	63.7	NS
	SEM	0.69	1.43	1.43	1.27	1.43	1.43	
OM	LSM	64.1	65.7	65.4	65.6	63.8	66.9	NS
	SEM	0.65	1.36	1.36	1.21	1.36	1.36	
CP	LSM	70.7	68.8	70.3	68.4	70.0	70.6	NS
	SEM	0.78	1.59	1.60	1.45	1.59	1.58	
GE	LSM	61.1	62.8	62.1	61.7	59.3	62.8	NS
	SEM	0.75	1.55	1.55	1.38	1.55	1.55	

There were no differences in digestible energy and metabolisable energy contents (MJ/kg DM) between the barley and wheat diets. DM intake (g/kg LW) was higher ($P<0.001$) for wheat than barley diets, but there were no differences between the different additives.

Conclusions The present findings do not indicate that the additives used have an important effect on the digestibility of whole crop barley or wheat silage.

Reference

SAS (1999). SAS/STAT User's Guide, version 8. SAS Institute Inc.:Cary, NC, USA.

Effects of varying dietary ratios of lucerne to maize silage on production and microbial protein synthesis in lactating dairy cows

G.A. Broderick and A.F. Brito
USDA, ARS, US Dairy Forage Research Centre and University of Wisconsin, 1925 Linden Dr. West, Madison, WI 53706, USA. Email: gbroderi@wisc.edu

Keywords: microbial yields, rumen metabolism, N-efficiency

Introduction Lucerne silage (LS) is high in total CP and rumen degraded protein (RDP) but low in fermentable energy, while maize silage (MS) is a good source of fermentable energy but low in RDP. Thus, these silages are complementary and feeding them at optimum ratio should increase nutrient efficiency in lactating cows. Dhiman & Satter (1997) observed greater milk yield when the dietary forage was 2/3 LS and 1/3 MS. The objective of this experiment was to optimise the dietary LS:MS ratio for production, microbial protein and N utilisation.

Materials and methods Twenty-eight (8 with rumen cannulae) multiparous Holstein cows were blocked by days-in-milk and assigned to replicated 4 x 4 Latin squares (28 d periods). The 4 diets were: A (51% LS, 43% high moisture shelled maize (HMSM), and 3% solvent soyabean meal (SSBM)), B (37% LS, 13% MS, 39% HMSM, and 7% SSBM), C (24% LS, 27% MS, 35% HMSM, and 12% SSBM), and D (10% LS, 40% MS, 31% HMSM, and 16% SSBM). Dietary CP was 17.2, 16.9, 16.6, and 16.3%, respectively. Intake and yield of milk and milk components were determined during the last 14 d of each period. Rumen digestion and metabolism, including microbial protein yields, were quantified using omasal sampling (Ahvenjarvi *et al.*, 2000).

Results Dry matter intake, yield of milk and fat, and milk fat content decreased linearly when MS replaced LS. Depressed fat yield may have resulted from lower rumen acetate. Milk protein content increased linearly with increasing MS; however, there was a quadratic effect of LS:MS on protein yield with maximum at 31% dietary LS. Nitrogen efficiency increased because N excreted in urine and faeces decreased linearly when MS replaced LS. Production was significantly depressed on LS:MS of 10:40 and microbial non-ammonia N (NAN) flow was lowest on this diet. A quadratic effect also was observed on microbial protein synthesis with maximum at 38% LS, suggesting that maximal microbial protein formation required a balance between fermentable energy and RDP supply.

Table 1 Effects of dietary ratios of lucerne silage to maize silage (LS:MS) on production and metabolism

Item LS:MS	51:0	37:13	24:27	10:40	SED	LS:MS
DM intake (kg/d)	26.8^a	26.5^a	25.4^b	23.7^c	0.44	R, L
Milk yield (kg/d)	41.5^a	42.0^a	41.5^a	39.5^b	0.86	R, L, Q
Milk fat (%)	3.81^a	3.58^{ab}	3.38^{bc}	3.34^c	0.12	R, L
Milk fat (kg/d)	1.56^a	1.51^{ab}	1.40^{bc}	1.33^c	0.06	R, L
Milk protein (%)	3.07^b	3.13^{ab}	3.14^{ab}	3.17^a	0.04	R, L
Milk protein (kg/d)	1.26	1.32	1.30	1.25	0.03	R, Q
Urinary N (g/d)	217^a	215^a	201^b	188^b	7.05	R, L
Faecal N (g/d)	275^a	263^a	230^b	211^b	10.2	R, L
Rumen ammonia N (mg/dl)	10.5^a	10.0^{ab}	8.72^b	6.19^c	0.92	R, L
Rumen acetate (mM)	88.6^a	84.8^{ab}	79.6^{bc}	74.0^c	3.46	R, L
Omasal flows						
RDP supply (% of DMI)	11.7^a	11.4^a	10.5^b	10.1^b	0.33	R, L
FAB NAN (g/d)	197	219	198	197	15	NS
PAB NAN (g/d)	268	260	261	225	18	NS
Total microbial NAN (g/d)	465^a	479^a	460^a	423^b	12	R, L

[a, b, c] Means in the same row with different superscripts differ ($P \le 0.05$).
[1] R, L, and Q = significant ($P < 0.05$) ratio, linear and quadratic effects; NS = non-significant; SED = standard error of the difference of least square means; FAB = fluid-associated bacteria; PAB = particle associated bacteria.

Conclusions The results of this study indicate that maximal milk protein yield and microbial protein supply occurred at dietary LS:MS ratios of 31:19 to 38:12.

References

Ahvenjarvi, S., A. Vanhatalo, P. Huhtanen & T. Varvikko (2000). Determination of reticulo-rumen and whole-stomach digestion in lactating cows by omasal canal or duodenal sampling. *British Journal of Nutrition*, 83, 67-77.

Dhiman, T.R. & L.D. Satter (1997). Yield response of dairy cows fed different proportions of alfalfa silage and corn silage. *Journal of Dairy Science*, 80, 3298-3307.

Effects of two different chopping lengths of maize silage on silage quality and dairy performance

K. Mahlkow and J. Thaysen
Agricultural Chamber of Schleswig-Holstein, 24101 Kiel, Holstenstr. 106-108, Germany Email: kmahlkow@lksh.de

Keywords: maize silage, chopping length, dairy performance

Introduction Maize silage harvested at dough stage contains less fibre structure than grass silage. Due to higher proportions of maize silage in dairy rations and due to the fact that with longer chopping length of maize silage the contribution of fibre structure increases. The objective of this study was to assess the effect of two chopping lengths in the silage on animal performance with respect to feed intake, milk yields and ingredients as well as physiological effects of digestion.

Materials and methods A mixture of two varieties of maize was harvested in early September 2003 at the experimental station 'Futterkamp'. One half of the maize silage was chopped with a length of 7 mm (CL 7), the other with a length of 22 mm (CL 22). The processing of the kernels during the chopping process was identical for both chopping lengths. Expecting a lower density of the stack with CL 22 mm the compaction weight was increased by nearly 90%. Both silages were consolidated identically with 0.19 t DM/m^3, which was 25% below the target value. After a storage period of 120 d the silages were fed in a total mixed ration consisting of 9 kg DM/d maize silage, 3.7 kg DM/d grass silages, 10 kg DM/d energy components and 280 g/d minerals to 2 x 32 cows in the stage of early lactation. The trial was conducted over 112 d. Feed intake of each cow was recorded daily, milk yield and ingredients once per week. Physiological parameters were assessed 3 x during the experiment.

Results Fermentation characteristics, aerobic stability and microbial composition indicated a medium quality for both silages, and a sufficient similarity for the dairy feeding trial (Table 1). Although the compression effort of the silage CL 22 mm was enhanced, the density was below target value. Except the feed intake all items of yield and physiology including weight change of the animals (results not shown) of the dairy feeding experiment were not significantly different (Table 2). During the experiment both rations containing the differently treated maize silages were analysed for homogeneity of particle size and fibre structure. It was proven, that the applied feeding technique guaranteed the ration with both chopping lengths effectively.

Table 1 Effect of 7 and 22 mm chopping length on mean silage parameters of maize silages[1]

Parameter	CL 7	CL 22
DM content (%)	36[a]	37[a]
pH value	3.9[a]	4.0[a]
Lactic/Acetic acid	7.4[a]	5.7[b]
NH$_3$ N/total N (%)	6[a]	6[a]
DM-losses (% DM)	7.5[a]	7.1[a]
Aerobic stability (days)	1.2[a]	1.0[a]
Stability losses (% DM)	11.8[a]	11.7[a]
Yeasts (log cfu/g FM)	6.5[a]	6.6[a]
Moulds (log cfu/g FM)	5.4[a]	5.6[a]

[1]different superscripts indicate statistical significant differences ($P>0.05$)

Table 2 Effect of 7 and 22 mm chopping length on mean parameters of dairy cows fed maize silages

Parameter	CL 7	CL 22
Feed intake (kg DM/day)	21.6	20.6
Milk yield (kg/day)	36.1	35.5
Milk fat (kg/day)	1.55	1.51
Milk protein (kg/day)	1.16	1.16
NSBA (mmol/L urea)	134	116
ß-hydroxybutyric acid (mmol/L serum)	0.70	0.72
Bilirubin (µmol/L serum)	3.29	3.17

Conclusions For maize silage of >35% DM content harvested at dough stage the enhancement of the chopping length up to 22 mm had no influence on dairy performance in terms of yield and animal health. This is in agreement with data from the literature (Onetti *et al.*, 2003).

References

Onetti, S.G., R.D. Shaver, S.L. Bertics & J.J. Grummer (2003). Influence of corn silage particle length on the performance of lactating dairy cows fed supplemental tallow. *Journal of Dairy Science* 86, 2949-2957.

Use of silage additives in ensiling of whole-crop barley and wheat - A comparison of round big bales and precision chopped silages

M. Knický and P. Lingvall
Swedish University of Agricultural Sciences (SLU), Department of Animal Nutrition and Management, Kungsängen Research Centre, SE – 753 23 Uppsala, Email: Martin.Knicky@huv.slu.se

Keywords: additives, barley, silage, wheat, whole-crop

Introduction An increasing use of whole-crop cereals, as supplementary feed, has increased interest in development of efficient ways of preserving these forages to achieve a high hygienic quality. It is known that ensiling of whole-cereals often results in silages with high concentrations of butyric acid (Weissbach & Haacker, 1988). Furthermore, problems with poor aerobic stability still persist despite the use of lactic acid bacteria (Filya *et al.*, 2000). Therefore, the objective of this study was to examine the effect of different types of additive mixtures on the fermentation process and aerobic stability of precision chopped and baled silages.

Materials and methods A spring variety of barley and autumn variety of wheat were harvested at the dough stage (42% DM). The crops were baled and wrapped with 12 layers of stretch film of white colour (width of 750 mm and thickness of 0.025 mm) or chopped to the particle size of 2 cm (4.5 l PVC silo). The following silage additives were used: C:control; KU:12% nitrite, 8% hexamine, 15.5% Na-benzoate, 5.5% Na-propionate; P1:42.5% formic acid, 20.5% propionic acid, ammonia; P2:22.4% formic acid, 41% propionic acid, ammonia; LAB1:lactic acid bacteria $5*10^5$cfu/g, saccharose; LAB2:lactic acid bacteria $5*10^5$cfu/g, cellulase 6500 IU/g, 7% Na-benzoate. The additives were applied at the dose of 4 l/t fresh forage. Each treatment included 3 replicates.

Results Additive treatments increased fermentation rate in both chopped silages, resulting in a significant pH drop compared with controls (Table 1). Butyric acid concentration was decreased by application of silage additives in both chopped silages as a consequence of restriction of clostridial growth. However, ensiling characteristics of chopped-barley silages were not reflected in the aerobic stability in contrast to chopped-wheat silages. Treatments LAB1 and LAB2 increased the fermentation rate in both barley and wheat bales, but only LAB2 silages had an enhanced aerobic stability. Treatments P1 and P2 improved the stability of wheat bales, although pH was significantly lower only in treatment P1. DM losses in all additive treatments were significantly reduced.

Table 1 Ensiling characteristics of barley and wheat silages. The stability was expressed as the number of days in ventilated silages until the CO_2 concentration in the out-going air reached 1% level

Forage			Treatment						$LSD_{0.05}$
			C	KU	LAB1	LAB2	P1	P2	
Barley	Silo	pH	5.0	4.5	3.9	4.0	4.3	4.3	*0.1*
		Butyric acid (%)	2.00	0.03	0.03	0.03	0.03	0.03	*0.07*
		Stability (d)	6.5	6.5	1.9	1.1	4.5	2.8	*0.8*
	Bale	PH	5.1	5.3	4.0	4.3	5.1	5.4	*0.3*
		Butyric acid (%)	0.05	0.03	0.03	0.03	0.06	0.03	*0.02*
		Stability (d)	1.3	4.8	1.2	4.8	1.7	2.2	*2.5*
Wheat	Silo	PH	4.7	4.4	3.9	3.9	4.1	4.3	*0.1*
		Butyric acid (%)	1.13	0.03	0.03	0.03	0.03	0.03	*0.1*
		Stability (d)	6.5	6.5	6.1	6.5	6.5	6.5	*0.5*
	Bale	PH	5.1	5.2	4.0	4.0	4.9	5.0	*0.2*
		Butyric acid (%)	0.03	0.03	0.03	0.03	0.03	0.03	*0.02*
		Stability (d)	2.6	4.7	2.3	5.8	6.1	5.6	*2.6*

Conclusions Silage additives were more effective in improving the silage quality in wheat forage. Bacterial inoculants were most effective in improving silage fermentation, mainly in baled silages, but only LAB2 containing Na-benzoate improved the stability of bales. The application of 4 l/t P1 and P2 was unsatisfactory in improving the quality of barley bales. KU appeared to be most applicable in both forages regardless of type of silage.

References
Filya, I., G. Ashbell, Y. Hen & Z.G. Weinberg (2000). The effect of bacterial inoculants on the fermentation and aerobic stability of whole crop wheat silage. *Animal Feed Science and Technology*, 88, 39-46.
Weissbach, F. & K. Haacker (1988). On the causes of butyric acid fermentation in silages from whole crop cereals. *Zeitschrift das wirtschaftseigene Futter* 3, 88-99.

Cob development in forage maize: influence of harvest date, cultivar and plastic mulch

E.M. Little[1,2,3], P. O'Kiely[1], J.C. Crowley[2] and G.P. Keane[3]
[1]Teagasc, Grange Research Centre, Dunsany, Co. Meath, Ireland, Email: elittle@grange.teagasc.ie, [2]Teagasc, Crops Research Centre, Oak Park, Carlow, Ireland, [3]University College Dublin, Belfield, Dublin 4, Ireland

Keywords: cob, starch, harvest date, mulch, cultivar

Introduction Forage maize grown for silage tends to be a compromise between reproductive and vegetative yield, and the cob component is the main driver of feeding value (Keane *et al.*, 2003). Thus the aim is to produce a well-developed crop of high dry matter (DM) and starch content reflecting large cobs of well-filled grains rather than crops with low DM and starch contents reflecting poorly developed (immature) cob components at harvest. The use of plastic mulch can increase total DM yields with the increase in cob yield accounting for 75% of the total yield increase (Easson & Fearnehough, 1997). In this experiment the composition of cob components (i.e. rachis plus kernel) of two cultivars of different maturity under Irish conditions grown with or without plastic mulch were monitored between the harvest dates of 10 September to 9 November.

Materials and methods Two forage maize cultivars of different maturity under Irish conditions (Tassilo: FAO 210 (early) and Benicia:FAO 270 (late)) were grown at Oak Park in 2002. Each plot consisted of 4 rows (70 cm spacing) of 5 m length sown in duplicate blocks either uncovered (NP) or under complete-cover clear polythene mulch (P; 6 micron; IP Europe Ltd) on 24 April using a Samco precision seed drill at a seed rate of 100,000 seeds/ha. Standard fertiliser (150 kg N, 50 kg P, 200 kg K/ha) and weed control (4.5 l atrazine/ha) was applied pre-sowing. At 10-d intervals from 10 September to 9 November the cob component was removed from plants from a one m length per plot. *In vitro* DM digestibility (DMD) was measured using the Tilley and Terry (1963) two-step method and acid detergent fibre (ADF) was measured using the Ankom fibre analyser. Data were analysed as a repeated measures analysis of variance using Genstat 7th Edition.

Results As harvest date was delayed cob starch content increased ($P<0.001$) and cob DMD and ADF generally decreased ($P<0.001$) (Figures 1-3). However cob ADF did increase initially in uncovered plants before decreasing.

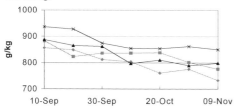

Figure 1 Cob in vitro DMD over time

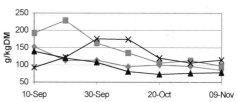

Figure 2 Cob ADF content over time

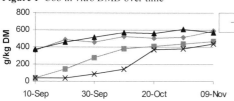

Figure 3 Cob starch content over time

Plastic mulch increased ($P<0.001$) cob starch content, this effect being most evident with Benicia ($P<0.01$) particularly in early September. It also reduced ADF ($P<0.01$) and DMD ($P<0.05$). Tassilo generally had a higher starch and lower DMD and a similar ADF to Benicia.

Conclusions A progressive rise in starch content was observed over time and was most evident in uncovered plants reflecting the greater maturity of covered plants. A corresponding decrease in ADF was observed as starch rose. The initial rise in cob ADF of uncovered plants could indicate the later, final stages of rachis development when compared to those under plastic cover. The temporal decline in cob *in vitro* DMD was quite large and may reflect decreasing degradability of starch and/or increasing indigestibility of the rachis as the cob matures.

References
Easson, D.L & W. Fearnehough (1997). Proceedings of the Agricultural Research Forum, p. 241-242.
Keane G.P, J. Kelly, S. Lordan & K. Kelly (2003). Agronomy factors affecting the yield and quality of forage maize in Ireland: effect of plastic film system and seeding rate. Grass and Forage Science, 58, 362-371.
Tilley, J.M.A. & R.A. Terry (1963). Journal of the British Grassland Society, 18, 104-111.

Yield and composition of forage maize: interaction of harvest date, cultivar and plastic mulch

E.M. Little[1,2,3], P. O'Kiely[1], J.C. Crowley[2] and G.P. Keane[3]
[1]Teagasc, Grange Research Centre, Dunsany, Co. Meath, Ireland, Email: elittle@grange.teagasc.ie, [2]Teagasc, Crops Research Centre, Oak Park, Carlow, Ireland, [3]University College Dublin, Belfield, Dublin 4, Ireland

Keywords: maize, cultivar, mulch, harvest date, yield, composition

Introduction Forage maize is established as a crop with the potential to consistently supply high yields of quality forage on some Irish farms. Despite its success, considerable variability in crop yield, quality and maturity at harvest can exist from year to year. These reflect differing prevailing weather conditions, particularly temperature during May to September. The use of plastic mulch has increased the likelihood of achieving higher yields of high quality crops and has permitted maize production to extend into areas once considered unsuitable for the crop. In this experiment two cultivars of differing maturity were grown with or without plastic mulch to examine how yield and composition altered during the harvest window of early September to early November.

Materials and methods Two forage maize cultivars of differing maturity under Irish conditions (Tassilo: FAO 210 (early) and Benicia: FAO 270 (late)) were grown at Oak Park in 2002. Each plot consisted of 4 rows (70 cm spacing) of 5 m length sown in duplicate blocks either uncovered (NP) or under complete-cover clear polythene mulch (P; 6 micron; IP Europe Ltd) on 24 April using a Samco precision seed drill at a seed rate of 100,000 seeds/ha. Standard fertiliser (150 kg N, 50 kg P, 200 kg K/ha) and weed control (4.5l atrazine/ha) were both applied pre-sowing. Crop samples (2x1 m per plot) were taken every 10 days from 10 September to 09 November. Data were analysed as a repeated measures analysis of variance using Genstat 7th Edition.

Results Plastic mulch increased ($P<0.001$) crop DM yield, the proportion of cob in crop DM, crop starch content and crop DM content for both cultivars, as shown in Figures 1-4. The late cultivar Benicia demonstrated the greater increase ($P<0.05$) in cob proportion and starch content when sown under plastic cover. As harvest date was delayed an increase ($P<0.001$) in crop DM content, cob proportion in DM and starch content was observed (Figures 2-4). Yields of DM increased initially but remained constant once peak yield was achieved which tended to be before mid October (Figure 1). Cultivar type did not influence ($P>0.05$) overall DM yield but the early cultivar Tassilo did have increased ($P<0.001$) crop DM, cob proportion and to a lesser extent increased ($P<0.05$) starch content compared to Benicia under both sowing regimes.

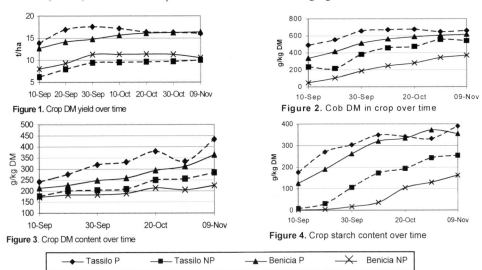

Figure 1. Crop DM yield over time

Figure 2. Cob DM in crop over time

Figure 3. Crop DM content over time

Figure 4. Crop starch content over time

Tassilo P   Tassilo NP   Benicia P   Benicia NP

Conclusions Plastic mulch increased crop DM yield, cob proportion and starch content and advanced crop ripeness (increased crop DM content) in both cultivars. Little yield benefit was obtained from prolonging harvest after 30 September, however starch content and cob content in DM of the plants, particularly those not grown under plastic (NP), continued to increase after this date. Tassilo (early) was about three weeks more advanced in terms of crop ripeness than Benicia (late) when grown under plastic. Benicia grown without plastic mulch did not mature to the desired level and remained very low in DM content.

Parameters of ensiled maize with biological and chemical additives

J. Grajewski[1], A. Potkański[2], K. Raczkowska-Werwińska[2], M. Twarużek[1] and B. Miklaszewska[1]

[1]Bygoszcz University of Kazimierz Wielki, Institute of Biology and Environmental Protection, Chodkiewicza 30, 85-064 Bydgoszcz, Poland, Email: jangra@ab-byd.edu.pl, [2]August Cieszkowski Agricultural University, Department of Animal Nutrition and Feed Management, Wołyńska 33, 60-637 Poznań, Poland

Keywords: mycotoxins, maize silage, fermentation, silage additives

Introduction The amount of maize grown in Poland has increased rapidly. Nowadays it takes about 600,000 ha, 40% of which is used as silage. Changing climate in Poland, with dry summer followed by wet autumn with ground frost causes extensive moulds contamination and high presence of the fusarium toxins in the maize during the harvest. The norms accepted in the EU concerning the acceptable level of deoxynivalenol (DON) and zearalenol (ZON) in feedstuffs for cattle require detailed examination of this problem as it decides on the health quality and production results. The aim of the study was to evaluate the effects of ensiling forage maize with microbiological additive and chemical preservative on the DON and ZON amount. The effects of the secondary fermentation after the silos were open (stability evaluation) were also examined.

Materials and methods The study samples taken from the maize type F70 (Flint type grain from Austria) were ensiled in mid September 2003 after grinding in three ways (5 microsilos each), i.e. 1 - control; 2 - control + 0.25% Kemisile 2000 preservative; 3 - control + 0.2% Lactacel L bacterial enzymatic substance (*Lactobacillus plantarum* 108 cfu/g + enzymes). Whole pieces of maize plant such as leaves, stem and cob were evaluated too. After 12 weeks of ensiling the quality of the silages, microbiological variables and the presence of Fusarium toxins were evaluated. Furthermore, the same analysis was carried out for the silage, which underwent 7-days oxygen exposure. The aflatoxins (AFLA), ochratoxin A (OTA) and ZON were determined by HPLC, DON by Elisa.

Results The results confirmed high moulds contamination of the harvested maize as well as high level of DON (7690 ppb). Ensiling significantly reduced DON and the number of fungal flora. However, the oxygen exposure of the silage increased the level of both mycotoxins and moulds (especially Aspergillus, Penicillium and Mucor genera). The additives Kemisile 2000 had a positive influence on the quality and stability of the silage. The selected parameters have been presented in the Table 1 (mean value from 3 repetitions). The AFLA and OTA were not detected in the raw material.

Table 1 Effect of additive addition on silage hygienic value

Parameters	Ensiling raw material	Silage Control after opening	Control after stability	Kemisile 2000 after opening	Kemisile 2000 after stability	Lactacel L after opening	Lactacel L after stability
Moulds (cfu/g)	1.1×10^7	1.0×10^2	5.6×10^5	1.3×10^3	5.0×10^7	2.1×10^2	1.7×10^7
Yeast (cfu/g)	1.1×10^8	1.1×10^4	6.3×10^8	1.0×10^3	2.5×10^8	6.2×10^4	7.9×10^8
LAB* (cfu/g)	2.5×10^8	2.4×10^8	2.1×10^8	1.9×10^8	1.7×10^8	2.6×10^8	2.2×10^8
Clostridia (cfu/g)	5.4×10^6	1.6×10^6	9.7×10^5	7.6×10^5	1.9×10^5	1.3×10^6	3.5×10^5
E. coli (cfu/g)	6.8×10^8	1.0×10^5	9.3×10^4	7.4×10^4	5.5×10^4	9.3×10^4	6.4×10^4

* LAB – Lactic acid bacteria

References

Grajewski, J., J. Böhm, W. Luf, B. Składanowska, K. Szczepaniak & M. Twarużek (1999). The effect of time on fermentation and fungal growth of two ensiled maize varieties in Austria. In: *Proceedings of XIIth International Silage Conference, Uppsala, Sweden*, pp. 135-136.

Oldenburg, E. & F. Hoppner (2003). Fusarium mycotoxins in forage maize-occurrence, risk assessment, minimisation. *Mycotoxin Research*, 19, 43-46.

Ensiling of tannin-containing sorghum grain

E.M. Ott, Y. Acosta Aragón and M. Gabel
University of Rostock, Faculty of Agricultural and Environmental Sciences, Institute of Farm Animal Sciences and Technology, Justus-von-Liebig-Weg 8, D-18059 Rostock, Germany, Email: edda.ott@uni-rostock.de

Keywords: sorghum, grain silage, tannin

Introduction Sorghum is known as important feed-stuff in tropical regions where rainfall is insufficient for the cultivation of maize. Furthermore, those sorghum cultivars rich in tannins are naturally protected to a certain extent against bird damage, insect pests and moulds. Nevertheless, tannins impair the feed quality. Thus, the objectives of this study were to investigate whether ensiling could be a suitable preservation method for sorghum grain originally rich in tannins and if it is possible to reduce tannin content during fermentation.

Materials and methods Bruised grain of sorghum variety CIAP 2E (with red coloured grain) was used for an ensiling study. After adjusting the DM content of the grist to 70% (using aqua dest.), 600 g of this material were filled in plastic bags (3 per treatment on opening) either pure or inoculated with lactic acid bacteria (LAB; 3×10^5 cfu/g fresh matter (FM)) and molasses (2% of FM). Air was evacuated. Bags were sealed and stored at 30°C for 1, 2, 7, 14 and 56 d. After incubation fermentation parameters (pH; lactic acid: HPLC; acetic acid, propionic acid, butyric acid, i-valeric acid, ethyl alcohol, 2,3 - butandiol: GC), content of condensed tannins (CT; extraction with butanol/HCl by Porter *et al.*, 1986), total phenol amount (TPA; staining with Folin-Ciocalteus-solution by Julkunen-Tiito, 1985) and content of non-tannin-phenols (NTP; Polyvinyl polypyrrolidone by Laurent, 1975 and Waterson & Butler, 1983) were analysed. The content of tannin-phenols (TP) was calculated by subtraction (TP=TPA-NTP). A t-test (Duncan) was performed to compare the results of treatments and control.

Table 1 pH-values, lactic acid contents and tannin fractions in silage from sorghum grain depending on additives and time of storage (means, n=3)

Parameter after time of storage (days)						
	0	1	2	7	14	56
Variants						
PH-value						
Control	6.45	5.84	5.97	5.71	5.48	4.74
LAB+mol	6.45	4.73	4.12	3.93	3.99	3.93
Lactic acid (g/kg DM)						
Control	0.00	0.00	0.30	2.10	3.30	8.20
LAB+mol	0.00	7.20	14.20	14.60	14.20	16.70
Condensed tannins (g/kg DM)						
Control	11.91[a]	3.02[a]	1.53[a]	0.80[a]	0.72[a]	0.41[a]
LAB+mol	11.91[a]	3.92[b]	3.14[b]	1.23[b]	0.96[b]	0.65[b]
Total phenol amount (g/kg DM)						
Control	19.55	9.45	8.54	8.82	9.20	9.83
LAB+mol	19.55	11.41	12.22	12.12	9.14	11.23
Non-tannin-phenols (g/kg DM)						
Control	6.43	2.26	2.17	1.91	2.36	2.94
LAB+mol	6.43	3.04	3.12	2.58	2.64	2.93
Tannin-phenols (g/kg DM)						
Control	13.17	7.22	6.46	7.05	6.83	6.94
LAB+mol	13.17	8.44	9.12	9.56	6.67	8.31

Control: without additives
LAB+mol: addition of lactic acid bacteria (*Lb. plantarum*) and molasses
[ab]: different letters show significant differences between variants

Results Organoleptic evaluation of the silages showed good quality in all variants tested, even in those variants exhibiting pH-values above 5. This could be approved by analytical determination of the fermentation parameters (e.g. no occurrence of butyric acid). Inoculation with LAB led to a rapid acidification and had a positive influence on silage quality (Table 1). Contents of acetic acid were low in all variants (0.08-0.10% of DM). Amounts of CT, TPA, NTP and TP decreased during storage, with the greatest decline after 1 d storage. Decrease was most evident for the fraction of CT, with a reduction to 25% after 1 d and to 3.5% after 56 d of storage (control, Table 1).

Conclusions Ensiling of bruised grain of sorghum (70% DM) results in anaerobically stable silages of good quality with lower contents of tannins than observed in the original material. To verify mechanisms of tannin-reduction further investigations are necessary. Conceivable explanations are, for instance:
- degradation of tannins during moist storage
- enzymatic degradation by LAB (Nishitani *et al.*, 2004)

References
Julkunen-Tiito, R. (1985). Phenolic constituents of leaves of Northern Willows: methods of analysis of certain phenolics. *Journal of the Science of Food and Agriculture*, 33, 213-217.
Laurent, S. (1975). Etude comparative de differentes methodes d'extraction et de dosage des tanins chez quelques pteridhytes. *Archive International Physiology and Biochemistry*, 83, 735-752.
Nishitani, Y., E. Sasaki, T. Fujisawa & R. Osawa (2004). Genotypic analyses of lactobacilli with a range of tannase activities isolated from human faeces and fermented foods. *Systematical Applied Microbiology*, 1, 109-117.
Porter, L.J., R.Y. Wong, H.F. Benson, B.G. Chang, V.N. Viswanadhan, R.E. Gandour & W.L. Mattice (1986). Conformational analysis of flavan. *Journal of Chemistry Research*, 86, 830.
Waterson, J.J. & G.L. Butler (1983). Occurrence of an unusual leucoanthocyanidin and absence of proanthocyanidins in sorghum leaves. *Journal of Agriculture and Food Chemistry*, 31, 41-45.

Fermentation characteristics of maize/sesbania bi-crop silage

M. Kondo[1], J. Yanagisawa[2], K. Kita[1] and H. Yokota[1]

[1]*Graduate School of Bioagricultural Sciences, Nagoya University, Aichi, 470-0151, Japan, Email: i021009d@mbox.nagoya-u.ac.jp, [2]Aichi Agricultural College, Aichi, 444-0802, Japan*

Keywords: maize, *Sesbania cannabina*, bi-crop, silage

Introduction Maize is one of the main forages for dairy production and is a suitable material for silage making because of high fermentable carbohydrates, high counts of lactic acid bacteria (LAB) and low buffering capacity (BC) (Nishino *et al.* 2003; McDonald *et al.* 1991). Whole crop maize silage is high in energy but low in crude protein (CP). On the other hand, legumes are high in CP but difficult to conserve because of their low water soluble carbohydrates (WSC) and high BC. It might be possible that maize/legume bi-crop silage compensate for their negative points. Sesbania is a legume originated in tropical area and might be suitable for inter-crops with maize. Therefore, we investigated the fermentation characteristics of bi-crop silage from maize and sesbania.

Materials and methods Maize (*Zea mays*), sesbania (*Sesbania cannabina*) and these bi-crops were sown on 27 May 2004 at Aichi Agricultural College. The bi-crops were produced from inter-cropped two lines of maize and one line of sesbania. The seeds rates were 36 kg/ha of maize alone, 42 kg/ha of sesbania alone, 24 kg/ha of maize and 14 kg/ha of sesbania for bi-crops. The forages were harvested with a maize harvester on 10 August and packed into glass bottle silos (900 ml capacity) in triplicates and then maintained at 30°C for 30 days.

Results Table 1 shows dry matter (DM) yield and chemical composition of ensiled forages. Maize had high WSC and LAB counts and moderate CP content. On the other hand, sesbania contained low WSC but high CP. Maize/sesbania bi-crop contained higher WSC than sesbania and higher CP than maize. Fermentation characteristics of the silages are given in Table 2. Maize/sesbania bi-crop silage showed similar pH and organic acid concentration to maize alone silage. However, sesbania alone showed higher pH and lower lactic acid than maize alone and bi-crop silage ($P<0.05$).

Table 1. DM yield and chemical composition of forages

	Maize	Maize-Sesbania	Sesbania
DM yield (t/ha)	13.3	11.3	4.4
DM (g/kg)	24.9	21.6	23.5
WSC (g/kg DM)	93.1	62.9	38.8
CP (g/kg DM)	85.1	105.8	157.7
Buffering capacity (meq/kg DM)	218	246	306
Lactic acid bacteria (log cfu/g FM)	6.26	6.22	4.69

Table 2. Fermentation characteristics of silages

	Maize	Maize-Sesbania	Sesbania	s.e.m.
DM (g/kg)	249 b	216 a	235 ab	6.8
pH	3.79 a	3.93 a	5.49 b	0.06
Lactic acid (g/kg DM)	74.4 b	76.3 b	1.6 a	2.4
Acetic acid (g/kg DM)	13.6 a	17.5 a	34.5 b	1.2
Propionic acid (g/kg DM	n.d.	n.d.	6.0	0.9
Butyric acid (g/kg DM)	n.d.	n.d.	14.5	1.5
NH$_3$-N (g/kg total N)	20.5 a	31.5 a	78.6 b	3.4
DM loss (g/kg DM)	61.9	77.1	66.4	8.5

Conclusion Maize/Sesbania bi-crop could be useful as one of the forage production systems. Feed intake and milk production of the bi-crop silage in dairy cattle should be further investigated.

References

llell
McDonald, P., A.R. Henderson & S.J.E. Heron (1991). *The Biochemistry of silage.* Chalcombe Publications.
Nishino, N., M. Yoshida, H. Shiota & E. Sakaguchi (2003). Accumulation of 1,2-propandiol and enhancement of aerobic stability in whole crop maize silage inoculated with *Lactobacillus buchneri*. *Journal of Applied Microbiology*, 94, 800-807.

The influence of crop maturity and type of baler on whole crop barley silage production

P. Lingvall, M. Knicky, B. Frank, B. Rustas and J. Wallsten
SLU, Kungsängen ResearchCentre, S-753 23 Uppsala, Sweden, Email: per.lingvall@huv.slu.se

Keywords: whole crop barley silage, stage of maturity, round baler, intake, liveweight gain

Introduction Bale ensiling is based on long cut forages. Earlier studies (Honig, 1984; 1987) have shown the importance of laceration and high density in preventing fungi growth and storage instability. On the other hand use of an efficient baling technology reduces the time between moving and wrapping of bale to less than 10 minutes with a combi-baler compared to two hours with a separate wrapper. Even during feeding late fermentation is restricted as the bale is fed within some hours after opening. Ensiling of whole crop cereals needs the addition of silage additives to avoid clostridial fermentation (Weissbach *et al.*, 1988). Late silage additive studies have shown the impact of using sodium benzoate in combination with sodium nitrite to baled crops (Knicky & Lingvall, 2002).

Materials and methods The spring variety of barley was harvested by a mover/conditioner Taarup 3028 at three stages of maturity; heading, milk stage and dough stage. The relation between ears and straw as % of forage dry matter (DM) were: at heading 23/77, at milk stage 37/63 and at dough stage 55/45. Two types of baler were used: Taarup BaleInOne (14 knives - 70 mm cut length and Welger 220 Profy (23 knives – 45 mm cut length). The forage was treated with Kofasil Ultra at the rate of 4 l/t during baling and were wrapped with 12 layers of white stretch film. Losses during moving, baling, conservation and storage were registered. Ensiled forage was used in feeding experiments with heifers.

Results The data in Figure 1 demonstrates a significant influence ($P<0.001$) of DM yield with increasing forage maturity. The losses during moving/conditioning of the crops increased ($P<0.001$) from nil at heading to 21 g/kg DM at milk stage and 75 g/kg DM at dough stage. Losses during baling also showed an increasing trend as forage matured ($P<0.001$), on average 30 g/kg DM at heading, 50 g at milk stage and 110 g at dough stage. A considerably higher loss ($P<0.001$) during baling was obtained from Welger 220 Profy at all stages of maturity in comparison with Taarup BaleInOne ($P<0.001$). The main loss consisted of ears. During conservation and storage, the losses were 28 g/kg DM both from silage produced at heading and milk stage but only 13 g/kg DM from dough stage ($P<0.03$). Preliminary results from the feeding trial with heifers (average live weight kg, 300 at start and 370 at the end) showed a daily gain of 0.8 kg/d, an average intake of 2.07 kg DM and 1.10 kg NDF per 100 kg live weigh independent on stage of maturity.

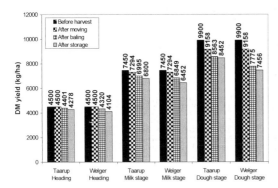

Conclusions Ensiling of whole-crop barley at dough stage resulted in the highest forage production per ha. The mover/conditioner gave high losses of ears at the dough stage. The Taarup baler with restricted number of knives gave lower losses than the Welger baler. The same indication was found comparing ensiling in bales and precision chopped forage in silos. In a "gas tight" bale laceration seems to increase fermentation and losses. No hygienic quality problems were found among the 300 bales produced.

Figure 1 Yield of whole-crop barley forage in relation to the stage of maturity and type of baler

References

Honig, H. (1984). Losses silage in stacks due to CO_2-flew off. *7th Silage Conference, Belfast.* September 1984.
Honig, H. (1987). Influence of forage type and consolidation on gas exchange and losses in silo. *8th Silage Conference, IGAP, Hurley.* September 1987.
Knicky, M. & P. Lingvall (2002). Possibilities to control the hygienic quality, nutrient losses and storage stability in clover/grass and whole-crop barley silages. *The XIIIth International Silage Conference*, Auchincruive, Scotland, September 2002, pp. 224.
Weissbach, F. & K. Macker (1988). Über die Ursachen der Buttersäuregährung in Silagen aus Getreideganzpflanzen. Zeitschrift " Das wirtschaftseigen Futter" Band 3, Heft 2:88.

Effect of stage of maturity on the nutrient content of alfalfa

Y. Tyrolova and A. Vyborna
Research Institute of Animal Production, Pratelstvi 815, 104 00 Praha 10, The Czech Republic

Keywords: alfalfa, maturity, nutrients

Introduction Alfalfa (*Medicago sativa*), one of the major agricultural crops in the Czech Republic, is grown on 15.8% of the arable land. Alfalfa silage forms a substantial part of diets for farm animals. It is very important to choose a suitable term of harvesting for ensilage from the view of optimal nutrient content. The stage of maturity at cutting has large effects on each component, except crude protein (Yu *et al.*, 2004). Alfalfa can be difficult to ensile due to a high buffering capacity and a low WSC content.

Materials and methods The objective of experiment was to examine changes in nutrient content and yield of alfalfa at four different maturity stages (small buds, big buds, bloom, after bloom). Alfalfa (*Medicago sativa*), cultivar Europe, was grown on the experimental field at the Research Institute of Animal Production in Prague (sugar beet growing region, 280 m above sea level). The area 15x15 m was marked out on the 10 ha alfalfa field. The four samples (every 1x1 m) were analysed in this area in all stage of maturity. Alfalfa was planted at seeding rate 18 kg/ha in 2002 with wheat as a foregoing crop and legume-cereal mixture as a cover crop. The plants were picked up from 12 May to 10 June 2003, during the first cut. The yield was recorded and the plants were analysed.

Results and discussion The results of chemical analyses are shown in the Table 1. Protein content of alfalfa decreased significantly after the big bud stage ($P<0.05$). The highest water soluble carbohydrate (WSC) content was observed in the stage of big buds. Thereafter the WSC content decreased while the fibre crude content increased ($P<0.05$). When the crop stand passes over to bloom stage, the nutrient content changes and the forage quality falls. It is due to the fact that some nutrients are transferred to the generative organs. Protein content in the stage of after bloom was reduced by about 30% in comparison with the stage of small buds.

Table 1 Nutrient content of alfalfa at four stages of maturity

	Unit	Small buds	SE	Big buds	SE	Bloom	SE	After bloom	SE
Dry matter	g/kg	165.3[a]	3.03	175.2[a]	4.61	207.8[b]	8.37	231.8[b]	1.15
Protein	g/kg	219.6[a]	4.30	203.1[a]	4.10	173.5[b]	2.67	154.2[c]	0.53
Crude fibre	g/kg	226.7[a]	4.81	235.4[a]	5.38	320.4[b]	9.08	312.7[b]	3.78
WSC	g/kg	37.4[d]	1.41	45.3[c]	1.71	27.6[b]	0.69	23.3[a]	0.57
Yield of DM	t/ha	2.9[a]	0.09	3.4[b]	0.14	5.5[c]	0.39	5.6[c]	0.05

[a,b,c,d,e,f] Means followed by the same letter on a row are not different (Tukey α = 0,05)

Conclusions It is recommended to ensile alfalfa at the maturity stage of big buds. In this stage the protein content is only about 7.5% less than in the small buds stage and in this stage the WSC content is highest and the crude protein content is still optimal. The yields of dry matter are highest in the maturity stage after bloom. In the after bloom maturity stage the dry matter yield increases but the nutritive value is reduced.

Acknowledgement Supported by the Ministry of Agriculture of the Czech Republic (Project No. MZE 0002701403).

References
Marley, C.L., R. Fychan, M.D. Fraser, A. Winters & R. Jones (2003). Effect of sowing ratio and stage of maturity at harvest on yield, persistency and chemical composition of fresh and ensiled red clover/lucerne bi-crops. *Grass and Forage Science* 58 (4), 397-406.
Yu, P., D.A. Christensen & J.J. McKinnon (2004). *In situ* rumen degradation kinetics of timothy and alfalfa as affected by cultivar and stage of maturity. *Canadian Journal of Animal Science*, 84 (2), 255-263.

Field beans and spring wheat as whole crop silage: yield, chemical composition and fermentation characteristics

L. Ericson, K. Arvidsson and K. Martinsson
The Swedish University of Agricultural Sciences, Forage Research Centre, Box 4097, S-904 03 Umeå, Sweden,
Email: Kjell.Martinsson@njv.slu.se

Keywords: field bean, fermentation, feed value

Introduction There has been an increasing interest in field beans (*Vicia faba L.*) in recent years because of its N-fixating ability. The objective of this study was to compare the yield, chemical composition and fermentation characteristics of field bean/spring wheat as whole-crop silage ensiled with and without an additive.

The crop was drilled on 27 May 2003 at a seed rate of 205 kg/ha field beans and 68 kg/ha spring wheat. The crop was harvested at four different growth stages (Zadoks *et al.*, 1974); end of blooming, (stage 69), when 50% of the pods had reached full length, (stage 75), pods fully formed, (stage 79) and when 10% of the pods are filled (stage 81). Yield and botanical composition were evaluated. Samples of the forage were analysed for dry matter (DM) and chemical composition at harvest. Forage, at stages 75, 79 and 81 was wilted overnight then chopped (20 mm) and ensiled in 10 kg silos. Half the forage was ensiled untreated (control). The other half was treated with PROENS (60-66% formic acid and 23-29% propionic acid, Perstorp Speciality Chemicals, Sweden) applied at a rate of 6 l/t fresh matter. The silos were incubated for a period of 90 d and then analysed for DM, chemical composition and fermentation characteristics.

Results and conclusions The yield and some chemical characteristics of the fresh crop are presented in Table 1. Chemical composition after ensiling is presented in Table 2. Delaying the harvest gave higher DM and CP yields and also higher starch content while the concentration of NDF was significantly decreased. The use of an acid additive gave a restricted fermentation resulting in a significantly decreased content of volatile fatty acids (VFA), lactic acid and ammonium-N. Also the proportion of soluble protein (SP) was decreased.

Table 1 Production and chemical composition of whole crop field beans/spring wheat cut at different stages of development. Four plots at each stage

	Stage 69	Stage 75	Stage 79	Stage 81	s.e.d	Significance level
Time of harvest	31 July	12 August	21 August	5 September		
CP (g/kg DM)	154	156	152	154	10	NS
Yield (kg DM/ha)	3800	4900	5000	5940	221	***
Field beans (% of total DM)	66	NC	75	75	2	**
WSC (g/kg DM)	NC	100	87	102	15	NS
Buffering capacity (meq/kg)	NC	203	195	179	2	***

* = $P<0.05$; ** = $P<0.01$; *** = $P<0.001$. NS = the difference was not significant. NC = the values were not calculated

Table 2 Chemical composition of whole crop field beans/spring wheat after ensiling. Four replicates of each treatment

	Stage 75		Stage 79		Stage 81		Signifi cance	Significance level	
	Control	Proens	Control	Proens	Control	Proens	s.e.d	Stage	Additive
DM (g/kg)	201	223	282	289	271	277	6	**	NS
NDF (g/kg DM)	490	522	466	501	451	442	18	**	NS
SP (% of CP)	49	42	49	46	49	47	2	NS	*
Starch (g/kg DM)	104	98	149	112	164	169	12	**	NS
VFA (1) (g/kg DM)	19.2	4.0	10.2	3.3	11.5	3.2	2.4	NS	***
Lactic acid (g/kg DM)	83.6	4.1	50.3	3.6	50.7	7.5	3.6	NS	***
Amm.-N (g/kg N)	56	40	55	40	66	44	7	NS	***

(1) Butyric acid was not detected. * = $P<0.05$; ** = $P<0.01$; *** = $P<0.001$. NS= the difference was not significant. There were no significant interactions

References
Zadoks, J.C., T.T. Chang & C.F. Konzak (1974). A decimal code for the growth stages of cereals. *Weed Research.* 14:6, 415-421.

Utilisation of whole-crop pea silages differing in condensed tannin content as a replacement for soya bean meal in the diet of dairy cows

K.J. Hart, R.G. Wilkinson, L.A. Sinclair and J.A. Huntington
*Animal Science Research Centre, Harper Adams University College, Newport, Shropshire, TF10 8NB, UK,
Email: k.j.hart@talk21.com*

Keywords: whole-crop pea silage, dairy cows

Introduction Adesogan *et al.* (2004), has demonstrated that ensiled pea wheat intercrops can reduce the amount of concentrate fed to dairy cows by 50% without affecting milk yield or composition. A limitation of forage peas for high yielding dairy cows is the low by-pass protein content, but it has been suggested by Broderick (1995), that feeding forages that contain low levels of condensed tannin can improve nitrogen utilisation. The objective of the current experiments was to evaluate the potential of whole-crop pea silage differing in condensed tannin content to replace soya bean meal in the diet of late lactation, pregnant dairy cows.

Materials and methods Whole-crop pea silage was produced from spring-sown crops of Racer (high tannin) and Croma (low tannin), which were cut 13 weeks after sowing, wilted for 30 h and ensiled with a bacterial inoculant. A conventionally managed crop of winter wheat (*cv.* Equinox) was harvested at target DM of 440 g/kg and ensiled (WCW). Grass silage (G) was cut from a predominantly perennial ryegrass sward, wilted for 24 h and ensiled. Eighteen multiparous pregnant, late lactation Holstein-Friesian dairy cows were randomly allocated to one of two experiments in a 3x3 latin square design. Each 28 d period consisted of a 21 d dietary adaptation period followed by a 7 d sampling period. The control diet was a 50:50 mix (DM basis) of G and WCW offered *ad libitum* and fed with 6.9 kg/d standard concentrate and 1.1 kg/d soya (GWS). The test diets were a 25:25:50 mix (DM basis) of G, WCW and one of the two pea silages offered *ad libitum*, fed with 6.9 kg/d standard concentrate. Cows in experiment 1 were offered the high tannin pea silage with the addition of either 1.1 kg/d soya (HTS) or 1.1 kg/d wheat (HTW), and those in experiment 2 were offered the low tannin pea silage with either 1.1 kg/d soya (LTS) or 1.1 kg/d wheat (LTW).

Results The high and low tannin pea silages had a similar chemical composition and only differed in condensed tannin content (93.4 and 47.2 g tannic acid equivalents/kg DM respectively), crude protein content (189 and 177 g/kg DM respectively) and starch content (58 and 87 g/kg DM respectively). Average daily forage intake, milk yield, milk composition and nitrogen efficiency (N. eff.) for experiments 1 and 2 are presented in Table 1. In experiment 1 N. eff. was higher ($P<0.05$) in cows fed GWS compared to those fed HTW, cows fed HTS had a higher ($P<0.05$) forage DM intake compared to those fed GWS. In experiment 2 cows fed GWS had a lower ($P<0.05$) forage DM intake compared to those fed LTS or LTW. There was no difference in milk protein concentration between cows fed either GWS or LTW. Cows fed GWS had the highest N. eff. and those fed LTS had the lowest N. eff. in experiment 2. There was no difference between any dietary treatment and milk yield or 4% fat corrected milk yield (FCM) in either experiment 1 or 2.

Table 1 Effect of diet on daily intake, milk yield, milk composition and nitrogen efficiency for milk production

	Experiment 1					Experiment 2				
	GWS	HTS	HTW	s.e.d.	*P*	GWS	LTS	LTW	s.e.d.	*P*
Forage intake (kg DM/d)	12.0	13.4	13.0	0.53	0.041	11.6	13.9	13.2	0.35	<0.001
Milk yield (kg/d)	23.1	23.8	22.0	0.86	0.123	24.2	25.4	23.8	0.43	0.299
4% FCM (kg/d)	24.7	25.6	24.1	0.89	0.264	26.3	25.5	26.9	0.64	0.108
Butterfat (g/kg)	43.1	43.1	43.9	0.87	0.669	43.6	41.9	45.7	1.04	0.011
Protein (g/kg)	35.7	35.4	36.1	1.15	0.847	35.4	34.4	34.7	0.42	0.112
N. eff. (%)	28.9	23.4	24.6	0.74	<0.001	30.0	23.4	25.9	0.79	<0.001

Conclusions The results from these two experiments demonstrate that the inclusion of whole-crop pea silage into the diets of late lactation dairy cows can replace 1.1 kg soya bean meal daily without affecting milk yield or composition. The forage mixes containing the high tannin forage had a lower nitrogen efficiency for milk production than those containing the low tannin forage.

References
Adesogan, A.T., M.B. Salawu, S.P. Williams, W.J. Fisher & R.J. Dewhurst (2004). Reducing concentrate supplementation in dairy cow diets while maintaining milk production with pea-wheat intercrops. *Journal of Dairy Science,* 87, 3398-3406.
Broderick, G.A. (1995). Desirable characteristics of forage legumes for improving protein utilisation in ruminants. *Journal of Animal Science,* 73, 2760-2773.

Ensiled high moisture barley or dry barley in the grass silage-based diet of dairy cows

S. Jaakkola, E. Saarisalo and R. Kangasniemi

MTT Agrifood Research Finland, Animal Production, FI-31600 Jokioinen, Finland, Email: seija.jaakkola@mtt.fi

Keywords: grass silage, ensiled barley, protein supplementation, dairy cow

Introduction Ensiling high moisture grain is based on a procedure similar to ensiling grass. Soluble carbohydrates of grain are partly fermented into acids and some protein is degraded. Thus the nutritional quality is modified by the preservation method. Two trials were conducted to compare the effects of dry barley (DB) and ensiled barley (EB) in a total mixed ration (TMR) on feed intake and milk production of dairy cows.

Materials and methods Sixteen cows were used in both replicated (n=4) Latin square experiments with four 21 day periods and a 2 x 2 factorial arrangement of treatments to evaluate the effects of barley preservation method and protein supplementation. EB was combine harvested at an earlier stage of maturity (dry matter (DM) 560 g/kg) and DB at full maturity (782 g DM/kg) from the same field. EB grains were rolled in a crimper (Murska 350 S2, Aimo Kortteen Konepaja Oy) and treated with a formic acid additive (3,3 l/t, AIV2000, Kemira Oyj), and then stored in a bunker silo and sealed beneath plastic sheeting. DB was dried to the final DM content of 890 g/kg. Timothy-meadow fescue grass was wilted, harvested with precision chopper, treated with formic acid additive (AIV2000, 5 l/t) and ensiled in bunker silos. Experimental treatments consisted of two TMRs (including ensiled or dry barley) offered *ad libitum* and supplemented with two levels of extra rapeseed meal (RSM) fed during milking. On DM basis TMR contained 55% grass silage and 45% concentrate, which included 64% barley, 22.5% RSM, 10.5% molassed sugar beet pulp (SBP) and 3% minerals. The crude protein content of concentrate and silage were 173 and 129 g/kg DM, respectively. The extra RSM levels were 0 or 1 kg/d (Trial 1) and 0 or 2 kg/d (Trial 2) fed as a mixture of RSM and SBP (3:1).

Results The fermentation quality of EB was good as indicated by low pH (3.86), ammonia-N (33 g/kg N) and VFA (11 g/kg DM) content. Similarly the quality of grass silage was good with low pH (4.03), ammonia-N (45 g/kg N) and VFA (25 g/kg DM) content. Lactic acid content in EB and grass silage was 40 and 59 g/kg DM, respectively. The preservation method of barley did not affect TMR intake in Trial 1 (Table 1). In Trial 2, daily intake of DB diet was higher than the intake of EB diet. No significant differences were observed between EB and DB diet in milk yield or milk composition in Trials 1 and 2. RSM supplementation decreased the intake of TMR but increased total DM intake with 0.75 (Trials 1) and 1.20 kg DM/d (Trial 2). In both experiments, RSM supplementation increased milk and energy corrected milk (ECM) yield, the response being 0.85 kg ECM/kg RSM. No significant effects of RSM on milk composition were observed. The effects of extra RSM on feed intake and milk production were the same with EB and DB diets except on milk protein content in Trial 2 (interaction barley x RSM, *P*<0.05).

Table 1 Effect of ensiled or dry barley and rape seed meal (RSM) on feed intake and milk production

Barley	Dry		Ensiled		SEM	Barley	RSM
Trial 1: RSM (kg/d)	0	1	0	1			
DM intake (kg/d)	22.8	23.2	22.5	23.6	0.21		
Milk (kg/d)	30.1	31.3	30.3	30.8	0.29		*
Fat (g/kg)	43.1	43.3	43.8	43.1	0.72		
Protein (g/kg)	33.9	34.0	33.6	34.2	0.24		
Lactose (g/kg)	47.9	48.1	48.1	48.2	0.30		
Milk N/feed N	0.290	0.283	0.298	0.277	0.0048		*
Trial 2: RSM (kg/d)	0	2	0	2			
DM intake (kg/d)	22.6	23.6	21.4	22.8	0.19	***	
Milk (kg/d)	32.6	34.5	31.8	34.1	0.31		***
Fat (g/kg)	43.0	42.2	44.4	42.8	0.63		
Protein (g/kg)	32.4	31.6	31.8	32.1	0.18		
Lactose (g/kg)	49.4	49.2	48.8	49.2	0.17		
Milk N/feed N	0.303	0.273	0.313	0.288	0.0035	**	***

* *P*<0.05, ** *P*<0.01, *** *P*<0.001

Conclusions The results suggest that ensiled, crimped barley has the same nutritive value as dry barley when fed for dairy cows given grass silage-based diet.

Effects of species, maturity and additive on the feed quality of whole crop cereal silage

E. Nadeau
Department of Animal Environment and Health, Swedish University of Agricultural Sciences, Box 234, 532 23 Skara, Sweden. Email: elisabet.nadeau@hmh.slu.se

Keywords: whole crop cereal silage, maturity, silage additive, feed quality

Introduction Chemical composition of whole crop cereals differ among species and maturity stages. These chemical differences create variations in silage quality (Bergen *et al.*, 1991). There is only limited information available on the effects of plant species and maturity on the use of additives for whole crop cereal silage. The objective of this experiment was to determine the effects of species, maturity, additive and their interactions on nutrient composition and fermentation characteristics of whole crop cereal silage.

Materials and methods Triticale, barley, spring wheat and oats were harvested as direct cut at early milk stage (73) and at early dough stage of maturity (83; Zadoks *et al.*, 1974) in 2002 and 2003. Material was ensiled for 100 d in 4-l silos with or without the use of additive. The additives were Proens™ (4 l/t of 2/3 formic acid and 1/3 propionic acid; Perstorp AB, Perstorp, Sweden) and Lactisil 200® NB (200 000 cfu of *Lactobacillus plantarum, Enterococcus faecium, Pediococcus acidilactici* and *Lactococcus lactis*/g fresh herbage plus cellulase and sodium benzoate; Medipharm AB, Kågeröd, Sweden). The experiment was conducted as a split-split-split-plot with three field replicates per year. Because most variables had no significant interactions with year, data are presented as averages across years.

Results Delayed harvest increased dry-matter (DM) (from 286 to 373 g/kg) and starch concentrations (from 32 to 192 g/kg DM) but decreased sugar (from 194 to 93 g/kg DM) and fibre (NDF) (from 563 to 495 g/kg DM) concentrations as well as *in situ* rumen degradability of NDF (from 65 to 57%). Triticale had the highest sugar content but the lowest starch content. Oats had the lowest sugar content and barley had the highest starch content. Barley and triticale had higher starch + sugar content (281 vs 222 g/kg DM), lower NDF content (511 vs 547 g/kg DM) and higher rumen degradability of NDF (66 vs 56%), resulting in higher organic matter digestibility in barley and triticale than in oats and wheat (79 vs 68%). Silage fermentation characteristics were improved by use of additives (Table 1). In contrast to inoculant treatment, acid treatment restricted fermentation and resulted in higher sugar concentration in the silage than in fresh material (174 vs 144 g/kg DM). The higher lactic acid concentration in inoculated silage resulted in the lowest pH (4.1 vs 4.2 (control) and 4.4 (acid)). Ammonia-nitrogen concentration was only slightly higher in acid-treated silage than in fresh material (105 vs 97 g/kg total nitrogen). Inoculated silage had less protein degradation than the control (135 vs 163 g NH₃-N/kg total N).

Table 1 Fermentation characteristics of whole-crop cereal silages averaged over two maturities and two years

Herbage treatment	Plant Species				Significance	
(g/kg dry matter)	Barley	Triticale	Oats	Spring wheat	*P*	s.e.m.
Sugar						
Control	33	66	22	50	<0.0001	4.1
Inoculant	85	138	37	88		
Acid	196	189	122	189		
Lactic acid						
Control	69	62	73	41	<0.0001	3.1
Inoculant	81	87	67	57		
Acid	17	28	26	4		
Acetic acid						
Control	20	24	11	14	<0.0001	0.8
Inoculant	7	8	6	5		
Acid	3	12	3	2		

Conclusions Barley and triticale have higher nutritive value and more available carbohydrates for ensiling than oats and spring wheat. Use of additives improved silage quality.

References

Bergen, W.G., T.M. Byrem & A.L. Grant (1991). Ensiling characteristics of whole-crop small grains harvested at milk and dough stages. *Journal of Animal Science*, 69, 1766-1774.
Zadoks, J.C., T.T. Chang & C.F. Konzak (1974). A decimal code for the growth stages of cereals. *Weed Research*, 14, 415-421.

Comparison of different maize hybrids cultivated and fermented with or without sorghum

Sz. Orosz, Z. Bellus, Zs. Kelemen, E. Zerényi and J. Helembai
Szent István University, Department of Nutrition, Gödöllő, PO B.3, H-2103, Hungary.   Email: Orosz.Szilvia@mkk.szie.hu

Keywords: sorghum, maize, silage, yield

Introduction In Hungary our key forage crop is silage maize, however, the joint growing of maize and sorghum is increasingly important in arid regions. The reason is, that sorghum varieties tolerate well the various ecological stresses (drought). The joint growing of maize and sorghum varieties has several advantages and disadvantageous in respect of yields, safety of production, fermentability of the crop and nutrient content of the silage. The basis of realising the complementary qualities of the two crops and of the successful joint growing and preservation is the suitable pairing of hybrid varieties.

Materials and methods The authors studied certain early maturing (290-350), medium maturing (FAO 350-450) and late maturing (> FAO 450) maize hybrids (12), grown alone and jointly (2x2) with Sucrosorgo silage sorghum (1 hectare per each hybrid). The preservation and storage were carried out in 3 shared plastic tubes (60 m long with diameter of 3.0 m and 0.27 mm thickness). The nutrient and energy content of the fresh maize and maize-sorghum mixtures and silages, the pH, the lactic - and volatile fatty acid content of silages (sampled on the days 14, 28 and 140 of fermentation) were analysed according to the Hungarian National Standards (Hungarian Feed Codex, 2004).

Table 1 Result of the different maize hybrids grown and ensiled alone or with sorghum

Maize hybrids	Green yield (t/ha)		DM (t/ha)		Nel (GJ/ha)		DM (g/kg)		NEl (g/kg DM)		CF (g/kg DM)		pH (28th day)		Lactic. acid % total acid (28th day)	
	M+S*	M**	M+S	M	M+S	M	M+S	M	M+S	M	M+S	M	M+S	M	M+S	M
LG2483	42.2	23.1	11.0	8.3	63.2	52.4	256.7	364.0	5.22	6.17	275.5	165.1	3.7	3.9	81.7	81.5
LG2470	44.4	27.4	11.1	10.4	61.9	67.9	251.3	392.1	5.50	6.27	309.1	178.6	3.7	4.0	82.4	84.0
GEYSER	41.5	24.4	11.3	10.2	62.1	66.7	269.2	397.6	5.50	6.22	260.0	178.2	3.8	3.9	86.0	83.7
VASALICA	40.5	27.2	11.0	11.3	61.0	75.5	293.4	424.6	5.46	6.30	266.8	164.2	3.7	3.9	83.0	84.4
CORALBA	41.6	29.2	10.8	10.7	60.6	71.2	266.8	373.0	5.58	6.32	273.3	174.1	3.8	3.8	85.6	83.9
DK 527	38.8	25.2	10.3	10.5	57.4	69.9	224.5	387.4	5.38	6.34	288.1	136.3	3.9	3.9	71.0	83.5
DK 523	39.4	23.2	10.5	9.5	59.3	63.5	276.0	424.9	5.45	6.24	278.2	160.5	3.8	3.9	82.1	80.1
DK 557	39.4	27.3	10.7	10.3	59.1	68.0	298.2	394.0	5.56	6.33	274.8	153.0	3.8	3.9	82.9	84.0
DK 366	39.2	17.5	10.8	7.2	60.3	47.7	265.0	372.7	5.51	6.22	242.9	168.7	3.7	4.1	85.3	80.6
MAXIMA	39.1	21.4	10.1	8.0	56.7	53.0	267.9	377.6	5.48	6.22	280.8	185.3	3.8	4.0	85.3	88.3
KÁMA	36.7	22.0	10.2	8.4	56.5	56.6	273.8	378.1	5.52	6.30	277.9	154.8	3.8	3.9	83.9	83.3
SZETC 465	33.3	23.5	9.0	11.0	49.5	73.2	247.3	436.6	5.57	6.20	291.8	158.4	3.8	4.0	82.7	83.2

Note*M+S: Maize grown with sorghum (2 x2). **M: maize grown alone. DM: dry matter. NEl: net energy for lactation. CF: crude fibre

Conclusions Where drought can be expected (during July-August in Hungary), the available acreage is limited or crop conditions are not ideal for any other reason, it is recommended to grow maize hybrids which perform well with sorghum and produce together a large green and dry matter yield safely (Vasalica, LG2470, Coralba, LG 2483 and Geyser). Where no such limitations exist (in good soil and weather conditions, with low risk of drought and high yielding dairy herd), higher energy yielding maize hybrids sown on their own are more important (Szetc 465, Vasalica, Coralba, DK 527 and DK 557). Where the high performance has priority the sowing of maize is highly recommended alone, because the sorghum decreased the dry matter (32%), net energy content (13%), while increasing the crude fibre content (68%) of the silages compared to maize silages.

References

Hungarian Feed Codex. III. (2004.). Laboratory methods and operations. National Institute for Agricultural Quality Control. Budapest. ISBN 963 86097 5 3, Hungary.

Utilisation of coffee grounds for total mixed ration silage

C. Xu, Y. Cai, N. Hino, N. Yoshida and M. Ogawa
National Institute of Livestock and Grassland Science, Nishinasuno, Tochigi 329-2793, Japan. Email: cai@affrc.go.jp

Keywords: coffee grounds, nutritive value, silage fermentation

Introduction In the beverage industry, wastes from coffee grounds are of particular importance given their rapid increase in recent years. Although a small part is converted into raw compost material, wastes generated from tea grounds are generally incinerated. There is increasing demand for efficient use of by-products due to economic and environmental concerns. Approximately 200,000 t of coffee grounds are produced annually in Japan. These grounds usually have high protein, fat, fibre, and nitrogen-free extract and possibly could be a source of nutrients for ruminant (Xu *et al.*, 2004). The objectives of this study were to evaluate the fermentation characteristics of silages prepared from coffee grounds mixed with various feeds and their nutritive values with sheep.

Materials and methods Three silages were prepared using commercial formula feed, timothy hay, lucerne hay, dried beet pulp and vitamin and mineral supplement and coffee grounds at 0%, 10% and 20% on dry matter (DM) basis. Wet coffee grounds were obtained from a local beverage factory. The silages were ensiled with an inoculant Chikuso-1 (*Lactobacillus plantarum*, Snow Brand Seed Co. Ltd.; 5 mg/kg fresh weight) in polyethylene bag silos (350 kg/bag). These silos were stored outdoors at -0.7 to 34.8°C for 225 days. Six 2-year-old Suffolk sheep (73.0 ± 3.2 kg) were used in a two 3×3 Latin square. A 7-d preliminary adjustment period was followed by 7-d period during which all faeces and urine were collected.

Results Three silages were well preserved, with low pH value and ammonia-nitrogen content, and high lactic acid content. The propionic acid and butyric acid were not detected. The DM, organic matter (OM) and crude protein (CP) contents of all silages were similar, but the ether extract (EE), acid detergent fibre (ADF) and neutral detergent fibre (NDF) contents were significantly ($P<0.05$) higher with the increase in the proportion of coffee grounds (Table 1). Increasing concentrations of coffee grounds in the silages decreased the digestibility of DM, CP, ADF, NDF and energy, and increased that of EE. Total digestible nutrients (TDN) and voluntary feed intake for the silage with 20% coffee grounds was significantly ($P<0.05$) lower than those with 0% and 10% coffee grounds (Table 2). With increases in coffee ground concentrations in silages, nitrogen intake was not different, but amount of nitrogen retained decreased as the faecal and urinary nitrogen losses increased.

Table 1 Chemical composition of total mixed ration silage (% of DM)

	The mixing ratio of coffee grounds		
	0	10	20
DM	42.6±0.75	43.1±0.78	43.1±0.71
OM	93.1±0.25	92.9±0.40	93.4±0.35
CP	14.5±0.21	14.7±0.15	14.8±0.25
EE	2.4±0.10[a]	4.1±0.12[b]	5.6±0.17[c]
ADF	24.1±0.41[a]	25.6±0.55[b]	27.1±0.21[c]
NDF	37.4±0.55[a]	39.2±0.45[b]	41.0±0.19[c]

Mean ± SD. (n = 3)

[a, b, c] Values with different superscript letters differ ($P<0.05$).

Table 2 Digestibility and nutrient content of total mixed ration silage (% of DM)

	The mixing ratio of coffee grounds		
	0	10	20
DM	71.9±0.88[a]	70.5±1.03[b]	67.2±1.36[c]
CP	72.7±0.42[a]	66.8±0.85[b]	61.6±0.61[c]
EE	71.2±1.24[c]	80.0±1.21[b]	83.5±1.26[a]
ADF	57.5±1.58[a]	54.5±1.51[b]	46.3±2.03[c]
NDF	59.9±2.25[a]	57.0±1.79[b]	54.4±1.07[c]
TDN	71.2±0.82[a]	71.5±0.79[a]	70.4±0.74[b]

Mean ± SD. (n = 6)

[a, b, c] Values with different superscript letters differ ($P<0.05$).

Conclusions These results suggest that all three silages with various concentrations of coffee grounds were well preserved, and the proportion of the coffee grounds in silages should not exceed 20% (DM basis) based on the voluntary feed intake and TDN content.

References

Xu, C., Y. Cai, T. Kida, M. Matsuo, H. Kawamoto & M. Murai (2004). Silage preparation of total mixed ration with green tea grounds and its fermentation quality and nutritive value. *Grassland Science*, 50, 40-46.

Forage preferences of horses

C.E. Müller
Swedish University of Agricultural Sciences, Department of Animal Nutrition and Management, Kungsängen Research Centre, SE-753 23 UPPSALA, SWEDEN, Email: Cecilia.Muller@huv.slu.se

Keywords: horse, silage, haylage, hay, free-choice

Introduction In the northern climates the forage fed to horses has by tradition been hay. However, hay is subjected to moulding unless it is stored dry. Mould spores together with actinomycetes are responsible for the condition Recurrent Airway Obstruction (RAO), which is the second largest reason for culling of warm-blood horses in Sweden (Wallin, 2001). Therefore, the possibility of replacing hay with haylage and silage in the feed rations of horses is interesting. The objective of this study was to investigate the preferences among horses for different types of conserved grass, to gain more knowledge about the suitability of haylage and silage as a horse feed.

Materials and methods Four mature horses of different age and breed were used in a preference test, in which the horses had simultaneous access to four forages harvested from the same grass ley (585-608 g neutral detergent fibre (NDF), 26-102 g water soluble carbohydrates (WSC) per kg DM) at the same date, but with different dry matter (DM) levels and different techniques. Silage with 350 g DM/kg, haylage with 550 g DM/kg, haylage with 700 g DM/kg and hay was baled with a high density hay baler. Silage and haylage bales were wrapped (eight layers of white 360 mm wide stretch film) and hay bales were put on a barn-drier. During the period of the preference test, the horses were kept on autumn pasture and were offered the experimental forages as an extra meal for two hours daily inside a stable. All forages were served in amounts of 1 kg DM in plastic containers of identical size, colour and square shape, and the order of the containers was different from day to day according to a randomised scheme, to avoid that a certain place was preferred instead of a certain forage. During the two-hour period, real time observations were made for every horse, and the activities were registered according to a predetermined protocol with certain definitions. The preference for eating place of each individual horse was also registered. The preference test was repeated for 24 days. All comparisons were done using SAS GLM procedure, and for a difference to be considered as statistically different $P<0.05$.

Results The horses chose to eat the silage with 350 g DM/kg first more often than any of the other forages (Figure 1, $P<0.001$). In the beginning of the experimental period, the horses spent more time smelling at the different forages before deciding to stay and eat from one particular forage, than at the end. The only forage that was never left in favour to any other forage was silage with 350 g DM/kg. In contrast, hay was the forage that was left in favour of the other forages most often (Figure 2, $P<0.001$). The horses did not have any preference for eating place in the study. The analytical results of the forages showed differences in content of NDF and WSC which were lowest in the 350 g DM/kg silage, as well as levels of VFA (volatile fatty acids) which were highest in silage with 350 g DM/kg (data not shown).

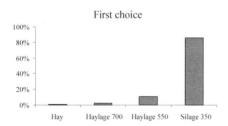

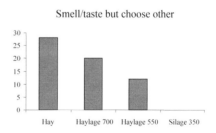

Figure 1 First choice of forages. The type of forage is indicated by the DM level on the x-axis

Figure 2 No. of times horses smelled at a forage but left it in favour for another

Conclusion The result of the present study shows that horses seem to make active choices of forages and that silage is the preferred forage when given the choice. WSC-content did not seem to be involved in preference, since the forage with the highest WSC-content was the least preferred.

References

Wallin, L. (2001). *Longevity and early prediction of performance in Swedish horses.* Agraria 288. PhD Thesis. Swedish University of Agricultural Sciences, Uppsala, Sweden.

Section 3

Developments in ensiling techniques

The effect of silage harvester type on harvesting efficiency

J.P. Frost and R.C. Binnie
Agricultural Research Institute of Northern Ireland, Hillsborough, Co. Down BT26 6DR, U.K. Email: peter.frost@dardni.gov.uk

Keywords: silage harvesting system

Introduction Choice of harvesting system can significantly influence production costs. Whether a tractor-powered or self-propelled forage harvester or a self-loading forage wagon system is used will depend on particular circumstances. However, in order to make an informed choice relevant information has to be available. A trial was commissioned by Landmec Pottinger (Ivybridge, UK) and Traynors (Clonmel, Ireland) at this Institute to investigate the performance of a self-propelled forage harvester system and a self-loading forage wagon system.

Materials and methods Wilted grass cut from a predominantly perennial ryegrass sward was rowed up and alternate swaths were harvested either by a John Deere 6850 self-propelled forage harvester (SPFH) or a Pottinger Torro 5100 self-loading forage wagon (SLFW) powered by a Fendt 716 tractor. The SPFH was serviced by 3 tractors with 12 t trailers and 1 tractor with a 10 t trailer; the standard harvesting team at this Institute. The sward was cut on the 1 June 2004 and harvested on 2 June. Transport distance from field to silo return was 3.4 km. Herbage from each system was ensiled in identical roofed concrete silos (80 t capacity). Representative samples of herbage taken from each load were used to determine DM concentration of the herbage. The Hillsborough Feeding Information Service was used to assess ensilability of herbages and quality of the resultant silages. Chop lengths of the herbage ensiled were determined by hand separating a 50 g sample from each load into 5 length categories (0–20, 21–40, 41–60, 61–80, 81–100 and >100 mm). The herbage in each length category was dried, weighed and the percentage distribution in each of the categories calculated. For each load of herbage the times taken to harvest, transport and turn-round at the silos were recorded. Also recorded were the times taken to fill and roll the herbage in the silos. Forward speeds of the two harvesters during harvesting were recorded, as was fuel consumption by all vehicles in both systems.

Results and discussion Herbage harvested averaged 23.4 t/ha and 286 g DM/kg. There was no treatment effect on the analyses of the herbages as ensiled or on the analyses of the resultant silages. Particle size distribution in the 21–80 mm range was similar for both systems being 66.6 and 66.2% for the SPFH and SLFW respectively. Particles in the 0–20 mm category were greatest in SPFH harvested herbage (22.1 vs. 6.6%) while particles >81 mm were greatest in herbage harvested by the SLFW (27.3 vs. 11.3%). Harvesting and transporting the herbage to the silos by the SPFH required 5 people for the 10½ loads compared with 1 person for the 8 loads with the SLFW. The quantity of herbage harvested and transported per person per hour with the SLFW system was more than double that of the SPFH system (Table 1). The fuel used to harvest and transport herbage to the silo with the SLFW was half of that required by the SPFH (0.67 vs. 1.32 l/t). Data relating to some of the other parameters measured are presented in Table 1. Factors influencing the choice of silage harvesting system for a particular farm include availabilities of labour, machinery, time and finance as well as transport distance. Potential outputs and resource requirements for the SPFH and SLFW systems for circumstances at this Institute are given in Table 1. These data should assist when choosing an appropriate silage harvesting system to suit different circumstances. For example, data in Table 1 indicate that 2 people, each with a SLFW, could harvest and transport almost as much herbage in a given time as 5 people with a SPFH system.

Table 1 Comparison of self-propelled forage harvester (SPFH) and self-loading forage wagon (SLFW) systems

	SPFH	SLFW		
Number harvesters/number of operators	1/5	1/1	2/2	3/3
Harvester power available (kW)	330	103	206	309
Transport power available (kW/unit)/number of units	95.5/4	103/1	103/2	103/3
Total power available (kW)	712	103	206	309
Output (t fresh herbage/h)	53.4	24.8	49.6	74.4
Output per person (t/h harvest and transport)	12.4	24.8	24.8	24.8
Output (t fresh herbage/10 h d)	534	248	496	744
Fuel used (l/t harvest and transport)	1.32	0.67	0.67	0.67
Weight herbage per load (t)	6.6	8.5	8.5	8.5
Average transport speed (km/h)	21.5	22.2	22.2	22.2

Conclusion Data presented indicate that, compared to SPFH silage harvesting systems, there is significant potential for SLFW silage harvesting systems to maximise output per person and improve fuel efficiency.

Harvesting silage with two types of silage trailer (feed rotor with knives and precision chop)

H. Arvidsson[1] and P. Lingvall[2]

[1]The Swedish University of Agricultural Sciences (SLU), Department of Agricultural Research for Northern Sweden, Box 4097, SE-905 96 Umeå, Sweden, Email: hans.arvidsson@njv.slu.se, [2]The Swedish University of Agricultural Sciences (SLU), Animal Nutrition and Management, Kungsängens forskningscentrum, SE-753 23Uppsala, Sweden, Email: per.lingvall@huv.slu.se

Keywords: harvesting silage, capacity, power need

Introduction Harvesting silage with a silage trailer that combines both a precision chopper and a trailer in the same machine is common in Sweden. A silage trailer with a feed rotor and knives has recently been put on the market. The objective of this study was to compare the two systems

Material and methods The crop was second harvest in second year ley 2 (timothy, red clover and meadow fescue). Two types of silage trailer were used, one with a feed rotor with knives (Pöttinger Jumbo 7200) and the other with a precision chopper (JF ES 3600). The same tractor was used for both trailers (CASE IH MXM190). The power required was measured at the PTO and the harvesting speed was measured with an extra ground wheel. The material from the two trailers was ensiled in separate bunker silos.

Results The dry matter (DM) of the herbage harvested was 35%. The herbage harvested by the trailer with rotor was longer (30% <40 mm) compared to that harvested by the trailer with precision chopper (84% <40 mm). Compared to the precision chopper trailer, the rotor trailer had a greater volume, a higher loading rate and required less energy (Table 1). There was no difference in density of silage or amount of bad silage. However, in the corners of the bunker with silage from the rotor trailer there was some more bad silage. The silage process had gone longer in the material from the chopper trailer.

Table 1 Power and capacity comparison between two types of silage trailers

		Precision chopper ES 3600 ProTec	Rotor with knives Jumbo 7200
Maximum speed in windrow	m/s	2.5	5.0
	km/h	9	18
PTO power at max speed	kW	112	104
PTO power at 2.5 m/s	kW	112	61
PTO power required	kW	43.3*m/s +3.6	17.2*m/s +18
at speed range	m/s	1-2	2-5
Loading time in windrow/load	s	399	407
Forage/load	kg	8061	13559
Average loading rate	kg/s	20	33
Density in trailer	kg/m³	224	307
Average unloading time	s	128	139
Fuel consumption diesel (harvest + transport)	l/ton	1.20	0.82

Conclusions The higher loading rate and lower power requirement of the rotor trailer compared to the chopper trailer might be explained by: longer cut of the material, different system of cutting, different system of transporting the material within the wagon and compaction in the wagon.

There was no difference in volume weight in the bunker silos between the systems. It was expected that density would have been less in the rotor silo since the material was longer. The quality of silage was the same in both bunkers. However, in both silages there were high levels of butyric acid that may have resulted from the low density and high DM content. If additives had been used (normally recommended) the result would most probably have been improved.

With a long material it is normally harder to get good silage. In the current work, there was more bad silage in the corners at the ends of the bunker with the silage made from the longer chop rotor material.

The effects of a new plastic film on the microbial and fermentation quality of Italian ryegrass bale silages

G. Borreani and E. Tabacco
Dipartimento di Agronomia, Selvicoltura e Gestione del Territorio – University of Turin, Via L. da Vinci, 44 10095 Grugliasco (TO) Italy, Email: giorgio.borreani@unito.it

Keywords: bale silage, plastic film, yeast, moulds, fermentation quality

Introduction Problems associated with big bale silage include the high permeability of plastic wrapping films to O_2, their low resistance to damage and the large amount of plastic that must be used to limit aerobic deterioration during conservation. Low permeability film, used in the packaging of food and recently proposed for bunker silos (Degano, 1999), could reduce fungal development in bale silage. The aim of this work was to compare the microbial and fermentation quality of big-baled silage, wrapped with commercially available plastic film and a new stretch film with low O_2 permeability, over different conservation periods.

Materials and methods Silage was made from a permanent pasture (80% Italian ryegrass) in the Po Valley, NW Italy. Herbage was field wilted for 2 days to a DM content of approximately 300 g/kg. Forage was baled (150 cm diameter) from alternate windrows and individually wrapped (6 layers) with one of two plastic films: a conventional polyethylene film (standard - ST) and a triple co-extruded stretch film with two outer layers of polyethylene and a central layer of polyamide (low permeability - LP), specifically produced for this experiment (IPM, S.p.A, Mondovì, Italy). The thickness of the two plastic films was 0.25 μm and O_2 permeability was 7120 and 400 cm^3/m^2 per 24 hours at 23°C for the ST and LP respectively. Six bales were allocated to each treatment. Bales were sampled after 2 and 4 months of conservation using a steel corer. Two cores were taken from each bale. One core was split into two portions: 1-20 cm (peripheral) and 21-55 cm (centre) which were analysed for yeast, mould and clostridia spores. The second core was analysed for fermentation quality.

Results All the silages were well fermented with no butyric acid and low ammonia concentration, with the exception of those wrapped with ST film after 4 months of conservation (Table 1). The yeast, mould and clostridia spores increased in the ST silage over the conservation period, especially in the peripheral zone of the bale, which is more prone to oxygen penetration (Table 2). No differences in microbiological quality were found after 4 months of conservation in the LP silage, showing the lower oxygen permeability of the wrapping.

Table 1 Fermentation characteristics of baled silages wrapped with low permeability and standard plastic films after two and four months of conservation

Plastic film	2 months of conservation					4 months of conservation				
	pH	lactic	acetic	butyric	NH_3-N	pH	lactic	acetic	butyric	NH_3-N
LP	4.1	47	16	n.d.	85	4.3	48	16	0.3	77
ST	3.8	52	15	n.d.	50	4.1	67	21	0.3	113

Lactic, acetic and butyric acids (g/kg dry matter); NH_3-N (g/kg total nitrogen); n.d. not detected.

Table 2 Mould and yeast colony forming units and clostridia spores in baled silages wrapped with low permeability and standard plastic films after two and four months of conservation

Plastic film	Bale zone	Yeast (log cfu/g)		Mould (log cfu/g)		Clostridia spores (log MPN/g)	
		2 months	4 months	2 months	4 months	2 months	4 months
LP	Peripheral	3.60bc	3.14bc	1.58a	1.06a	1.96a	1.52a
	Centre	<1.00a	1.74ab	<1.00a	1.11a	1.57a	1.55a
ST	Peripheral	2.60bc	4.26c	<1.00a	2.72b	1.71a	3.12b
	Centre	3.13bc	3.12bc	<1.00a	1.24a	1.71a	2.82b

Values with a different letter are significantly different within each parameter for $P<0.05$; values lower than 1 were set to 0.7 for statistical analysis.

Conclusions The results showed that the experimental stretch film with low oxygen permeability was able to reduce moulds and clostridia development during conservation and to maintain high quality silages for longer period than conventional polyethylene film.

References
Degano, L. (1999). Improvement of silage quality by innovative covering system. *13th International Silage Conference, 296-297.*

This work was funded by the Regione Lombardia, Direzione Generale Agricoltura, Project MARINSIL.

Section 3

Developments in ensiling techniques

A. Silage fermentation

Influence of different alfalfa-grass mixtures and the use of additives on nutritive value and fermentation of silage

P. Lättemäe and U. Tamm
The Estonian Research Institute of Agriculture, 75501 Saku, Estonia, Email: paulioma@hot.ee

Keywords: alfalfa-grass mixture, legumes, fermentation, silage quality, additive

Introduction Legumes have a high nutritive value but they are known to be difficult to ensile and often result in poorly fermented silage. This is usually due to high buffering capacity and low available sugar concentration. However, the results have shown that silage quality can considerably be improved by using additives or when legume-grass mixtures are ensiled (Lättemäe & Tamm, 2002). Different legume-grass mixtures differ in their ensiling properties and also may affect the fermentation. The objective of this experiment was to study the effect of alfalfa-grass mixtures and the use of additives on nutritive value and fermentation of silage.

Materials and methods In this ensiling study alfalfa cvs. "Jõgeva 118" and "WL 252 HQ" were used in a mixture with timothy cv. "Goliath". In order to obtain different proportions of alfalfa and timothy in the mixture, the seeding rates of both alfalfa were 18 kg/ha and timothy 2 and 6 kg/ha respectively (mixture 1 and 2 with alfalfa "Jõgeva 118"; mixture 3 and 4 with alfalfa "WL 252 HQ"). Nitrogen fertiliser was not used and fertilisers (P16K66 kg/ha) were added in autumn. The grass was mown, chopped to 4-8 cm and ensiled in 3 l glass jars. The following chemical additive treatments were used: untreated control, Niben treated 5 l/t fresh matter (FM) and AIV-2000 treated 5 l/t FM. Niben is based on sodium benzoate and AIV-2000 on formic acid. The spoilage micro-flora prior to ensiling was also used. The silos were sealed with plastic film and kept in room temperature 18-25°C for 130 days.

Results The results of the chemical analyses are presented in Tables 1 and 2. The lowest fermentation quality had untreated silage and when mixtures 1 and 2 were ensiled. Both additives reduced clostridial fermentation, proteolysis and dry matter losses. There were also interaction effects when using additives and different alfalfa mixtures (data not shown). The nutritive value of silages slightly varied. Alfalfa "Jõgeva 118" resulted in higher CP and CF concentrations in the silage.

Table 1 The effect of using additives on the chemical composition and dry matter losses in ensiled alfalfa-grass mixture. The values are averaged across the mixtures

Indicators	Untreated	Niben 5 l/t FM	AIV-2000 5 l/t FM	LSD$_{0,05}$
Dry matter (g/kg)	161	181	183	17.6
Crude protein (g/kg DM)	153	165	173	19.7
Crude fibre (g/kg DM)	310	285	287	24.8
pH	5.7	5.1	5.1	0.3
Ammonia N (% total N)	24.1	10.0	11.5	6.4
Butyric acid (g/kg DM)	11.1	4.2	7.1	3.8
Dry matter losses (%)	15.2	5.6	7.5	2.6

LSD$_{0,05}$- Least significant difference at the 5% probability level, n=8

Table 2 The effect of ensiling different alfalfa-grass mixtures on chemical composition and dry matter losses

Indicators	Mixture 1	Mixture 2	Mixture 3	Mixture 4	LSD$_{0,05}$
Dry matter (g/kg)	170	173	174	182	15.4
Crude protein (g/kg DM)	175	165	170	150	17.2
Crude fibre (g/kg DM)	310	324	276	292	27.2
pH	5.5	5.5	5.0	5.1	0.4
Ammonia N (% total N)	22.7	16.6	8.0	12.9	7.3
Butyric acid (g/kg DM)	9.8	8.0	7.0	5.0	4.0
Dry matter losses (%)	11.3	9.1	7.3	10.1	3.2

LSD$_{0,05}$- Least significant difference at the 5% probability level, n=6

Conclusions The fermentation quality of silage was dependent on the use of additive, species of alfalfa and its mixture with timothy. Both additives improved fermentation whereas Niben was more effective. The nutritive value of silage made of different alfalfa-grass mixtures varied slightly.

References

Lättemäe, P. & U. Tamm (2002). The improvement of lucerne silage quality by using additives and lucerne-grass mixtures. *Journal of Agricultural Science*, 6, 337-341.

The effect of neutralising formic acid on fermentation of fresh and wilted grass silage

E. Saarisalo and S. Jaakkola

MTT Agrifood Research Finland, Animal Nutrition, FI-31600 Jokioinen, Finland, Email: eeva.saarisalo@mtt.fi

Keywords: grass silage, additive, formic acid, ammonium formate

Introduction Rapid drop in pH is essential for minimising proteolysis and successful ensiling. Use of acid additives typically reduces protein degradation and restricts fermentation. The effects of acid additive depend on application rate and type of herbage. Corrosiveness and risks in handling formic acid (FA) can be reduced by using salts of FA like ammonium formate (AF). Increasing proportions of AF replacing FA were applied into grass at two dry matter (DM) contents to evaluate the effects of neutralised FA on silage pH and fermentation.

Materials and methods First cut timothy-meadow fescue grass was wilted for 1.5 (Fresh) or 21 h (Wilted) prior to chopping and ensiling in mini silos (120 ml). Herbage DM was 210 and 406 g/kg, crude protein 172 and 180 g/kg DM, and water soluble carbohydrates (WSC) 151 and 137 g/kg DM for Fresh and Wilted, respectively. Six additive treatments consisted of untreated (UT), and AF:FA (w:w %) 0:85 (AF0); 10:75 (AF10); 20:65 (AF20); 30:55 (AF30) and 40:45 (AF40). In addition, all contained 15% water. The application rate was 6 g/kg grass. Two silos per treatment were opened after 1, 3, 7, 21 and 97 days of ensiling. The data were tested for each DM separately with a GLM model using SAS. Sum of squares for treatment effect was further separated using orthogonal contrast into single degree of freedom comparisons.

Results The increasing proportion of AF affected the drop in pH at the beginning of ensiling (Figure 1). Still, pH was lower with all AFs compared to UT. In Fresh UT pH dropped below that of AFs between days 3 and 7 while in Wilted UT pH remained higher than in AFs until day 97 (Table 1). The fermentation quality of all silages, including UT, in either DM was good. Based on ammonia-N (NH_3-N, g/kg N), after subtracting the amount applied in additive, all AFs restricted proteolysis compared to UT.

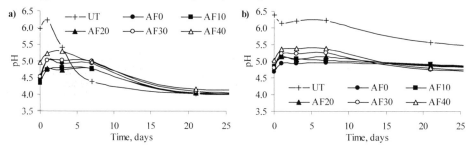

Figure 1 Effect of additives on silage pH at the beginning of ensiling in (a) Fresh and (b) Wilted grass silage

Table 1 Effect of additives on silage quality after 97 days of ensiling. For NH_3-N: a analysed, b analysed-added

				Treatment						Statistical significance			
FRESH	UT	AF0	AF10	AF20	AF30	AF40	SEM	UT vs AF	AF lin	AF quad	AF cub	AF quar
pH	4.03	4.07	4.00	4.04	4.08	4.16	0.008	***	***	***	*	
WSC (g/kg DM)	17	25	27	26	27	21	1.0	***	*	**		
Lact.acid (g/kg DM)	121	71	66	71	71	91	1.1	***	***	***	*	**
VFA (g/kg DM)	35	25	24	25	24	34	4.4					
NH_3-N (g/kg N) (a)	61	29	35	48	64	95	1.1	***	***	***	*	
NH_3-N (g/kg N) (b)	61	29	15	7	4	11	1.5	***	***	***		
WILTED												
pH	4.46	4.20	4.27	4.25	4.23	4.27	0.015	***			*	
WSC (g/kg DM)	15	32	39	35	22	19	3.9	**	*			
Lact.acid (g/kg DM)	69	54	50	56	57	62	2.9	**	*			
VFA (g/kg DM)	17	16	16	16	17	18	0.5					
NH_3-N (g/kg N) (a)	29	17	24	33	43	49	1.4		***			
NH_3-N (g/kg N) (b)	29	17	15	13	13	9	1.6	***	**			

Conclusions In this experiment silage quality was not compromised by replacing a part of FA with AF. Only the highest level of AF decreased the restrictive effect of FA on fermentation especially in Fresh silage. Still AF40 restricted fermentation compared with UT when recommended application rate (5 l/t) was used.

Effects of inoculation of LAB on fermentation pattern and clostridia spores in easily ensilable grass silages

J. Thaysen[1], G. Pahlow[2] and E. Mathies[3]
[1]Agricultural Chamber of Schleswig-Holstein, D-24783 Osterrönfeld, Am Kamp 9, Germany, Email: jthaysen@lksh.de, [2]Institute of Grassland and Forage Research, Bundesallee 50, D-3811 Braunschweig, Germany, [3]IS-Forschung GmbH, An der Mühlenau 4, D-25421 Pinneberg

Keywords: grass silage, lactic acid bacteria (LAB), clostridia spores

Introduction Clostridia can damage the protein quality of grass silages. They cause high gas losses during the fermentation process and quality problems in dairy products like semi-hard cheeses. In comparison to the effect of chemicals such as nitrite on undesirable clostridia in grass silages the respective inhibitory mechanism of LAB requires further investigation. The objective of this experiment was to study under laboratory conditions novel isolates of lactic acid bacteria (LAB), selected for their inhibitory effect on clostridia in grass silages.

Materials and methods Eight grass swards from the first cut 2004 were used for fermentation trials comparing a control (C) without additive to silage treated with a commercial LAB-inoculant (LAB). The grass crops were slightly wilted and treated with a mixture of *Clostridium sporogenes* (DSM 795, DSM 633), *Clostridium tyrobutyricum* (DSM 2637) and *Clostridium butyricum* (DSM 10702) at a rate of 1.0×10^2 spores/g FM at the time of ensiling. The trials were conducted according the guidelines of the German silage additive approval scheme (DLG 2000). The inoculant, consisting of *Lactobacillus paracasei* (DSM 16245) *Lactococcus lactis* (NCIMB 30160) *Pediococcus acidilactici* (DSM 16243), was applied as a suspension at a rate of 5.0×10^5 cfu/g of grass. The silages were stored for 14, 49 and 90 days at 25°C. Silage parameters, yeasts, moulds and clostridia spores were analysed according official methods (LUFA). Aerobic instability and corresponding DM losses were assessed by measuring the temperature rise above ambient according to Honig (1990). Fermentation losses were monitored by frequent weighing of the 1.5 l jars.

Results The mean sward composition on a dry matter basis at ensiling was as follows: 31% of FM, 16.4% of crude protein, 13.5% WSC and BC of 6.4 g lactic acid per 100 g DM. The mean fermentability coefficient (FC) was 48. The material was free of nitrate and contained 4.5 log MPN/g FM of clostridia spores. In Table 1 selected fermentation parameters, aerobic stability as well as numbers of LAB and clostridia spores are shown. Extending the storage period up to 90 days increased the amount of butyric acid (BA) in the untreated control. The pH decline during the ensiling process reduced the clostridia spores on average by one logarithmic unit in the untreated silages. Addition of LAB improved the fermentation profile by lowering pH, BA, NH_3-N and gas losses. However, aerobic instability and DM losses were increased in the treated silages, as well as the numbers of yeasts and moulds. Clostridia spores were reduced at all stages of storage period.

Table 1 Mean silage parameters and clostridia spores of grass silages at 14, 49 and 90 days of storage

Item		14 days		49 days		90 days	
Treatment	unit	C	LAB	C	LAB	C	LAB
DM	%	32.9	33.0	32.0	32.2	31.9	32.1
pH		4.4	3.9	4.1	3.8	4.0	3.8
BA	% FM	0.1	0.0	0.14	0.0	0.25	0.01
NH_3-N	% total N	8.0	6.0	10.0	8.0	10.0	7.0
DM-losses	% DM	7.5	6.5	8.8	6.8	8.0	6.7
Aerobic stability	days	-	-	3.1	1.9	-	-
Stability losses	% DM	-	-	9.7	15.2	-	-
Yeasts	log cfu/g FM	-	-	4.4	5.3	2.4	3.2
Moulds	log cfu/g FM	-	-	2.4	3.1	2.4	2.5
Clostridia spores	log cfu/g FM	3.2	2.4	3.9	3.0	3.3	2.9

Conclusions The results of the study showed that inoculation with LAB significantly improved the fermentation quality of easy ensilable grass silages and reduced the number of clostridia spores. However, the aerobic stability was reduced as a consequence of lower concentrations of butyric acid in the inoculated silages. Further analysis of DOM will be conducted.

References

DLG. (2000). DLG-Guidelines for testing of silage additives for compliance with DLG quality symbol requirements. Unpublished paper.
Honig, H. (1990). Evaluation of the aerobic stability. In: *Proceedings of the Eurobac Conference, Swedish University of Agricultural Sciences*, Uppsala/Sweden, Special Issue.

Effect of biological additives in red clover – timothy conservation

A. Olt[1], H. Kaldmäe[1], E. Songisepp[2] and O. Kärt[1]
[1]Estonian Agricultural University, Animal Nutrition, Kreutzwaldi 1, 51014 Tartu, Estonia, Email: slabor@eau.ee, [2]University of Tartu, Institute of Microbiology, Ravila 19, 50411 Tartu, Estonia

Keywords: silage, fermentation, additives, red clover, timothy

Introduction Red clover at early flower bud formation is difficult to ensile. For efficient improvement of the quality of leguminous silages, chemical additives are used. The present research focuses on the effectiveness of biological additives with different composition on the fermentation and quality of clover silage.

Materials and methods The chemical composition, nutritive value and quality of silage, prepared from red clover-timothy mixture (50% red clover variety 'Jõgeva 433') at early bud formation stage, were investigated. For this purpose test silages were prepared. Raw material for silage was cut at the height of 5 cm, wilted for 24 h, chopped into 2 cm pieces, supplemented with chemical (AIV) or 4 variants of biological additives containing combinations of 4 *Lactobacillus* sp. belonging into facultative and obligatory heterofermentative groups and conserved into 3 l jars. In 90 days the jars were opened. Silage was analysed for dry matter (DM), crude protein (CP), crude fibre (CF), neutral detergent fibre (NDF), acid detergent fibre (ADF), pH and ammonia-N in total N, according to generally accepted methods; ethanol, volatile fatty acids, lactic acids (LA) were analysed with gas chromatograph. Biological additives were the following: I-1, I-2, I-3 and I-4.

Results DM losses during fermentation were the lowest in the silage treated with AIV additive (4.8%), in silages treated with biological additives these values were I-1 5.2%, I-2 10%, I-3 11.6%, I-4 9.7%, respectively and in untreated silage (17.4%). The chemical compositions of silages are presented in Table 1. The quality of silages, treated with biological and chemical additives, was high, compared to that of the untreated silage. Figure 1 illustrates the content of lactic acid (LA), acetic acid (AA), butyric acid (BA) and ethanol (E). Lactic acid concentration was higher in the silages inoculated with biological additives (within the range of 96.4 and 127.5 g/kg in DM) compared to that in the AIV treated silages (44.3 g/kg in DM) or the untreated silages (73.7 g/kg in DM). The content of AA in studied silages remained between 8.15 to 9.7 g/kg in DM. The content of BA in DM was 23.9 g/kg in the untreated silage and below 0.4 g/kg in the remainder of the silages. Biological additives favoured lactic acid fermentation by increasing LA content. However, as to the nutrition of ruminants, such a high LA content does not play a significant role in animal metabolism, as normally the ruminal LA content is low (McDonald *et al.*, 1991).

Table 1 Chemical composition of silages

Item	Untreated	I-1	I-2	I-3	I-4	AIV
DM (g/kg)	256	294	279	274	279	295
Crude protein (g/kg DM)	178	167	170	172	174	172
Crude fibre (g/kg DM)	209	204	195	192	189	185
Nitrogen-free extract (g/kg DM)	467	496	497	497	499	506
ME (MJ/kg DM)	10.0	10.1	10.2	10.1	10.2	10.1
NH$_3$–N (% total N)	6.3	1.5	1.6	1.6	1.7	3.9
pH	5.0	4.2	4.2	4.2	4.2	4.8

Conclusions The use of biological additives in ensiling pre-wilted material, rich in red clover, improved fermentation and silage quality, also decreased DM losses.

References

McDonald, P., A.R. Henderson & S.J.E. Heron (1991). The biochemistry of silage. Chalcombe publication, UK, 340 pp.

This work was funded by ESF G4985.

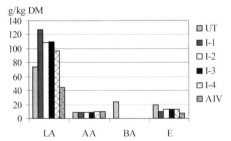

Figure 1 The content of lactic-, acetic- and butyric acids, and ethanol in the test silages

Application of a new inoculant "Chikuso-1" for silage preparation of forage paddy rice

Y. Cai, C. Xu, S. Ennahar, N. Hino, N. Yoshida and M. Ogawa
National Institute of Livestock and Grassland Science, Nishinasuno, Tochigi 329-2793, Japan. Email: cai@affrc.go.jp

Keywords: forage paddy rice, lactic acid bacteria, inoculant, silage

Introduction Forage paddy rice is currently one of the most important silage crops in Japan. In fact, the use of paddy rice culture for silage production has been steadily increasing in recent years, not only because this represents a new way towards achieving self-sufficiency in animal feed, but also because of the interest of combining crop cultivation and livestock farming as a more effective use of idle paddy fields that often remain unused. However, the preparation of quality silage from paddy rice and its long-term storage are often challenging (Cai *et al.*, 1999, 2003). In this study, a new bacterial inoculant was developed and its application for silage preparation of forage paddy rice was examined.

Materials and methods Two paddy rice cultivars grown in a farm field (Saitama, Japan), were harvested at the ripe stage. A new lactic acid bacterial inoculant Chikuso-1 (*Lactobacillus plantarum*, Brand seed Ltd., Sapporo, Japan) was used at 1.0×10^5 cfu/g of fresh matter (FM) for silage fermentation. Silage was prepared by using a round bale system. Untreated controls and Chikuso-1-inoculated samples from three round bale silages per treatment were then monitored for silage quality through microbiological and chemical analyses. Data were subjected to one-way analysis of variance and treatment means were compared by Tukey multiple range test SAS Institute Inc. 1988.

Table 1 Fermentation quality and microbiological analysis of silage

	Hamasari cultivar		Kusahonami cultivar	
	Control	Chikuso-1	Control	Chikuso-1
Ferementation quality				
pH	5.35[b]	3.72[a]	5.28[b]	4.05[a]
DM (%)	31.37	30.32	31.11	32.24
Lactic acid(% FM)	0.35[a]	1.33[b]	0.24[a]	0.95[b]
Acetic acid(% FM)	0.44	0.32	0.46	0.40
Butyric acid(% FM)	0.38	nd	0.56	nd
Propionic acid(% FM)	0.06	nd	0.08	nd
Ammonia N(g/kg FM)	0.66	0.27	0.82	0.33
Microorganism composition (log colony-forming units per gram of FM)				
Lactic acid bacteria	4.56[b]	6.86[a]	5.20[b]	7.05[a]
Aerobic bacteria	4.55[b]	3.20[a]	5.24	nd
Clostridia	3.22	nd	3.67	nd
Bacilli	4.50	4.02	4.80	3.20
Yeast	3.80	4.20	4.20	4.80
Mould	3.32	nd	4.60	nd

FM, fresh matter; nd, not detected. Chikuso-1: *Lactobacillus plantarum* ; Silage ensiled for 300-days, [a, b] Values are means of three silage samples.

Results Results overall showed counts of 10^6 (cfu/g of FM) aerobic bacteria, 10^3 coccus-shaped lactic acid bacteria, 10^5 molds and 10^4 yeasts in the two sets of silage samples. *Lactobacilli* were too few to be detected in any of the samples analysed. The silages treated with Chikuso-1 were well preserved; had significantly lower pH values, butyric acid, propionic acid and ammonia N concentrations, gas production, and dry matter losses; and higher contents of lactic acid compared with the control silages after 300 d of fermentation. During silage fermentation, the control silages displayed growth of clostridia and molds, whereas in Chikuso-1-inoculated silages, these were at or below the detectable levels (Table 1).

Conclusions These results suggest that the inoculation of paddy rice silage with the inoculant Chikuso-1 results in beneficial effects by promoting the propagation of lactic acid bacteria and inhibiting the growth of clostridia, aerobic bacteria and molds, as well as improving the overall fermentation quality.

References

Cai, Y., Y. Fujita, M. Murai, M. Ogawa & N. Yoshida (2003). Application of lactic acid bacteria (*Lactobacillus plantarum* Chikuso-1) for silage preparation. *Grassland Science,* 49, 477-485.
Cai, Y., S. Kumai, M. Ogawa, Y. Benno & T. Nakase (1999). Characterisation and identification of *Pediococcus* species isolated from forage crops and their application for silage preparation. *Applied Environmental Microbiology,* 65, 2901-2906.

Synergism of chemical and microbial additives on sugarcane (*Saccharum officinarum* L.) silage fermentation

T.F. Bernardes, G.R. Siqueira, R.P. Schocken-Iturrino, A.P.T.P. Roth and R.A. Reis
Universidade Estadual Paulista, FCAV, Jaboticabal, SP, Brazil,14.884-900. Email: tfbernardes@yahoo.com

Keywords: *L. buchneri*, losses, nutritive value

Introduction Sugarcane has a high productive potential (30 t DM/year) and it is commonly used in its fresh form. The ensiling of sugarcane is increasing but little research has been carried out to reduce nutrient losses during fermentation.

Material and methods The tested cultivar was SP70-1143. The production observed at 15 months of vegetative growth was 80 t/ha with 16% of pol (sucrose in sugarcane juice). Following factorial scheme with three inoculations (control, *Propionibacterium acidipropionici* (cepa MS 01) (PROP) + *Lactobacillus plantarum* (Cepa MA 18/50) and *Lactobacillus buchneri* (Cepa NCIMB 40788) (BUCH)) and four chemical additives (control, urea (1.5% DM), sodium benzoate (0.1% DM) and sodium hydroxide (1% DM)), with three replications was evaluated. *In vitro* dry matter digestibility (IVDMD) was estimated as amount of residual digestible DM in relation to digestible DM ensilaged, and dry matter recovery (DMR) was determined. This work aimed to evaluate quantitative and nutritional losses during the fermentative process associated with sugarcane ensilage.

Results Silage treated with BUCH in relation to those inoculated and silages treated with NaOH compared to those treated with chemical additives had higher IVDMD. There were synergic effects between BUCH and NaOH (Table 1). Higher IVDMD means that the silage nutritive value was maintained in relation to original forage. The yeast activity is intense during sugarcane ensilage (Alli *et al*., 1983 and Pedroso *et al*., 2002) promoting high soluble sugar consumption and, consequently, reduction in DM and IVDMD. Inoculation with BUCH controlled yeast population, probably because of their capacity for acetic acid production, which reduced the quantitative and qualitative losses in sugarcane ensilage.

Table 1 Recovering of DM (DMR) and IVDMD in relation to chemical and microbial additives, expressed in percentage of sugarcane dry matter ensilaged

	DMR (%)				IVDMD (%)			
	Control	PROP	BUCH	Mean	Control	PROP	BUCH	Mean
Control	67.5 Bb	66.4 Bc	80.8 Ac	71.6 d	45.0 Bc	39.8 Bc	74.3 Ab	53.0 d
Urea	72.8 Ba	75.0 Bb	79.7 Ac	75.8 c	51.6 Bb	64.0 Ab	63.7 Ac	59.8 c
Benzoate (0.1%)	74.8 Ba	74.8 Bb	87.2 Ab	78.9 b	53.2 Bb	58.5 Bb	80.6 Aab	64.1 b
NaOH (1%)	76.1 Ca	86.5 Ba	93.7 Aa	85.4 a	68.5 Ca	76.0 Ba	84.8 Aa	76.4 a
Mean	72.8 C	75.7 B	85.3 A	77.9	54.6 C	59.6 B	75.9 A	63.3
*CV (%)				2.17				4.59

Mean followed by the same capital letter in line and small letter in column are statistically similar by Tukey test (*P*<0.05).
* Coefficient of variation

Conclusions The *L. buchneri,* even in an isolated action, was efficient for controlling quantitative and qualitative losses during sugarcane ensilage. Association of microbial inoculums aiming to control losses provoked by yeast and NaOH seems to be an alternative for enhancing the effects of inoculums.

References

Alli, I., R. Fairbairn, B.E. Baker (1983). The effects of ammonia on the fermentation of chopped sugarcane. *Animal Feed Science and Technology*, 9, 291-299.

Pedroso, A.F., L.G Nussio, S.F. Paziani, D.R.S. Loures, M.S. Igarasi, L.J. Mari, R.M. Coelho, J.L. Ribeiro, M. Zopollatto, & J.Horii (2002). Bacterial inoculants and chemical additives to improve fermentation in sugar cane (*Saccharum officinarum*) silage. *Proceedings of the XIII International Silage Conference, Auchincruive, Scotland*. p-66.

The influence of the application of a biological additive on the fermentation process of red clover silage

Ľ. Rajčáková[1], R. Mlynár[1] and M. Gallo[2]

[1]The Research Institute of Animal Production, Nitra, Institut of Animal Nutrition, Research Station Poprad, SNP 2/1278, 058 01 Poprad, Slovak Republic, Email: rajcak@rspp.vuzv.sk, [2]Biofaktory, s.r.o., Černyševského 26, 851 01 Bratislava, Slovak Republic

Keywords: red clover, silage, fermentation

Introduction In Slovakia, mainly in the submontane and mountainous regions, growing of red clover is an important source of proteinous feeds. It is grown on 3.0% of arable land. It was the aim of this work to verify the possibilities of using a biological additive in red clover silage conservation.

Materials and methods The experiments were carried out with crops of tetraploid red clover from the second cut, wilted for 48 hours (Table 1). The cut crops were homogenised and filled into 1.7 l silos. The filled silos were placed in a dark room at 22^0C. For treatment the following biological additive was used, consisting of *Lactobacillus plantarum* (DSM 3676, 3677) and *Propionic bacterium* (DSM 9576, 9577). The application rate was 4 l additive/t feed. After 180 days of incubation the samples were examined for nutrient content and basal characteristic of the fermentation process.

Results Application of silo additive proved to have a positive effect on the fermentation process. Untreated samples have lower pH, higher content of acetic and butyric acid and lower content of NH_3-N of total N. Differences in nutrient levels between silages were minimal. Treated silage had a higher content of crude protein and fat and a lower content of crude fibre and ash. The digestibility of dry matter and organic matter was also higher in treated silage. There were also minimal differences between silages. Silage treated with biological additive appeared slightly better. Observations similar to this study were reported by Hetta (1999) during grass and clover-grass crops conservation with lactic acid bacteria. The author reported decreased pH, acetic acid and alcohol levels as well as decreased NH_3-N of total N in treated clover-grass silage. From the point of view of nutrient levels the author observed a decrease in crude fibre content and in fractions in clover-grass silage which corresponded with our findings. Earlier research (Gallo *et al.*, 2001 and Gallo *et al.*, 2002) also showed that the application of a biological silage additive produced a positive effect in the conservation of red clover.

Table 1 Nutrient composition and parameters of the fermentation red clover silage in g/kg DM

Parameter n = 6	Fresh matter	Untreated		Treated		Statistical significance of differences	
	$\bar{x}$	$\bar{x}$	s	$\bar{x}$	s	$P<0.05$	$P<0.01$
Dry matter	279	263	2.47	265	2.24		
Crude protein	195	192	0.94	188	2.40		**
Crude fibre	287	289	4.86	282	4.59	*	
Fat	24	29	0.91	30	2.41		
Ash	81	87	0.97	85	0.41		**
pH	-	4.45	0.02	4.14	0.05		**
Lactic acid	-	58	6.32	87	12.14		**
Acetic acid	-	13	1.10	8	1.69		**
Butyric + isobutyric acid	-	0.4	0.35	0.3	0.02		
NH_3 - N of total N	-	92	2.2	67	9.4		**
Digestible DM	-	603	0.60	605	0.76		
Digestible OM (g/kg)	-	568	0.66	571	0.75		

Conclusions Application of biological additive showed its positive effect on the quality of fermentation process and nutrient levels in clover silage with a low content of dry matter.

References

Gallo, M., V. Jambor, R. Mlynár & Ľ. Rajčákova (2002). Effect of the application of different silage preparations upon the fermentation process in red clover. In: *Proceedings of the XIIIth International Silage Conference, Great Britain, Auchincruive: SAC*, pp. 110–111.

Gallo, M., R. Mlymár & Ľ. Rajčákova (2001). The effect of the combination of biological and biological-enzymatic additive with sodium benzoate upon the fermentation process in red clover silages. In: *The Xth International Symposium Forage Conservation, Brno* : MZLU Brno, pp. 100–101.

Hetta, M. (1999). Ensiling during difficult conditions of two direct cut forages, with different botanical composition. In: *Proceedings of the XIIth International Silage Conference, Uppsala*, pp. 94–96.

Inoculant effects on ensiling and *in vitro* gas production in lucerne silage

R.E. Muck[1], I. Filya[2] and F.E. Contreras-Govea[1]
[1]*U.S. Department of Agriculture, Agricultural Research Service, U.S. Dairy Forage Research Center, Madison, Wisconsin 53706-1108 U.S.A. Email: remuck@wisc.edu* [2]*Department of Animal Science, Uludag University, Bursa, Turkey*

Keywords: inoculant, lucerne, *in vitro* fermentation

Introduction Inoculants are the most common additives used in making silage. While inoculant effects on fermentation and dry matter (DM) recovery are understood, animal performance effects are often greater than expected. *In vitro* analyses may help uncover how inoculants affect rumen fermentation and ultimately dairy cattle performance. Our objective was to study how inoculation of lucerne silage affected *in vitro* gas production.

Materials and methods Lucerne was ensiled in two trials (first (48% DM) and second cutting (39% DM)) in 1-l and 500-ml Weck® jars, respectively. Each trial had fifteen treatments (Table 1), four silos per treatment. Eight inoculants were commercial products, the others were single strains provided by two companies. All inoculants were applied at 10^6 colony-forming units (cfu)/g crop (not label rates). Silages were stored for a minimum of 30 d at 22°C. *In vitro* gas production was measured on 1-g samples of the wet-ground (Büchi mixer) silage in 160-ml serum bottles. *In vitro* analysis was carried out at 39°C, and gas pressure was measured at 3, 6, 9, 24, 48 and 96 h (Weimer *et al.*, 2005). At 96 h, the bottles were opened, and pH was measured. Treatment differences were determined using the MIX procedure of SAS®; silage characteristics (pH, lactate, acetate, ethanol, fibre fractions, crude protein) were correlated with gas production using the CORR procedure.

Results In first cutting, the epiphytic lactic acid bacteria population at ensiling was 1.5×10^5 cfu/g, and all inoculants except *Enterococcus faecium* C reduced pH relative to that of the control (Table 1). The commercial homofermentative inoculants produced the largest reductions in pH. In second cutting, the epiphytic population (2.7×10^7 cfu/g) at ensiling was high. The commercial homofermentative inoculants were the only treatments producing lower pH values than the control. We expected inoculants would improve *in vitro* DM digestibility and increase gas production. Surprisingly inoculants had either no effect on 96-h gas production per g DM or gas production was reduced (Table 1). In first cutting, gas production was correlated positively with silage pH (0.62, $P<0.02$), negatively with lactic acid concentration (-0.56, $P<0.04$). In second cutting, gas production was correlated positively with acid detergent lignin (0.62, $P<0.02$) and hemicellulose concentrations (0.57, $P<0.03$).

Table 1 Silage pH and *in vitro* gas production (96 h) for the first and second cuttings (Buch - commercial *L. buchneri* inoculants, Comm - commercial homofermentative inoculants)

	First Cutting		Second Cutting	
Treatment	pH	Gas (ml/g DM)	pH	Gas (ml/g DM)
Control	5.081	190	4.422	205
Buch A	4.825	186	4.642	203
Buch B	4.899	188	4.651	198
Comm A	4.497	185	4.336	208
Comm B	4.429	186	4.399	201
Comm C	4.335	182	4.287	205
Comm D	4.511	178	4.418	187
Comm E	4.377	184	4.318	192
Comm F	4.507	184	4.397	198
E. faecium C	5.144	190	4.470	195
E. faecium Q	4.578	185	4.445	192
L. pentosus	4.657	178	4.464	197
L. plantarum	4.462	176	4.425	189
P. pentosaceus A	4.569	184	4.463	198
P. pentosaceus E	4.577	189	4.459	197
S.E.M.	0.017	3.24	0.018	3.63

Conclusions These results indicate that inoculation of lucerne at ensiling affects *in vitro* fermentation, but differences in gas production could not be consistently explained by fermentation products or fibre fractions.

Reference
Weimer, P.J., B.S. Dien, T.L. Springer & K.P. Vogel (2005). *In vitro* gas production as a surrogate measure of the fermentability of cellulosic biomass to ethanol. *Applied Microbiology and Biotechnology* (In press).

Effects of stage of growth and inoculation on fermentation quality of field pea silage

G. Borreani[1], L. Cavallarin[2], S. Antoniazzi[2] and E. Tabacco[1]
[1]Dip. Agronomia, Selvicoltura e Gestione del Territorio, Università degli Studi di Torino, Via Leonardo da Vinci, 44, 10095 Grugliasco, Italy, Email: giorgio.borreani@unito.it, [2]Istituto di Scienze delle Produzioni Alimentari, CNR, Via L.da Vinci44, 10095 Grugliasco, Italy

Keywords: legume, field pea, lactic acid bacteria, stage of development

Introduction Field peas (*Pisum sativum* L.) are a short-term catch crop with a high crude protein content, which provides a high forage yield in a short growing period. Since field peas are a succulent crop and are difficult to field cure, it is preferable to directly ensile them to prevent weather damage and excessive grain losses. The onset of lodging is delayed in field pea varieties, since the crop is supported by the tendrils in a more erect manner, and this allows easy harvesting without soil contamination even at advanced stages of maturity (Koivisto *et al*., 2003). To our knowledge, no information is available on the ensiling of peas in Southern Europe. The aim of the study was to investigate the effect of the stage of maturity and inoculant application on the quality of silage produced from directly-cut field peas in the Po Valley, NW Italy.

Material and methods Stands of semi-leafless cv. Baccara field pea were sown on 21 March 2001. Herbage was harvested 4 times at progressive morphological stages (end of flowering, I; beginning of pod filling, II; advanced pod filling, III; beginning of ripening, IV) over the period 1-21 June. The herbage was chopped and directly ensiled in 2-litre laboratory glass silos with an inoculant (I) (*Lactobacillus plantarum)* and without an inoculant (C). Silages were analysed for fermentation quality after 60 days of conservation.

Results High levels of ethanol and volatile fatty acids, especially lactic and acetic acid, were observed in all the silages. Despite the low pH values, all the silages showed detectable levels of butyric acid. The silages prepared from forage harvested at the IV stage had a significantly lower lactic acid content than silages made from forage at the three previous stages. Ethanol content significantly increased with increasing forage maturity. The inoculation treatment affected the pH and lowered the ethanol and ammonia concentrations in all the silages, with the exception of the first stage. The occurrence of butyric acid in the silages is likely to have been the result of the ensilability characteristics of the herbage. Legumes have a two-fold disadvantage in being both highly buffered and having a low WSC content, and as a consequence clostridia tend to dominate the fermentation of these crops. The high buffering capacity of the herbage (data not shown) especially in the first stages, probably caused a slow drop in pH and the fermentation proceeded on different pathways. At the same time the high levels of WSC at ensiling (130, 189, 198, 111 g/kg DM for stage I, II, III, IV, respectively) led to high levels of lactic acid in the silages.

Table 1 Composition of control (C) and inoculated (I) pea silages at four stages of growth

Stage of growth	I		II		III		IV				
Inoculation	C	I	C	I	C	I	C	I	S[A]	I	S x I
DM (g/kg)	143	148	163	158	188	198	212	209			
pH	4.2	4.4	4.2	3.9	4.2	3.9	4.8	4.1	*[B]	*	NS
Lactic acid (g/kg DM)	160	120	140	185	138	162	63	99	***	NS	*
Acetic acid (g/kg DM)	28	49	20	22	17	19	14	21	***	**	*
Butyric acid (g/kg DM)	5.1	14.2	4.0	4.7	8.7	2.7	26.9	2.3	NS	NS	NS
Propionic acid (g/kg DM)	5.7	4.6	0.0	1.5	2.0	2.0	3.2	2.2	NS	NS	NS
Ethanol (g/kg DM)	5.0	2.0	8.5	6.2	17	15	26	13	***	***	**
WSC (g/kg DM)	13	8.0	42	15	26	25	4.0	5.0	***	***	***
Total N (g/kg DM)	39	39	35	36	32	32	33	32	***	NS	NS
NH3-N (g/kg TN)	137	134	116	52	131	68	132	105	***	***	*

[A] S = stage of growth, I= inoculation
[B] NS = $P>0.05$; * = $P<0.05$; ** = $P<0.01$; *** = $P<0.001$.

Conclusions The data show that field peas can be successfully directly ensiled at advanced stages of maturity with the aid of LAB inoculum.

References
Koivisto, J., L. Benjamin, G. Lane & W. Davies (2003). Forage potential of semi-leafless grain peas. *Grass and Forage Science*, 58, 220-223.

A novel bacterial silage additive effective against clostridial fermentation

E. Mayrhuber[1], M. Holzer[1], W. Kramer[1] and E. Mathies[2]
[1]Lactosan Starterkulturen GmbH. & Co. KG, Industriestr. West 5, 8605 Kapfenberg, Austria Email: mayrhuber@lactosan.at [2]Union-Agricole Holding AG, An der Mühlenau 4, D-25421 Pinneberg, Germany.

Keywords: butyric acid, clostridia spores, grass silage, lactic acid bacteria (LAB), silage additive

Introduction Silage quality is determined by factors including the content of butyric acid and ammonium-N. These parameters have to be restricted especially in lightly wilted silages due to a higher risk of clostridia contamination. In this study a novel silage additive was tested in grass silages of low dry matter content. The objective of this experiment was to explore the effect of the silage additive on quality parameters in comparison to an untreated control.

Materials and methods Ensiling: In three trials grass (each first cut of permanent grassland of different farms) was ensiled in 6.5 l laboratory silos and analysed after 90 d of storage at 20°C. In each trial a control without additive was compared with silages treated with the new biological silage additive (three homofermentative lactic acid bacteria: *Lactobacillus paracasei*, *Pediococcus acidilactici* and *Lactococcus lactis* treatment). The inoculant was applied at a dosage of 2.5×10^5 cfu/g grass.
Analyses: An aqueous silage extract (50 g silage extracted with 250 g distilled water) was prepared and the following chemical analyses were performed. Sugars and organic acids were analysed by HPLC (Agilent 1100, Column: Transgenomic ICSep ICE-ION 300). Ammonia-N was determined by distillation (Gerhardt). Clostridia spores were determined on silage samples by MPN–method on a microtiter plate scale on RCM-bouillon with D-cycloserine and neutral red (Jonsson, 1990; Kaufmann & Weaver, 1959).

Results The dry matter range of grass ensiled was between 19 and 26%. The lowest pH value reached in the untreated controls was 4.8, whereas the inoculated silages reached significantly lower pH values (3.8 and 4.1). Due to the high production of lactic acid (mean value 12.7 g/100 g DM) an improved conservation effect was achieved in the treated samples. This is demonstrated by low concentrations of butyric acid and ammonia-N and by lower counts of clostridia spores. The strong lactic acid fermentation minimised weight losses by inhibiting activity of spoilage organisms (Control 3.1 g/100 g DM; Treatment 1.2 g/100 g DM).

Table 1 Silage parameters of trials with grass silages after 90 days of storage

Laboratory silage d 90	Trial 1		Trial 2		Trial 3		Mean values		Standard error		Signifi-cance
Groups	C	T	C	T	C	T	C	T	C	T	
DM (g/100 g FM)	25.0	26.1	20.1	20.8	22.8	19.0	22.6	22.0	2.4	3.7	-
pH – value	4.8	3.8	5.2	3.8	5.2	4.1	5.0	3.9	0.2	0.1	**
Ammonia-N (g/100 g TKN)	11.6	7.4	26.4	7.7	14.9	8.9	17.6	8.0	7.8	0.8	-
Weigth losses (g/100 g FM)	2.9	1.6	3.1	0.6	3.2	1.3	3.1	1.2	0.2	0.5	**
Lactic acid (g/100 g DM)	3.3	12.2	0.4	15.8	0.7	10.1	1.5	12.7	1.6	2.9	**
Butyric acid (g/100 g DM)	2.0	0.3	7.0	0.3	4.3	0.5	4.4	0.3	2.5	0.1	*
Cl spores (log (cfu/g FM))	6.2	4.9	6.3	3.9	6.8	4.9	6.5	4.7	0.3	0.6	**

C = Control; T = Treated; TKN = Total Kjeldahl nitrogen; FM = fresh matter; DM = dry matter; * $P<0.05$, ** $P<0.01$

Conclusions In summary it was observed that the additive improved silage quality. The main effects were the increased production of lactic acid (highly significant) connected with a highly significant reduction of pH value. Additionally the activity of clostridia was reduced as indicated by lower butyric acid formation and lower counts of clostridia spores.

References

Jonsson A. (1990). Enumeration and Confirmation of Clostridium tyrobutyricum in silages Using Neutral Red, D-Cycloserine, and Lactate Dehydrogenase Activity. *Journal of Dairy Science*, 73, 719-725.
Kaufmann L. & R.H. Weaver (1959). Use of Neutral Red for the Identification of Colonies of Clostridia. *Journal of Bacteriology*, 79, 292-294.

In vitro gas production and bacterial biomass estimation for lucerne silage inoculated with one of three lactic acid bacterial inoculants

F.E. Contreras-Govea[1], R.E. Muck[1], I. Filya[2], D.R. Mertens[1] and P.J. Weimer[1]
[1]*U.S. Department of Agriculture, Agricultural Research Service, U.S. Dairy Forage Research Center, Madison, Wisconsin 53706-1108 U.S.A. Email: fecontre@wisc.edu* [2]*Department of Animal Science, Uludag University, Bursa, Turkey*

Keywords: inoculant, lucerne, *in vitro* fermentation

Introduction Silages inoculated with microbial inoculants frequently have a lower pH than non-inoculated crops. Less often inoculated crops have a positive effect on milk production (Weinberg & Muck, 1996). One hypothesis is that bacterial inoculants produce a probiotic effect that could enhance animal performance (Weinberg & Muck, 1996). Our objective was to use the method of Blümmel *et al.* (1997) to study differences in *in vitro* fermentation among lucerne silages inoculated with three microbial inoculants.

Material and methods Lucerne was ensiled in Weck® jars in two trials (48 and 39% DM) with four treatments (Table 1). The lucerne silages (1-g samples, wet-ground in a Büchi mixer, frozen until analysed) were incubated in sealed 160 ml serum bottles. *In vitro* gas kinetics was carried out at 39°C, and gas pressure was measured at 3, 6, 9, 24, 48 and 96 h (Weimer *et al.*, 2005). At 9 h, 4 bottles of each treatment were opened, pH measured, and microbial biomass yield estimated using the method of Blümmel *et al.* (1997). Statistical differences between treatments were determined using the MIX procedure of SAS®.

Results On average, the gas production increased linearly during the first 9 h of fermentation and was greater in control than inoculated silage. Although harvests were not statistically compared, greater GP, IVTD, and MBY were observed on second cut than first cut (Table 1). Even though treatment differences within harvests were not always significant, the trend among treatments was the same, lower GP and GE on Ecosyl MTD1 and *L. pentosus* (Agri-King, Inc.) than control and Pioneer 1174. In addition, lucerne inoculated with Ecosyl MTD1 had consistent trends toward higher IVTD and MBY than those of the control. Methane production was different among treatments, but trends were not consistent between cuts.

Table 1 Gas production (GP), *in vitro* true digestibility (IVTD), microbial biomass yield (MBY), gas efficiency (GE), methane produced (Methane), and methane efficiency (Meth E) of lucerne silage inoculated with one of four microbial inoculants. Values are means of two incubations at 9 h. Significance at $P<0.05$

Inoculant	GP (ml/g DM)	IVTD (mg/g DM)	MBY (mg/100 mg TD)	GE (ml/100 mg TD)	Methane (ml/g DM)	Meth E (ml/100 mg TD)
First cut						
Control	130a	715	25b	18a	6.8a	0.95a
Pioneer 1174	120b	713	26b	17a	6.2b	0.87b
Ecosyl MTD1	115c	723	29a	16b	5.7c	0.79c
L. pentosus	111c	704	25b	16b	5.5c	0.79c
s.e.m.	2.0	NS	1.53	1.11	0.15	0.023
Second cut						
Control	148b	788	32	19	6.4b	0.80b
Pioneer 1174	151a	786	33	19	7.7a	1.00a
Ecosyl MTD1	147b	804	34	18	7.9a	0.98a
L. pentosus	142c	806	33	18	7.8a	0.96a
s.e.m.	1.5	NS	NS	NS	0.39	0.045

Conclusions The results indicate that microbial inoculants, particularly Ecosyl MTD1, produced silages that shifted *in vitro* rumen fermentation toward less gas production and more microbial biomass than untreated silages.

References
Blümmel, M., H.P.S. Makkar, & K. Becker (1997). In vitro gas production: a technique revisited. Journal of Animal Physiology and Animal Nutrition, 77, 24-34.
Weinberg, Z.G., & R.E. Muck (1996). New trends and opportunities in the development and use of inoculants for silage. FEMS Microbiological Review. 19, 53-68.
Weimer, P.J., B.S. Dien, T.L. Springer, & K.P. Vogel (2005). In vitro gas production as a surrogate measure of the fermentability of cellulosic biomass to ethanol. Applied Microbiology and Biotechnology (In press).

Correlation between epiphytic microflora and microbial pollution and fermentation quality of silage made from grasses

B. Osmane and J. Blūzmanis
Latvia University of Agriculture, Research Institute of Biotechnology and Veterinary Medicine "Sigra", 1 Instituta Street, Sigulda LV – 2150, Latvia, E – mail: sigra@lis.lv

Keywords: correlation, microflora, fermentation quality, green material, silage

Introduction Grass forage in Latvia is the main and inexpensive cow feed, however its composition and nutritive value differ during the growth period of grasses. The traits of grasses, their natural ensilage capacity, count of epiphytic microflora, the timing of harvest and ensilage making technology affecting the quality of grass silage are important issues to be studied. Silage making for the winter period is the treatment of green material to minimise the breakdown of nutrients being the results of biochemical and microbiological processes. The aim of the research was to clarify the methodologies to reduce the count of epiphytic microflora and CFU count of microorganisms in grass silage and improve fermentation quality (Woolford, 1998; Wilkinson, 1999).

Materials and methods We investigated the fresh material and silage (2001-2003) from: Perennial ryegrass (*Lolium perenne L.),* Meadow fescue *(Festuca pratensis L.),* Timothy (*Phleum pratense)* and others at three stages of maturity (branching – stage 1, shooting – stage 2, blooming – stage 3). We analysed samples biochemically and microbiologically. Fermentation coefficient (FC) – calculated according to Weissbach:
FC = DM% + 8 WCS/BC. We used correlation analysis and three-factorial dispersion analysis in statistical data processing with SPSS computer program, GLM model.

Results and discussion Results of green material analysis showed different chemical composition, buffer capacity (BC), FC in grass during its development, which characterised green mass ensilage capacity. Total microorganisms colony forming units (CFU) count greatly varied in different grasses green material during growth. Count of microorganisms had a tendency to increase during growth in all green material of the studied grasses (except lactic acid bacteria). Count of silage microflora tended to decrease during growth in all silages of the studied grasses (except moulds) (Table 1). The count of epiphytic microflora influenced the fermentation quality in grass silage (P<0.01). The count of butyric bacteria was greater at blooming stage of maturity in fresh material of grasses, but it was higher at the branching stage in the ensiled mass. Grasses with a dry matter content of 250-300 g/kg resulted in good fermentation (i.e. stable pH 4-4.2) thus preventing the development of non desirable microorganisms. A negative correlation (r = 0.66) existed between butyric bacteria CFU count in fresh material and grass silages fermentation quality. The highest FC in fresh material (characterising ensilaging ability of mass) resulted in the lowest count of undesirable microorganisms in silage. In our investigations correlations were not observed between CFU count of yeast fungi in different grasses fresh material and in silages made from them.

Table 1 Changes of microflora in timothy fresh grass and silage (Lg from count of CFU in 1 ml susp.) (n=6)

Stage of maturity	Fresh grass			Silage		
	Lactic acid bacteria	Butyric acid bacteria	Moulds	Lactic acid bacteria	Butyric acid bacteria	Moulds
Branching	6.000	4.699	7.650	6.492	6.492	6.489
Shooting	5.802	5.627	8.401	5.514	5.925	6.798
Blooming	3.451	6.238	8.397	4.277	5.542	7.924

Conclusion The count and composition of microflora changed significantly (P<0.01) during growth in grasses fresh material and in silages made from it. The count of butyric bacteria correlated negatively with the count in fresh grass material and in silages (r = – 0.66). The CFU count of moulds in fresh grass material correlated positively with its count in silages during growth. The negative correlation existed between FC and count of lactic acid bacteria in fresh grass material, but positively correlated between FC and silage quality.

References
Wilkinson, I.M. (1999). Silage and health. Proceedings of the XII International Silage Conference, Uppsala, 67-83.
Woolford, M.K. and G. Pahlow (1998). The silage fermentation. In: Microbiology of Fermented Foods. London: Blackie, B.J.B. Wood, (Ed.), 73-102.

Hygienic value and mycotoxins level of grass silage in bales for horses

A. Potkański[1], J. Grajewski[2], K. Raczkowska-Werwińska[1], B. Miklaszewska[2], A. Gubała[1], M. Selwet[3] and M. Szumacher-Strabel[1]
[1]August Cieszkowski Agricultural University, Department of Animal Nutrition and Feed Management Wołyńska 33, 60-637 Poznań, Poland, Email: potkansk@jay.au.poznan.pl, [2]Bygoszcz University of Kazimierz Wielki, Institute of Biology and Environmental Protection, Chodkiewicza 30, 85-064 Bydgoszcz, Poland, [3]August Cieszkowski Agricultural University, Department of Agricultural Microbiology, Wołyńska 35, 60-637 Poznań, Poland

Keywords: mycotoxins, grass silage, bales, horses

Introduction Mycotoxins are secondary metabolites of moulds which have adverse effects on humans, animals, and crops and result in illnesses and economic losses. The toxins may occur in storage under conditions favourable for the growth of the toxin-producing fungus or fungi. The highest forage concentration of toxins was found in horizontal storage methods such as bunker silos and feed piles, which were left open to oxygen. In any fermentation storage system, temperature and the presence of moisture is sufficient for toxin production. In a plastic covered storage system, oxygen penetration is slowed but not eliminated. The longer silage is stored, the greater the opportunity for significant fungus growth and toxin contamination. Although the effects of mycotoxins on horses are not well documented in scientific literature, in many situations mycotoxin problems appear to be significant e.g. colic, neurological disorders, paralysis and brain lesions. The aim of this study was to determine the level of mycotoxins in grass silage prepared in bales for horses.

Materials and methods Grasses were conserved in two ways as a hay and silage. Grasses were ensiled in bales, which were opened after six and sixteen weeks. Samples for each treatment from six bales were taken for chemical and microbiological analysis of the basic nutrients, the total amount of fungi, yeast, LAB bacteria, mycotoxins – AFLA, OTA, DON, ZON.

Results Spores of *Aspergillus niger* that belongs to the allergens of respiratory system (*Cladosporium, Alternarium*) were dominated in hay. ZON (Zearalenol) was also detected in hay. Hay silage after 4 months of fermentation contained lower level of moulds than hay. Simultaneously, the level of lactic acid bacteria increased. Mycotoxins AFLA, OTA, ZON, DON were not detected.

Table 1 Hygienic value of used feeds

Material	Grass	Hay	Hay-silage 6 weeks	Hay-silage 16 weeks
Total fungi CFU/1gram	2.6×10^6	2.4×10^6	8.1×10^4	5.7×10^5
Total moulds CFU/1gram	3.2×10^5	7.1×10^5	2.9×10^3	9.3×10^3
Total yeast CFU/1gram	2.3×10^6	1.7×10^6	7.8×10^4	5.6×10^5
The dominant moulds	Cladosporium Aureobasidium Alternaria Mucorales Fusarium	Cladosporium Aureobasidium Alternaria Mucorales Fusarium Dematiaceae Endomyces	Acremonium Aureobasidium Mucor Humicola Penicillium	Fusarium Penicillium Endomyces
The total amount (30^0C) CFU/1gram	5.0×10^7	6.6×10^7	8.4×10^6	2.0×10^7
The total amount of lactic fermentation CFU/1gram	9.4×10^3	1.8×10^3	4.8×10^4	2.2×10^7
AFLA (B1, B2, G1, G2) ppb	NS	NS	NS	NS
OTA ppb	NS	NS	0,1	NS
ZON ppb	< 5	< 10,1	< 5	< 5
DON ppb	< 222	< 222	< 222	< 222

Conclusions Hay-silage had better hygienic value than hay and seems to be safer feed for horses.

Polyphenol oxidase activity and *in vitro* proteolytic inhibition in grasses

J.M. Marita, R.D. Hatfield and G.E. Brink
USDA-Agricultural Research Service, U.S. Dairy Forage Research Center, 1925 Linden Drive West, Madison, Wisconsin 53706 USA, Email: jmarita@wisc.edu

Keywords: polyphenol oxidase (PPO), *o*-diphenols, proteolysis, grasses, caffeic acid

Introduction Harvesting and storing high quality forage in the cool humid regions remains a challenge due to the potential for protein degradation during ensiling. Red clover is an exception as high protein levels are maintained during ensiling. Decreased proteolytic activity in red clover is due to polyphenol oxidase (PPO) activity and appropriate *o*-diphenol substrates (Jones *et al.*, 1995, Sullivan *et al.*, 2004). This project was undertaken to determine if PPO activity is present in a range of grasses and the potential role in proteolytic inhibition in the presence of the *o*-diphenol caffeic acid.

Methods Fifteen grass species were established in the greenhouse under a 14/10 h (day/night) lighting regime. Leaf blades were harvested from plants at the vegetative stage, frozen in liquid nitrogen and stored at -80°C. Individual samples were processed and analysed following modified methods of Sullivan *et al.* (2004). The PPO activity was determined in duplicate using *o*-diphenol substrates (caffeic acid, chlorogenic acid and catechol; 2mM final concentration). To determine the potential impact of PPO and *o*-diphenols on proteolytic activity leaf blades were prepared similar to above. Two samples were prepared with one processed through a G-25 sephadex spin column to remove low molecular weight materials and the other clarified by centrifugation and removal of the supernatant. At time zero, caffeic acid (3 mM final concentration) was added to the eluted protein (spin column). Aliquots were removed from each sample at specific time intervals (t_0, t_1, t_2, t_4 and t_{24} h) and soluble amino acids and small peptides were quantified to assess the degree of proteolysis.

Results and discussion Both species and the specific type of *o*-diphenol significantly altered PPO activity (Table 1). Orchardgrass, meadow fescue, ryegrass, and smooth bromegrass exhibited the highest PPO activities. Chlorogenic acid and/or caffeic acid were the preferred substrates, although there were differences among the most active grasses as to which was the best utilised. This suggests potential differences among the individual PPO enzymes. Generally, the addition of caffeic acid to isolated grass extracts resulted in proteolytic inhibition in grasses with substantial PPO activity. Such results suggest that several important grass species contain PPO activity, but may lack the appropriate *o*-diphenol substrates to effectively inhibit proteolysis. Initial results suggest that proteolytic inhibition can be achieved with the addition of caffeic acid.

Table 1 PPO activity in 15 grass species in the presence of three representative *o*-diphenol substrates and the percent reduction in *in vitro* proteolysis with the addition of caffeic acid after 24 h

Grass Species	Polyphenol Oxidase Activity (μmoles/μg/min)			Reduction in proteolysis
	Caffeic Acid	Chlorogenic Acid	Catechol	
Tall fescue (soft)	3.2 E-4	2.9 E-4	8.9 E-5	5%
Meadow fescue	3.5 E-3	4.1 E-3	9.0 E-4	90%
Timothy grass	1.5 E-4	2.2 E-4	1.1 E-4	6%
Smooth bromegrass	8.2 E-4	4.1 E-3	8.7 E-4	99%
Quackgrass	1.0 E-4	1.6 E-4	2.1 E-4	20%
Tall fescue	1.1 E-4	5.6 E-5	4.5 E-5	55%
Reed canary grass	5.1 E-5	3.4 E-5	3.1 E-5	49%
Ryegrass	2.6 E-3	1.3 E-2	7.9 E-4	78%
Spring wheat	2.1 E-5	1.0 E-5	2.6 E-5	8%
Winter wheat 1	2.5 E-5	1.8 E-5	2.5 E-5	16%
Winter wheat 2	1.1 E-5	1.0 E-5	2.1 E-5	15%
Rye	2.6 E-5	1.7 E-5	4.2 E-5	0%
Oat	2.6 E-5	1.7 E-5	1.9 E-5	0%
Orchardgrass	1.1 E-2	2.6 E-2	1.7 E-3	60%
Kentucky bluegrass	6.0 E-5	3.3 E-5	7.0 E-5	24%

References
Jones, B.A., R.E. Muck & R.D. Hatfield (1995). Red clover extracts inhibit legume proteolysis, *Journal of the Science of Food and Agriculture*, 67, 329-333.
Sullivan, M.L., R.D. Hatfield, S.L. Thoma & D.A. Samac (2004). Cloning and characterization of red clover Polyphenol oxidase cDNAs and expression of active protein in Escherichia coli and transgenic alfalfa, *Plant Physiology*, (accepted July 2004).

The effect of dry matter content and inoculation with lactic acid bacteria on the residual water soluble carbohydrate content of silages prepared from a high sugar grass cultivar

D.R. Davies, D.K. Leemans, E.L. Bakewell and R.J. Merry
Institute of Grassland and Environmental Research, Plas Gogerddan, Aberystwyth, SY 23 3EB, UK, Email: david.davies@bbsrc.ac.uk

Keywords: water soluble carbohydrate, grass silage

Introduction The introduction of new perennial ryegrass cultivars bred for high water soluble carbohydrate (WSC) content has created opportunities for improving the quality of grass silage, by not only providing adequate WSC for a good fermentation, but also sufficient to leave a higher residual level of WSC in the mature silage. High WSC silages have the potential to provide readily available energy during the early stages of rumen fermentation to balance energy and nitrogen supply and optimise rumen microbial growth. (Merry *et al.* 2002). The aim was to examine the effect of wilting and silage inoculants on the residual WSC content of grass silage.

Materials and methods A first cut of perennial ryegrass (cv. Aberdart) was mown, chopped and wilted for different periods of time to achieve dry matter (DM) levels of approximately 200, 250, 300 and 350 g DM/kg FM. The wilted herbages were ensiled in glass laboratory silos, either untreated or after application of 2 different silage inoculants (In 1) Pioneer (10^5 cfu/g FM containing homo- and heterofermentative lactic acid bacteria [LAB]) or (In 2) Powerstart, (10^6 cfu/g FM, containing only homo-fermentative LAB). Three replicates were prepared for each treatment and the silages were stored at 18-20°C for 90 days, before destructive sampling and analysis of chemical composition. Statistical analysis was carried out using the Anova package in Genstat.

Results After wilting DM levels of 223, 250, 282 and 358 were achieved, with respective WSC contents of 272, 285, 287 and 267 g/kg DM and a mean N content of 15.3 (0.57) g/kg DM (± SE). The chemical composition of the 20%, 30% and 35% DM silages is shown in Table 1. The 25% DM silage has been omitted but the full data set will be presented at the conference. pH values were significantly lower ($P<0.001$) and lactic acid concentrations higher ($P<0.001$) for Inoculant 2 silage than for the control or Inoculant 1 silages at all DM levels. There were significantly lower ammonia-N ($P<0.001$) concentrations in all of the Inoculant 2 silages compared to other treatments. Residual WSC content was high and >60 g/kg DM in both the control and Inoculant 2 silages at all DM levels and as DM increased above 243 g/kg FM values increased for all treatments, but were significantly higher ($P<0.001$) for Inoculant 2, with over 60% of the original WSC retained in the silage. The lowest WSC concentrations were observed with Inoculant 1 at all DM levels, with only 11% of WSC remaining.

Table 1 Chemical analysis of silages harvested at different DM contents and ensiled with or without inoculants

Target DM	20% DM			30% DM			35% DM			s.e.d.
	C	In 1	In 2	C	In 1	In 2	C	In 1	In 2	
pH	3.91	4.15	3.50	3.99	3.97	3.57	4.38	4.14	3.67	0.020
Dry Matter (g/kg FM)	201	197	205	251	250	257	330	336	334	
Lactic acid	76.4	27.7	119.7	64.4	40.1	97.5	39.8	23.7	78.1	3.80
Acetic acid	8.4	72.0	5.7	2.5	66.0	4.9	2.9	52.5	4.6	5.98
Butyric acid	6.2	ND	0.3	6.6	0.3	0.5	2.4	ND	0.6	0.83
Ammonia-N (g/kg TN)	98.6	84.4	45.5	88.1	72.4	37.3	67.8	85.7	41.3	7.22
WSC	68.2	15.4	70.2	113.8	37.8	160.5	67.0	31.5	174.9	9.49

All values are in g/kg DM unless otherwise stated. C = Untreated control; In 1= Inoculant 1; In 2 = Inoculant 2. WSC = Water soluble carbohydrate. DM = Dry matter.

Conclusions A high quality grass silage with exceptionally high residual WSC content can be prepared using a combination of wilting (to >23% DM) and treatment with a homo-fermentative lactic acid bacterial inoculant. Such high sugar silages have the potential to increase N use efficiency by the rumen microbial population.

Acknowledgements This work was funded jointly by The Department for Environment, Food and Rural Affairs and by a Framework V grant from the European Union.

References

Merry, R.J., D.K. Leemans & D.R. Davies (2002). Improving the efficiency of silage-N utilisation in the rumen through the use of grasses high in water soluble carbohydrate content. In: *Proceedings of The XIIIth International Silage Conference*, Auchincruive, Scotland eds. L.M. Gechie and C. Thomas.

Using the red clover polyphenol oxidase gene to inhibit proteolytic activity in lucerne

R.D. Hatfield, M.L. Sullivan and R.E. Muck
USDA-Agricultural Research Service, U.S. Dairy Forage Research Centre, 1925 Linden Drive West, Madison, WI 53706 USA, Email rdhatfie@wisc.edu

Keywords: polyphenol oxidase, *o*-diphenols, proteolysis, red clover, lucerne

Introduction Preserving high quality forage in cool humid regions of agricultural production remains a challenge due to potentially high levels of protein degradation during ensiling. Red clover is an exception maintaining its high protein levels during ensiling. Decreased proteolytic activity in red clover is due to polyphenol oxidase (PPO) activity and appropriate *o*-diphenol substrates (Jones *et al.*, 1995, Sullivan *et al.*, 2004). This work highlights potential strategies for utilising PPO as a means of decreasing proteolytic degradation during the ensiling of lucerne and other forages.

Methods Three red clover PPO genes (PPO1, PPO2 and PPO3) were cloned from a leaf cDNA library. Red clover PPO cDNAs under control of a constitutive promoter were expressed in lucerne. Leaves from PPO transformed lucerne were ground in liquid nitrogen and extracted with 50mM MES buffer (pH 6.5). Extracts were allowed to incubate at 37°C and subsamples removed at time intervals to determine the extent of proteolysis. Lucerne extracts were also incubated with or without the addition of *o*-diphenols to determine the impact of PPO activity and type of *o*-diphenol on proteolysis.

Results and discussion Since lucerne does not produce *o*-diphenols, PPO-transformed lucerne (PPO-Luc) is an excellent system for defining factors critical for PPO-mediated proteolytic inhibition. The amount of PPO activity measured in transformed lucerne (PPO-Luc) was at levels similar to that found in red clover using the *o*-diphenol caffeic acid as the primary PPO substrate. Addition of caffeic acid to PPO-Luc extracts resulted in approximately 80% inhibition of proteolytic activity (Table 1). Caffeic acid and chlorogenic acid *o*-diphenols were the most rapidly utilised substrates with other *o*-diphenols showing considerably less activity. However, a wide range of *o*-diphenols were effective in decreasing proteolysis over incubation times of 4 h or longer (Table 2). Addition of PPO-Luc extracts to oat leaf extracts along with caffeic acid resulted in a 50% reduction in proteolysis. Preliminary experiments ensiling PPO-Luc in mini-silos with the addition of *o*-diphenols resulted in a 25% decrease in proteolysis over control silos. These results indicate that strategies can be developed to utilise the PPO system to decrease protein losses during ensiling of forges.

Table 1 Amino acids released (mmol/mg total protein) from lucerne extracts incubated with or without 3 mM caffeic acid (+CA, -CA). Extracts were from Control-Luc (lucerne transformed with vector only) or from PPO-Luc (lucerne transformed with red clover PPO 1 gene)

Incubation hour	Control-Luc-CA	Control-Luc+CA	PPO-Luc-CA	PPO-Luc+CA
0	0.00±0.00	0.00±0.00	0.00±0.0	0.00±0.00
1	0.18±0.02	0.21±0.00	0.22±0.03	0.06±0.01
2	0.36±0.03	0.39±0.02	0.40±0.03	0.11±0.01
3	0.51±0.03	0.56±0.01	0.54±0.06	0.10±0.02
4	0.66±0.04	0.70±0.02	0.68±0.07	0.15±0.02

Table 2 Impact of different *o*-diphenols (3 mM) on proteolytic inhibition in lucerne extracts. Amino acids released after 4 h incubation at 30°C. CA= caffeic acid, HCA= hydrocaffeic acid, CGA= chlorogenic acid

	Amino Acid Released (mM/mg protein)				
Substrate	Luc-Control	Luc-PPO	Substrate	Luc-Control	Luc-PPO
None	0.71±0.05	0.72±0.12	CGA	0.80±0.06	0.26±0.06
CA	0.75±0.08	0.12±0.03	Catechol	0.71±0.10	0.37±0.05
HCA	0.70±0.07	0.25±0.07	Epicatichin	0.64±0.08	0.25±0.08

References

Jones, B.A., R.E Muck & R.D. Hatfield (1995). Red clover extracts inhibit legume proteolysis. *Journal of the Science of Food and Agriculture* 67, 329-333.

Sullivan, M.L., R.D. Hatfield, S.L. Thoma & D.A. Samac (2004). Cloning and characterization of red clover Polyphenol oxidase cDNAs and expression of active protein in Escherichia coli and transgenic alfalfa. *Plant Physiology* 136,3234-3244.

New results on inhibition of clostridia development in silages

E. Kaiser, K. Weiß and I. Polip
The Institute of Animal Science, Humboldt-University of Berlin, Invalidenstraße 42, D-10115 Berlin, Germany
Email: ehrengard.kaiser@agrar.hu-berlin.de

Keywords: clostridia, fermentation quality, fermentation process, butyric acid

Introduction The prevention of clostridial activity in silages is one of the most important aims in silage making. Clostridial activity in silages is especially expressed as the occurrence of butyric acid and as increased content of clostridial spores. A rapid reduction in the pH value at the beginning of fermentation process is considered as the most important factor for inhibition of clostridial development. It is assumed, that, if the "critical pH value" will be quickly achieved, clostridial activity in silages can be stopped. In experiments concerning the fermentation process it was found that the effect of acidification and dry matter content on the clostridial activity is different in ensiling material, containing nitrate, and in nitrate-free material. The object of the present paper was to clarify the conditions for clostridial development during the fermentation process, including examination of factors such as dry matter content, acidification and nitrate content.

Materials and methods The fermentation process in *Dactylis glomerata*, first growth and nitrate-free, was examined at four levels of dry matter (DM) content (217 g/kg, 325 g/kg, 416 g/kg, 515 g/kg) of the ensiling material under laboratory conditions, in each case without and with additives. To improve the acidification, lactic acid bacteria (LAB) was used in a concentration of 10^5 cfu g/FM and LAB (10^5 cfu g/FM) + glucose (2% of fresh matter) respectively. The additive of nitrate was 4.4 as well as 6.6 g NO_3/kg DM. The silages were analysed after 3, 7, 14, 28, 56 and 180 days of storage period under temperature constant conditions (25°C). To determine effects of DM, acidification and nitrate on clostridia development the ensiling material was contaminated with additives of clostridia spores (approximately 10^4 MPN g/FM).

Results Although a rapid setting of the critical pH value was achieved, significant butyric acid (BA) formation was observed in silages, which was earlier (on DM of 217 g/kg from d 7) and more comprehensive with lower DM content (Table 1). Butyric acid-free silages were only estimated in the fermentation process of ensiling material with the level of 515 g/kg DM. At lower levels of DM, the BA formation was only restricted, but not suppressed by acidification. In contrast the formation of BA was also suppressed at lower levels of DM by nitrate additive. The addition of LAB and LAB + glucose had no positive effect on the development of clostridia spores at lower DM levels 217 (Table 1) and 325 g/kg. Indeed the content of clostridia spores decreased after commencement of fermentation. The content of clostridia spores increased with commencement of butyric acid formation and reached an extent, which is higher than the content of ensiling material. With nitrate as an additive (Table 1) the clostridia spore content was reduced to nearly zero at two levels of DM. In contrast at high DM contents the effects on clostridia have been observed, whereas the effect of nitrate was not so clearly pronounced.

Table 1 Effects of DM content, acidification and nitrate on clostridia development during fermentation process

	After 56 days of storage period (DM g/kg)				After 14 days of storage period (DM content 217 g/kg)		
	217	325	416	515	Control	LAB	Nitrate
pH	4.3	4.4	4.6	4.6	4.4	3.6	4.
BA (g/kg)	36	16	12	0	14	6	0
Clostridia spores (MPN g/FM)	2.3×10^6	6.8×10^5	2.0×10^5	9.3×10^3	2.5×10^6	1.2×10^4	1.8×10^3

Conclusions The decrease in pH value is not the major inhibiting parameter against clostridia activity. At low DM, a directly acting clostridia inhibitor is necessary in addition to acidification. At high DM content the reduced availability of water (decreased a_w-value) is effective as an inhibiting parameter, whereas the importance of acidification decreased. Acidification, a_w-value and nitrate complement are inhibiting factors for clostridia development. This synergistic effect is very different at varying DM levels.

References

Kaiser, E., I. Polip & K. Weiß. (2001). Einfluss des Trockensubstanzgehaltes auf die Clostridiendynamik in Silagen. 113. VDLUFA- Kongress, Berlin, 107.

Polip, I. & E. Kaiser. (1998): Untersuchung zur Clostridienentwicklung in Silagen. Proc. 110.VDLUFA-Kongreß, Giessen, 194.

Polip, I. (2001): Untersuchungen zur Unterbindung von Buttersäuregärung und Clostridienaktivität in Silagen aus nitratarmem Grünfutter. Dissertation, Humboldt- Universität zu Berlin.

Ensilability and silage quality of different cocksfoot varieties

U. Wyss
Agroscope Liebefeld-Posieux, Swiss Federal Research Station for Animal Production and Dairy Products (ALP), 1725 Posieux, Switzerland

Keywords: cocksfoot varieties, ensilability, fermentation quality

Introduction Various factors determine the ensilability of plant crops. In addition to dry matter content, sugar and protein content, buffering capacity, plant structure, soil contaminations and epiphytic microflora are also important. In Switzerland, various mixtures and also pure swards are continuously revised on the basis of results from variety testing programs. The main parameters for these tests are the yield and nutrient contents. However, in terms of ensilability and silage quality, no systematic tests have been carried out. For this reason, we have tested the ensilability of different cocksfoot varieties as well as the quality of the silages.

Materials and methods In a trial five different cocksfoot varieties were tested, which were part of the testing program of the Swiss Federal Research Station for Agroecology and Agriculture in Zürich-Reckenholz. Forage was ensiled of the first as well as of the third cut. The forage was pre-wilted, short chopped and ensiled in laboratory silos each having a volume of 1.5 litres. Chemical parameters were analysed before ensiling and after a storage period of five months. Fermentation acids, ethanol, ammonia and pH were also analysed in the silage. Furthermore, the fermentability coefficient was calculated in the green forage. This parameter summarises the potential effects of dry matter as well as the ratio of sugar content and buffering capacity on the fermentation.

Results The dry matter contents in the green forage varied between 220 and 270 g/kg for the first cut and between 310 and 340 g/kg for the third cut. Concerning the nutrient contents, the forage of the first cut had higher protein contents as well as lower crude fibre contents in comparison to the third cut. The ash and sugar contents were only slightly higher for the forage of the first cut. Within the five varieties there were some differences. The fermentability coefficients varied between 31 and 38 for the first cut and between 40 and 45 for the third cut. There were significant variety and also cut number effects. One reason for the higher fermentability coefficients of the third cut is the higher pre-wilting degree. Forage with fermentability coefficients below 35 is reputed to be difficult to ensile. All silages contained butyric acid and only low lactic acid contents. The butyric acid contents were significantly different between the varieties. The relation between the dry matter content and butyric acid content is shown in Figure 1. According to the DLG evaluation scheme developed by Weissbach and Honig (1997) the silages attained scores between 24 and 42 for the first cut and between 45 and 54 for the third cut, out of a maximum of 100. There is a strong correlation between the fermentability coefficient and the DLG scores (Figure 2).

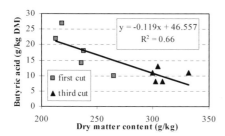

Figure 1 Relation between dry matter content and butyric acid

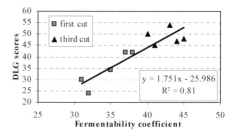

Figure 2 Relation between fermentability coefficient and DLG scores

Conclusions There were some differences concerning the nutrient contents and the fermentation parameters for the five cocksfoot varieties. However, the cut number and especially the dry matter content proved to be the major influencing factors for the ensilability as well as for the silage quality.

References

Weissbach, F. & H. Honig (1997). DLG-Schlüssel zur Beurteilung der Gärqualität von Grünfuttersilagen auf der Basis der chemischen Untersuchung. *Tagung des DLG-Ausschusses für Futterkonservierung vom 2*. Juli 1997 in Gumpenstein.

Ensiling characteristics of sudangrass silage treated with green tea leaf waste or green tea polyphenols

M. Kondo, K. Kita and H. Yokota
Graduate School of Bioagricultural Sciences, Nagoya University, Aichi, 470-0151, Japan,   Email: i021009d@mbox.nagoya-u.ac.jp

Keywords: green tea waste, green tea polyphenol, lactic acid, silage

Introduction   Green tea waste (GTW), emitted from beverage companies manufacturing tea drinks, contains high crude protein (CP) and polyphenols. Kondo *et al*. (2004) showed that GTW addition to forage ensiling enhanced lactic acid fermentation and decreased pH value. Ishihara *et al*. (2001) showed that high counts of *Lactobacillus* species were maintained and the counts of clostridia were decreased in the intestinal microflora of animals fed the diet containing green tea polyphenols (GTP). It is hypothesised that GTP might activate lactic acid bacteria and enhance silage fermentation. This study was conducted to evaluate the potential of GTW and GTP as silage additives and explored the mechanisms of enhanced lactic acid fermentation by GTW.

Materials and methods   Silages were made from sudangrass (*Sorghum sudanese*) harvested at heading stage on 2 August 2001. Soon after the harvest, the forages were chopped into about 2 to 3 cm in length using a forage cutter and then 400 g of the forages were mixed with fresh GTW, GTP, *Lactobacillus rhamnosus* or cell-wall degrading enzymes prior to ensiling. Silages were prepared in triplicates and stored at 25°C for 30 days.

Results   Dry matter (DM) contents of sudangrass and GTW at ensiling were 312 and 184 g/kg respectively. Water soluble carbohydrate contents of sudangrass and GTW were respectively 102.0 g/kg DM and 7.0 g/kg DM. The number of lactic acid bacteria associated with the grass and GTW were 5.90 and 7.96 $\log^{10}$ cfu/g fresh matter (FM) respectively. GTW contained 91.4 g/kg DM of polyphenols. Silages treated with GTW at 50 and 200 g/kg FM and *L. rhamnosus* decreased pH and increased lactic acid content ($P<0.05$) (Table 1). GTP and cell-wall degrading enzyme treatments did not increase lactic acid content compared with the control silage.

Table 1 Chemical composition of Sudangrass silage after 30 days of ensiling

Treatment	Rate (g/kg FM)	pH	DM (g/kg)	Lactic acid (g/kg DM)	Acetic acid (g/kg DM)	Butyric acid (g/kg DM)
Control		5.04a	298abc	22.4d	1.7c	8.3b
GTW	50	3.99d	298abc	75.7c	8.6b	0.0d
	200	3.86e	277dc	89.0b	10.5a	0.0d
GTP	0.2	4.90c	290bdc	24.1d	1.4c	11.3a
	1	4.94bc	291bdc	24.5d	1.6c	7.4b
	4	5.11a	288d	20.4d	1.3c	6.5b
L. rhamnosus		3.67f	304ab	107.8a	2.2c	0.0d
Enzyme		4.99b	303a	27.2d	1.6c	5.5c
s.e.m.		0.02	0.41	3.4	0.5	2.5

GTW: green tea waste, GTP: green tea polyphenols.   Values in the same column with different letters are significantly different ($P<0.05$).

Conclusions   GTW addition to grass at ensiling increased lactic acid content by a similar amount as *L. rhamnosus* inoculation. GTP treatment had no affect on lactic acid production. Enhancement of lactic acid fermentation by GTW would not be caused by its polyphenol content.

References

Ishihara, N., D.C. Chu, S. Akachi & J.R. Juneja (2001). Improvement of intestinal microflora balance and prevention of digestive and respiratory organ diseases in calves by green tea extracts. *Livestock Production Science*. 68, 217-229.
Kondo, M., K. Kita & H.Yokota (2004). Effects of tea leaf waste of green tea, oolong tea, and black tea addition on sudangrass silage quality and *in vitro* gas production. *Journal of the Science of Food and Agriculture*. 84, 721-727.

Effects of silage preparation and microbial silage additives on biogas production from whole crop maize silage

M. Neureiter, C. Perez Lopez, H. Pichler, R. Kirchmayr and R. Braun

BOKU - University of Natural Resources and Applied Life Sciences, Vienna - Department for Agrobiotechnology, IFA-Tulln, Konrad Lorenz Str. 20, A-3430 Tulln, Austria, Email: markus.neureiter@boku.ac.at

Keywords: maize whole crop silage, lactic acid bacteria, anaerobic digestion, methane yield

Introduction Biogas applications based on the production of energy from renewable resources have emerged in the past years due to several countries setting quotas for bioenergy thus promoting anaerobic digestion for heat and electricity generation. Maize is one of the most common substrates for biogas production based on energy crops because of the high yields per hectare with ensiling as the preferred method for storage. Experiments were performed to investigate whether conditions during the silage fermentation and the addition of starter cultures can affect the biogas yields.

Materials and methods Laboratory silage experiments were performed as described by Danner *et al.* (2003) with whole crop maize as raw material. Six treatments were tested; improperly compressed material, inoculation with homofermentative lactic acid bacteria (LAB), inoculation with homo- and heterofermentative LAB, addition of amylase, and inoculation with *Clostridium tyrobutyricum* after adjusting the dry matter contents to a lower level (29%) and adding $CaCO_3$ as a buffer. All experiments were made in triplicate. After 44 d the silages were chemically analysed and a portion was used as a substrate for anaerobic batch fermentation tests. Methane production was measured over a period of three weeks.

Results The chemical analyses of the silages are shown in Figure 1. Well ensiled silages exhibited high lactic acid concentrations in a comparable range, with the exception of the silage inoculated with heterofermentative lactic acid bacteria that showed lower lactic acid and elevated acetic acid formation. The silages prepared under unfavourable conditions showed lower lactic acid concentrations. High concentrations of butyric acid were only detected in the silage that was inoculated with *Clostridium tyrobutyricum*.

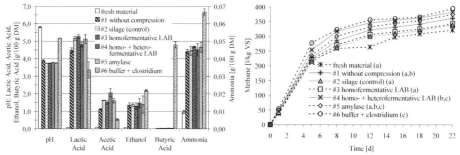

Figure 1 pH and concentration of metabolites in fresh and ensiled whole crop maize

Figure 2 Methane production from fresh and ensiled whole crop maize

Methane production based on volatile solids (VS) obtained during labscale batch tests is shown in Figure 2. Treatments that do not differ significantly (*P*<0.05, LSD) in their methane yields after 22 d are marked with same letters in parentheses. The silages generally show higher methane production from VS compared with fresh substrate. The highest productions after 22 d were obtained with addition of *Clostridium tyrobutyricum* (395 l/kg VS) and with the starter containing both homo- and heterofermentative LAB (389 l/kg VS).

Conclusions The results of the present study show that ensiling generally can improve methane production from whole crop maize. Obviously spoiled silages also showed good methane yields with respect to volatile solids. The highest methane productions were obtained by inoculating with *Clostridium tyrobutyricum* and with a starter culture containing both homo- and heterofermentative lactic acid bacteria, however differences in methane yield are not very distinct. Fermentation losses during the ensiling process may lower the overall yield from spoiled silages after longer storage periods.

References

Danner, H., M. Holzer, E. Mayrhuber & R. Braun (2003). Acetic acid increases stability of silage under aerobic conditions. *Applied and Environmental Microbiology*, 69 (1), 562-567

A 16S rDNA-based quantitative assay for monitoring *Lactobacillus plantarum* in silage

M. Klocke[1], K. Mundt[1], C. Idler[1], P. O`Kiely[2], S. Barth[3]
[1] *Institute of Agricultural Engineering Bornim e.V. (ATB), Dept. Bioengineering, Max-Eyth-Allee 100, D-14469 Potsdam, Germany, Email: mklocke@atb-potsdam.de,* [2] *Teagasc, Grange Research Centre, Dunsany, Co Meath, Ireland,* [3] *Teagasc, Oak Park Research Centre, Carlow, Co Carlow, Ireland*

Keywords silage, *Lactobacillus plantarum*, 16S-rDNA-based marker, quantitative PCR, Q-PCR

Introduction Ensilage of herbaceous biomass can be enhanced by applying pre-selected fermentative bacteria, however insufficient is known about the population dynamics of such starter cultures under a range of ensiling conditions. Classical methods for species-specific quantification of bacteria are labour intensive. An alternative approach is the detection of bacteria based on molecular markers for species-specific regions within their genomic DNA (e.g. the 16S rDNA sequence). In this study, a quantitative marker assay using the real-time PCR technique (Q-PCR) is described for *Lactobacillus plantarum*, a bacterium often used for silage starter cultures.

Materials and methods Based on a variable region in the 16S rDNA of *L. plantarum* (Chagnaud *et al.*, 2001), the following PCR-primers were developed: forward primer Lplan-vreg1-F 5'-TTACATTTGAGTGAGTGGCG AACT, reverse primer Lplan-vreg1-R 5'-AGGTGTTATCCCCCGCTTCT, TaqMan® probe Lplrh-vreg1-T 5'-VIC®-GTGAGTAACACGTGGGWAACCTGCCC-TAMRA®. Q-PCR was performed in triplicate using an ABI 7000 and the following conditions: reaction mixture 1x JumpStart™ *Taq* ReadyMix™, 900nM forward primer, 300 nM reverse primer, 200 nM TaqMan® probe, 0.25 µl internal dye ROX®, *ad* 25 µl H_2O_{dd}; cycle regime 1: 120s 94°C, 2: 15s 94°C, 3: 60s 59°C, 4: 60s 72°C, 2-4 were repeated 40x. Plasmid pATB875 containing a 1500 bp fragment of *L. plantarum* 16S rDNA was used as a standard. Grass was inoculated with equal concentrations of *L. plantarum* ATB-8 and *L. rhamnosus* ATB-14 and sampled on days 2, 13, 20 and 40 of ensilage. The genomic DNA was prepared from paddled samples (Rheims & Stackebrandt, 1999).

Results Using the pATB875 plasmid as a standard, an optimised Q-PCR protocol was developed. From a triplicate dilution series an equation for the species-specific estimation of the copy number of *L. plantarum* 16S rDNA sequences in unknown samples was calculated (Figure 1A, B). Applied to grass silage samples, the assay monitored the rise of the *L. plantarum* population during the first two weeks of ensiling. Due to the increased acidification, the population decreased during prolonged ensiling (Figure 1C).

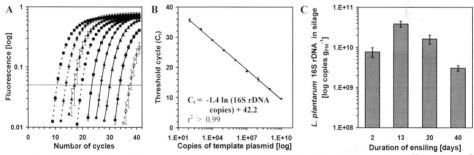

$$C_t = -1.4 \ln (16S\ rDNA\ copies) + 42.2$$
$$r^2 > 0.99$$

Figure 1 (A) Representative Q-PCR amplification curves of pATB875 10 fold dilution series (10^{10} (···■···) to 10^3 copies (—●—) and no-template control (○). (B) The mean C_t values of triplicates were plotted against the copy number of pATB875. (C) Determination of *L. plantarum* 16S rDNA copy number within grass silage samples.

Conclusions This Q-PCR assay is a first attempt at a direct DNA-based quantification of *L. plantarum* within silages. The Q-PCR assay enables the analyses of population dynamics of starter cultures containing *L. plantarum*. Similar approaches should also be applicable for monitoring other fermentative bacteria species.

Acknowledgements This work was co-financed by the European Molecular Biology Organization (220.00-04).

References
Chagnaud, P., K. Machinis, L.A. Coutte, A. Marecat & A. Mercenier (2001). Rapid PCR-based procedure to identify lactic acid bacteria, application to six common *Lactobacillus* species. *Journal of Microbiological Methods*, 44, 139-148.
Rheims, H. & E. Stackebrandt (1999). Application of nested polymerase chain reaction for the detection of as yet uncultured organisms of the class *Actinobacteria* in environmental samples. *Environmental Microbiology*, 1, 137-143.

A comparison of the efficacy of an ultra-low volume applicator for liquid-applied silage inoculants with that of a conventional applicator

G. Marley[1], G. Pahlow[2], H.-H. Herrmann[1] and T.R. Owen[1]
[1]Ecosyl Products Ltd, Ellerbeck Way, Stokesley, North Yorkshire, TS9 5QT, UK, Email: gordon.marley@ecosyl.com, [2]Institute of Crop and Grassland Science, Federal Agricultural Research Centre, Bundesallee 50, D-38116 Braunschweig, Germany

Keywords: silage inoculant, applicator, ultra-low volume

Introduction Liquid-applied silage inoculants are normally sprayed onto forages cut for ensiling at application rates from 1 to 3 l/t. Applicator tanks can require frequent re-filling, especially with large self-propelled forage harvesters having harvest rates in excess of 1000 t/d. This can be an issue for fields remote from the farm, for areas with restricted water availability and for contractors paid by the area harvested. This study was conducted to assess the efficacy of inoculant distribution on the crop using a simple, ultra-low volume (ULV) applicator compared with a conventional liquid-applied silage inoculant applicator.

Materials and methods A series of 10 experiments was conducted in the UK and Germany. Forage crops were harvested using self-propelled forage harvesters and forage samples were collected either untreated or treated with 'Ecosyl' silage inoculant, formulated to apply 10^6 cfu/g forage, using an 'Ecosyler' ULV applicator at 10, 15 or 20 ml/t or a 'Magnum' conventional applicator at 1500 ml/t. The 'Ecosyler' applicator comprises a small tank, a peristaltic pump and an electronic speed control box. It relies on the air speed (41-68 m/s) at the exit from the harvester accelerator fan, where the inoculant was applied, to atomise the inoculant suspension. Both applicators were calibrated prior to use. Twenty samples of each of the untreated or treated forages were taken for microbial counts to assess the homogeneity of distribution of the inoculant lactic acid bacteria (LAB) applied to the treated forages.

Results Mean LAB counts on the untreated and treated forages in each of the 10 experiments are shown in Table 1, together with the coefficients of variation (CV%) for each set of counts to express homogeneity of distribution of inoculant LAB with the different applicators.

Table 1 Distribution of LAB on forages

Experiment	Crop	ULV rate (ml/t)	Mean LAB count log 10 cfu/g, (CV%)		
			Untreated	Magnum	Ecosyler
1	Grass	10	3.60 (17.6)	5.82 (5.1)	5.83 (4.4)
2	Grass	10	3.96 (4.5)	5.89 (2.9)	5.76 (2.8)
3	Grass	10	3.95 (3.5)	5.87 (2.6)	5.94 (2.6)
4	Grass	10	4.60 (4.0)	5.96 (1.7)	5.84 (3.9)
5	Grass	20	5.11 (3.4)	6.18 (2.2)	6.35 (3.5)
6	Triticale	15	4.95 (2.0)	5.91 (2.6)	6.03 (2.3)
7	Wheat	15	4.90 (2.7)	5.70 (1.4)	6.00 (2.0)
8	Wheat	15	4.92 (1.8)	5.43 (3.3)	6.17 (3.4)
9	Wheat	15	5.92 (4.8)	6.46 (1.5)	6.32 (1.6)
10	Maize	15	4.56 (2.4)	5.83 (3.0)	6.00 (1.8)

The CV% for the counts from the 'Magnum' and 'Ecosyler' ranged from 1.4 to 5.1 (mean 2.8) and 1.6 to 4.4 (mean 2.6), respectively. Analysis of variance showed no statistically significant difference between the mean CV% for the 2 applicators (P=0.53, SEM 0.15) nor between the mean LAB counts achieved (Magnum 5.91, Ecosyler 6.02, P=0.20, SEM 0.042).

Conclusions The distribution of the inoculant LAB on the treated forage crops was similar for both the ULV applicator at 10-20 ml/t and the conventional applicator at 1500 ml/t. The air speeds of the forage harvesters used were sufficient to atomise the ULV-applied inoculant to give effective crop coverage at low liquid application rates. The variation in LAB application numbers achieved with both applicators is believed to be a result of the inherent inaccuracy of harvest rate estimates used to set applicator flow-rates.

Section 3

Developments in ensiling techniques

B. Aerobic stability

Improving the aerobic stability of whole-crop cereal silages

I. Filya, E. Sucu and A. Karabulut

Uludag University Agricultural Faculty, Department of Animal Science, 16059 Bursa, Turkey. Email: ifilya@uludag.edu.tr

Keywords: silage, propionic acid bacteria, aerobic stability

Introduction Whole-crop cereal silages, such as wheat, sorghum, and maize are susceptible to aerobic deterioration, especially in warm climates. This is because aerobic yeasts are the most active at 20-30°C (Ashbell *et al.*, 2002). Therefore, it is very important to find suitable additives that inhibit fungi and protect the silage upon aerobic exposure. *Propionibacterium acidipropionici* is propionic acid bacteria (PAB), which produce propionic and acetic acid in silage. Results with these micro-organisms in laboratory studies were promising with regard to aerobic stability. The purpose of the present work was to study the effects of PAB, lactic acid bacteria (LAB) and combinations of PAB + LAB on the fermentation and aerobic stability of whole-crop cereal silages.

Material and methods Wheat at the early dough stage (366 g/kg DM), sorghum at the milk stage (272 g/kg DM) and maize at the one-third milk line stage (358 g/kg DM) were harvested and chopped to about 1.5 cm and ensiled in 1.5-l glass jars (Weck®, Wher-Oflingen, Germany) equipped with a lid that enables gas release only. At the end of the ensiling period, 60 d, the silages were sampled for chemical and microbiological analysis and were subjected to an aerobic stability test, which lasted 5 d, in a "bottle" system developed by Ashbell *et al.* (1991). The following microbial additives were applied to fresh forage at the levels recommended by the manufacturer: control (no additives); *P. acidipropionici; Lactobacillus plantarum;* combination of *P. acidipropionici* and *L. plantarum* (final application rate of 1.0 x 10^6 cfu/g of fresh forage).

Results *P. acidipropionici* increased the concentrations of propionic acid of the silages. This was evident from propionic acid production. The higher amount of acetic acid in the *P. acidipropionici*-inoculated silages was expected because acetic acid is a co-metabolite of the fermentation of carbohydrates and lactic acid by *P. acidipropionici*. However, production of acetic and propionic acid in the *P. acidipropionici*-inoculated silages decreased all yeasts and moulds counts. Propionic and acetic acid are fungicidal agents, and high concentrations of propionate and acetate inhibit yeasts and moulds growth. Table 1 gives the results of the aerobic exposure test of the silages. Silage deterioration indicators are pH change, CO_2 production and an increase in yeast and mould numbers. The *P. acidipropionici*-inoculated silages had significantly higher levels of acetic and propionic acid than the *L. plantarum* or *P. acidipropionici* + *L. plantarum*-inoculated silages (*P*<0.05). Therefore, yeast activity was impaired in the *P. acidipropionici*-inoculated silages. As a result, *P. acidipropionici* decreased CO_2 production and improved aerobic stability of wheat, sorghum, and maize silages. However, the combination of *P. acidipropionici* + *L. plantarum* did not improve aerobic stability of the silages.

Conclusions The *P. acidipropionici* was very effective in protecting the wheat, sorghum, and maize silages exposed to air under laboratory conditions. The use of *P. acidipropionici*, as a silage inoculant can improve the aerobic stability of silages by inhibition of yeast activity. The combination of *P. acidipropionici* and *L. plantarum* do not look promising in protecting wheat, sorghum, and maize silages upon aerobic exposure.

References

Ashbell, G., Z.G. Weinberg, A. Azrieli, Y. Hen & B. Horev (1991). A simple system to study the aerobic deterioration of silages. *Canadian Agricultural Engineering* 34, 171-175.

Ashbell, G., Z.G. Weinberg, Y. Hen & I. Filya (2002). The effects of temperature on the aerobic stability of wheat and corn silages. *Journal of Industrial Microbiology and Biotechnology* 28, 261-263.

Table 1 The results of the aerobic stability test (5 days) of the silages

Forage type	Treatment	pH	CO$_2$ (g/kg DM)	Yeasts (log cfu/g)	Moulds (log cfu/g)
Wheat	Control	5.2^{ab}	14.8^b	5.2	4.0
	PAB	4.9^b	4.1^c	<2.0	<2.0
	LAB	5.3^a	33.7^a	8.6	4.4
	PAB + LAB	4.9^b	17.6^b	4.7	2.8
	SE	0.245	0.157		
Sorghum	Control	4.8^a	20.4^b	5.8	4.1
	PAB	4.2^{bc}	6.7^c	<2.0	<2.0
	LAB	4.8^a	38.3^a	8.0	4.3
	PAB + LAB	4.4^b	19.7^b	5.6	2.9
	SE	0.208	0.135		
Maize	Control	4.4^{ab}	25.6^b	6.1	4.5
	PAB	4.1^b	5.8^c	<2.0	<2.0
	LAB	4.7^a	44.5^a	8.3	4.8
	PAB + LAB	4.2^b	31.9^b	5.3	3.0
	SE	0.187	0.112		

Within a column and forage type means followed by different letter differ significantly (*P*<0.05).

Aerobic stability and nutritive value of low dry matter maize silage treated with a formic acid-based preservative

I. Filya, E. Sucu and A. Karabulut
Uludag University Agricultural Faculty, Department of Animal Science, 16059 Bursa, Turkey. Email: ifilya@uludag.edu.tr

Keywords: silage, formic acid, aerobic stability, nutritive value

Introduction Aerobic stability is one of the major problems of the ensiling process, especially in warm climates. Ashbell *et al.* (2002) have shown that at 30°C, the development of aerobic yeast and moulds in silages is most intensive. In Turkey all silages are susceptible to air penetration during storage and unloading with a large proportion of the silage spoiled and in extreme cases all the silage is spoiled. The purpose of the present work was to study the effects of formic acid-based preservative (FAB; Kemisile® 2000, Kemira Oyj-Industrial Chemicals, Finland) on the aerobic stability and nutritive value of maize silage.

Material and methods Maize was harvested at the milk stage (218 g/kg DM). FAB was applied at 1.0, 1.5, 2.0, 2.5, 3.0, 3.5 and 4.0 g/kg to 1.5-2.0 cm chopped fresh maize. Fresh material was ensiled in 1.5 l glass jars (Weck®, Wher-Oflingen, Germany) equipped with a lid that allowed gas to be released. The jars were stored at 26±2°C in laboratory conditions. At the end of a 90 d ensiling period, silages were subjected to a 5 d "bottle" system aerobic stability test (Ashbell *et al.*, 1991). Rumen dry and organic matter degradability of the silage was determined by the *in situ* nylon bag technique developed by Mehrez and Ørskov (1977).

Results FAB decreased lactic, acetic and butyric acid concentrations of maize silage. However, FAB prevented proteolysis in maize silage. Data obtained from the aerobic stability and ruminal degradability of maize silage are given in Tables 1 and 2 respectively. FAB inhibited yeast and mould growth and reduced CO_2 production of silage (Table 1) and improved the aerobic stability of maize silage. At the higher levels of application FAB increased both the dry and organic matter of maize silage (Table 2).

Table 1 The results of the aerobic stability test (5days) of maize silage

Treatment	pH	CO_2 (g/kg DM)	Yeasts	Moulds
		log CFU/g		
Control	4.4±0.1^a	6.9±0.5^a	4.0±0.3^a	4.7±0.3^a
1.0	4.1±0.1^{bc}	7.0±0.2^a	3.9±0.2^a	3.9±0.3^{ab}
1.5	4.0±0.1^{bcd}	7.1±0.4^a	3.9±0.2^a	3.2±0.3^{bc}
2.0	3.9±0.1^{cd}	7.0±0.2^a	3.8±0.3^a	2.5±0.2^{cd}
2.5	3.8±0.1^{de}	5.6±0.0^b	2.0±0.2^b	1.8±0.3^{def}
3.0	3.6±0.1^{ef}	4.7±0.4^b	1.6±0.2^b	1.5±0.3^{ef}
3.5	3.5±0^f	4.7±0.3^b	1.8±0.2^b	1.1±0.2^{fg}
4.0	3.4±0^f	4.6±0.4^b	1.6±0.1^b	0.5±0.2^g

Within a column means followed by different letter differ significantly (*P*<0.05)

Table 2 Rumen degradability characteristics of maize silage

Treatment	DM (g/kg)	OM (g/kg)
Control	462±19^c	482±02^b
1.0	464±23^c	484±15^b
1.5	468±06^c	496±09^b
2.0	469±18^c	500±03^b
2.5	476±09^{bc}	504±12^b
3.0	479±04^{bc}	503±05^b
3.5	510±03^{ab}	530±03^a
4.0	537±33^a	540±20^a

Within a column means followed by different letter differ significantly (*P*<0.05)

Conclusions Aerobic stability is an important factor and care should be taken to minimise losses associated with it. The data presented indicate that FAB can improve aerobic stability and nutritive value of low dry matter maize silage.

References
Ashbell, G., Z.G. Weinberg, A. Azrieli, Y. Hen & B. Horev (1991). A simple system to study the aerobic deterioration of silages. *Canadian Agricultural Engineering* 34, 171-175.
Ashbell, G., Z.G. Weinberg, Y. Hen & I. Filya (2002). The effects of temperature on the aerobic stability of wheat and corn silages. *Journal of Industrial Microbiology and Biotechnology,* 28, 261-263.
Mehrez, A.Z. & E.R. Ørskov (1977). A study of the artificial fibre bag technique for determining the digestibility of feeds in the rumen. *Journal of Agricultural Science,* 88, 645-650.

Microbial changes and aerobic stability in high moisture maize silages inoculated with *Lactobacillus buchneri*

R.A. Reis, E.O. Almeida, G.R. Siqueira, T.F. Bernardes, E.R. Janusckiewicz and M.T.P. Roth
Universidade Estadual Paulista, FCAV, Jaboticabal, SP, Brasil, 14.884-900, Email: rareis@fcav.unesp.br

Keywords: inoculant, yeast, losses

Introduction Oxygen exposure changes microbial and chemical profiles of silages after silo opening. Yeast and fungi are the main active microorganisms responsible by consumption of nutrients and fermentative residual products, which increases temperature in the ensilaged mass. The *Lactobacillus buchneri (L. buchneri)* has been considered an aerobic stability controller. This work aimed to evaluate the effect of *L. buchneri* in ensilage process of high moisture maize.

Material and methods High moisture maize was harvested with 70% of DM. Five doses of *L. buchneri* (NCIMB 40788) (in cfu per g of fresh maize) were used: control, LB1 (5×10^4), LB2 (1×10^5), LB3 (2×10^5) and LB4 (4×10^5). Ensilage period was 68 days. At silo opening was determined: % DM, DM recovering (DMR), pH and ammonia level in relation to total nitrogen (N). After silo opening, aerobic stability was determined (h) according to Taylor & Kung (2002) and after 5 and 10 days of aerobic exposure yeast and fungi were counted (Taylor & Kung, 2002). Four silos per treatment (plastic bucket with 20.0 cm height, 21.6 cm diameter, 5.0 kg of silage and 1,000 kg/m^3 density) were used. Means were compared by Tukey test ($P<0.05$).

Results No DMR and pH changes and small alterations in % DM and ammonia:N ratio were observed during fermentation with *L. buchneri* (Table 1). During aerobic exposure, from LB2 dose no significant increments were observed in relation to temperature increase (Figure 1). This effect was proved by the control of yeast and fungi. In relation to yeast counting, at day 5 of aerobic exposure (120 h) the control and LB1 silages showed 6.6 and 6.2 log cfu/g of silage, respectively. However, other treatments showed a lower number (4.4 log cfu/g silage). Only at day 10 (240 h) silages LB2, LB3 and LB4 had yeast counting around 6 log cfu/g silage. Fungi were identified in LB1 silages from day 5 and in all silages from day 10. Taylor & Kung (2002) related the yeast and fungi control to acetic acid production by *L. buchneri*.

Table 1 Parameters evaluated in silages of high moisture maize inoculated with *L. buchneri*

Item	Control	LB1	LB2	LB3	LB4	SEM
DM (%)	66.5^a	65.3^{bc}	65.6^{abc}	64.9^c	65.9^{ab}	0.22
DMR (%)	98.3^a	98.4^a	98.5^a	98.3^a	98.7^a	0.59
Ammonia/N	9.3^{ab}	7.4^b	8.5^{ab}	8.3^{ab}	9.7^a	0.45
pH	4.1^a	4.1^a	4.2^a	4.2^a	4.2^a	0.03
Yeast (0)[1]	2.7^a	3.5^a	4.0^a	3.4^a	3.9^a	0.33
Yeast (5)	6.6^a	6.2^{ab}	4.2^{ab}	4.4^{ab}	1.9^b	0.99
Yeast (10)	6.6^a	6.8^a	6.8^a	6.3^a	6.1^a	0.47
Moulds (0)	ND	ND	ND	ND	ND	-
Moulds (5)	ND	3.5^a	ND	ND	ND	-
Moulds (10)	5.2^a	5.2^a	5.4^a	3.2^a	3.9^a	0.93

[a,b,c] Means with different letters in rows are different *(P<0.05)*.
[1]log cfu/g silage at silo opening and after 5 and 10 days of aerobic exposure.

Figure 1 Effect of *L. buchneri* on aerobic stability (h)

Conclusions The *L. buchneri* (from dose 1×10^5 cfu/g fresh maize) was efficient for controlling yeast and fungi growth and, consequently, temperature during aerobic exposure.

References
Taylor, C.C. & L. Kung Jr. (2002). The effect of *Lactobacillus buchneri* 40788 on the fermentation and aerobic stability of high moisture corn in laboratory silos. *Journal of Dairy Science*, 84,1149–1155.

Effect of residual sugar in high sugar grass silages on aerobic stability

G. Pahlow[1], R.J. Merry[2], P. O'Kiely[3], T. Pauly[4] and J.M. Greef[1]
[1]Institute of Crop and Grassland Science, Federal Agricultural Research Centre (FAL) Bundesallee 50, D-38116 Braunschweig, Germany, Email: Guenter.Pahlow@fal.de, [2]Institute of Grassland and Environmental Research (IGER), Plas Gogerddan, Aberystwyth, SY23 3EB Ceredigion, Wales, UK, [3]Teagasc, Grange Research Centre, Dunsany, Co. Meath, Ireland, [4]Swedish University of Agricultural Science (SLU), Uppsala, Sweden

Keywords: *Lolium perenne*, silage, water soluble carbohydrates (WSC), aerobic stability

Introduction New varieties of *Lolium perenne*, bred for high sugar content, can contain up to 30% of water soluble carbohydrates (WSC). Only a fraction of such high contents are metabolised during a normal fermentation and the high residual sugar content (RSC) of these silages can improve the efficiency of use of nitrogen by ruminants. However, these RSC at opening for feed-out could be preferentially metabolised relative to fermentation products by all aerobically growing fungi and bacteria present on the forage. A high RSC thus can increase the risk of aerobic deterioration over that of extensively fermented silages, containing predominantly organic acids, which are initially utilised by certain yeasts. The objective of this study was to assess the relationship between RSC and aerobic stability of silages prepared with either optimal ensiling conditions or with a defined air challenge treatment to make them prone to aerobic deterioration. The latter is a useful method to test the efficacy of aerobic stability improving silage additives, requiring unstable controls (Pahlow *et al.*, 1999).

Materials and methods *Lolium perenne* from experimental plots was mown, wilted, chopped and ensiled in laboratory scale silos. Ten silages (3 reps.) were stored at 25°C for 90 days in gas tight containers (Group A) or for only 49 days with a defined air challenge treatment for 24 hours after 4 and 6 weeks of storage respectively (Group B). The silages were analysed for RSC and for yeasts on lactate agar. The aerobic stability was measured over 7 days by recording the days to persistent temperature rise by 3°C above 20°C ambient (Honig, 1990).

Results In Figures 1 and 2 mean RSC and corresponding aerobic stability are given for 10 silages respectively in Groups A and B, both arranged according to increasing RSC. For both groups stability varied with no noticeable relationship to RSC. The coefficients of correlation were r = -0.07 and -0.03 respectively. In Group A completely stable silages were observed with both low and high RSC. In contrast, Group B silages were 100 times higher in yeasts (>10^5 vs. 10^3 colony forming units per gram fresh matter) and had considerably shorter stability irrespective of their RSC, ranging from 5 up to 110 g WSC/kg.

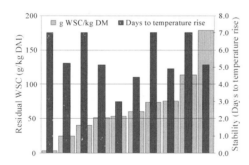

Figure 1 Stability and sugar content of 10 silages after 90 days under optimal gastight storage conditions

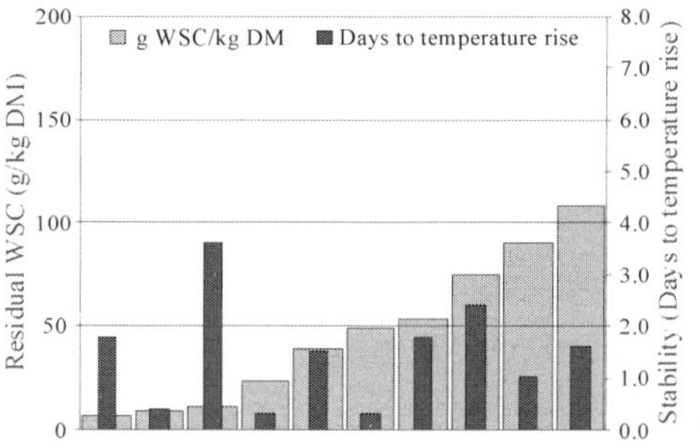

Figure 2 Stability and sugar content of 10 silages after 49 days storage with an air challenge treatment

Conclusion Optimal, gastight ensiling conditions allow the production of aerobically stable silages irrespective of their residual sugar content. Ingress of air during storage leads to aerobic instability caused by yeasts. In both cases (Groups A and B) there was no direct relationship between RSC and aerobic stability after silo opening.

References

Honig, H. (1990). Evaluation of aerobic stability. In: Lindgren, S. and Lundén Petterson. K. (eds) Proceedings of the EUROBAC Conference, Uppsala, 1986 Grovfoder Grass and Forage Reports, 3, 76-82.

Pahlow, G., B. Ruser & H. Honig (1999). Inducing aerobic instability in laboratory scale silages. In: Th. Pauly et al. (eds), Proceedings of the 12th International Silage Conference, Uppsala, 253-254.

This work was part of a Framework V project (QLK5-CT-2001-04980) funded by the European Union

An *in vitro* study on the influence of residual sugars on aerobic changes in grass silages

S.D. Martens, G. Pahlow and J.M. Greef
Institute of Crop and Grassland Science, Federal Agricultural Research Centre (FAL), D-38116 Braunschweig, Germany, Email: siriwan.martens@fal.de

Keywords: aerobic deterioration, pH, yeasts, residual water soluble carbohydrates (WSC), *in-vitro* method

Introduction How do residual sugars in high dry matter grass silages influence microbial metabolism? To answer this question a simple laboratory method was developed using pH as main indicator for aerobic changes.

Materials and methods Three grass silages made from *Lolium perenne* (35-40% dry matter (DM)) prone to aerobic deterioration were investigated. The silages were extracted with sterile water in the ratio 1:4 (g:ml) for 5 min in a Stomacher. The extracts served as a complex medium in an agitated batch culture system. Fungi and bacteria were provided by the indigenous microflora in the silage extracts. Fructose was added (0, 3 or 6% in fresh matter (FM)) to investigate the effect of different levels of residual sugar. Three replicate samples consisting of 40 ml aliquots of silage extract in each Erlenmeyer flask (100 ml) covered with aluminium foil were agitated on an orbital shaker at 175 rpm at 25°C for 2 d. pH was measured at the beginning (0 h) and after 21, 34, 45 and 51 h. Volatile fatty acids were analysed by HPLC after 0, 21 and 45 hours. Numbers of yeasts and lactic acid bacteria (LAB) were counted on modified malt extract and Rogosa agar.

Results The 3 silages used for the extracts contained: yeasts 6-7 log cfu/g FM, LAB 5-6 log cfu/g FM, lactate 8.8-11.7 g/kg FM, WSC 47-69 g/kg FM (fructose, glucose, sucrose and not including fructans) and had a pH in the range 4.6-4.8. During the first 21 h there was a pH decline with all sugar levels. Lactic acid was produced within that time in 6 out of 7 cases. Acetic acid contents rose in all media. Further changes in pH were directly affected by WSC content. The lower the WSC content the faster and steeper was the pH rise. After 45 h the five WSC groups differed significantly from each other by at least 0.09 pH units (Tukey test) (descending): 47, 59-69, 77, 89-99, 119 (g/kg FM initial WSC content). At the same time acetic acid contents rose further in general contrast to values for lactic acid content.

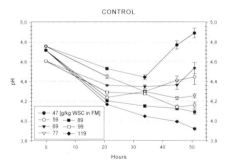

Figure 1 Changes of pH in silage extracts with different WSC contents (error bars = s.d.)

Table 1 Lactic and acetic acid contents (mg/ml) of media with different WSC levels after 0, 21 and 45 h

WSC	Lactic acid			Acetic acid		
(g/kg FM)	0 h	21 h	45 h	0 h	21 h	45 h
47	2.9	3.1	1.6	0.9	1.3	3.3
59	2.2	3.0	1.9	0.8	1.3	2.1
69	2.4	2.6	2.4	0.9	1.0	1.6
77	2.9	2.7	2.6	0.9	1.3	5.5
89	2.2	2.9	2.4	0.8	1.5	3.3
99	2.4	2.5	2.6	0.9	1.2	2.4
119	2.2	2.8	2.5	0.8	2.2	3.9

The regression analysis for the whole measurement period showed a significant correlation between pH and WSC x time (t): $pH_t = 5.19-0.024*t [h]+0.0003*t^2-0.06*WSC [\% \text{ in FM}]$, ($r^2 = 0.88$).

Conclusions In this defined *in vitro* system it was shown that the higher the residual WSC content of grass silage the higher was the tendency to increase the H^+ ion concentration in the initial phase of exposure to air. This was caused partly by lactate production from LAB metabolism within the first 21 h and partly by acetate production from LAB and yeasts under aerobic conditions during the 2 d of measurement (Condon, 1987; Flikweert, 1999). The method described can easily be applied and adapted to study microbial processes and their manipulation in many types of silages that have undergone a natural fermentation.

References
Flikweert, M.T. (1999). *Physiological roles of pyruvate decarboxylase in Saccharomyces cerevisiae*. PhD thesis, Delft University of Technology, Delft.
Condon, S. (1987). Responses of lactic acid bacteria to oxygen. *FEMS Microbiology Reviews*, 46, 269-280.

The effects of the growth stage and inoculant on fermentation and aerobic stability of whole-plant grain sorghum silage

E. Tabacco and G. Borreani
Dip. Agronomia, Selvicoltura e Gestione del Territorio – University of Turin, Via L. da Vinci, 44 10095 Grugliasco (TO) Italy, Email: ernesto.tabacco@unito.it

Keywords: grain sorghum, stage of development, lactic acid bacteria, nutritional quality, aerobic stability

Introduction Grain sorghum (*Sorghum bicolor*) is well adapted to environments with limited rainfall and low soil fertility. Today, on dry land, improved grain sorghum hybrids may be a valid alternative to maize silage and they may give DM yields and digestible energy that are comparable to maize, but at lower production costs (Legarto, 2000). Harvesting crops for silage at an early stage of maturity (low DM content) may result in silage with a higher acid content and low nutritional quality, while harvesting crops at a later stage of maturity may make the forage more difficult to chop and pack. Furthermore, drier silage could be more aerobically unstable during the feed-out phase. The aim of this work was to determine the optimum stage of development for silage purposes and to evaluate the effect of maturity and lactic acid bacteria (LAB) inoculant on the fermentation, nutritional quality and aerobic stability of whole-plant silage produced by grain sorghum grown without irrigation in the Po valley, NW Italy.

Materials and methods White grain sorghum (cv. Kalblanc), sown on 25 May 2002 and grown without irrigation, was harvested at 4 progressive morphological stages (early soft dough, I; late soft dough, II; early hard dough, III; and late hard dough, IV) about every 12 d from 9 August to 24 September. Sorghum was chopped and directly ensiled in 30-l laboratory silos with an inoculant (I) (*Lactobacillus plantarum)* and without an inoculant (C). The silages were analysed for the fermentation and nutritional quality after 150 d of conservation. After opening, change in temperature was measured to evaluate the aerobic stability.

Results The mean fermentation and quality characteristics of whole-plant grain sorghum silages are reported in Table 1. All silages were well fermented with no butyric acid and low contents of ammonia. Fermentation was restricted as the DM content at ensiling increased. The silages prepared from forage harvested at the early soft stage had a significantly higher value of lactic acid than the silages made at the three subsequent stages. No effect of LAB inoculant was observed on fermentation. The starch content increased, while the NDF and ADF contents decreased with advancing maturity as a result of the increase in grain content. The aerobic stability was in access of 86 h in the first 3 stages of maturity, while it dropped to less than 55 h in the IV stage.

Table 1 Composition of control (C) and inoculated (I) grain sorghum silages at four stages of growth

Stage of growth	I		II		III		IV				
Inoculation	C	I	C	I	C	I	C	I	S[1]	I	S x I
DM (g/kg)	208	221	265	255	264	294	295	299	***	NS	NS
PH	3.7	3.8	4.0	4.1	4.0	4.1	4.0	4.0	***	**	*
Lactic acid (g/kg DM)	87	86	59	57	55	48	49	51	***	NS	NS
Acetic acid (g/kg DM)	14	9	14	16	15	14	10	9	***	NS	**
Butyric acid (g/kg DM)	0	0	0	0	0	0	0	0	-	-	-
NH₃-N (g/kg total N)	95	75	102	105	95	95	84	66	***	**	*
CP (g/kg DM)	94	97	91	81	72	78	75	79	***	NS	***
Starch (g/kg DM)	57	25	266	223	257	282	318	333	***	NS	NS
NDF (g/kg DM)	592	604	430	475	463	419	417	376	***	NS	*
ADF (g/kg DM)	368	364	266	303	296	261	261	238	***	NS	*
Aerobic stability (h)	86	110	>166	>166	96	>166	54	55	***	***	***

[1] S = stage of growth, I= inoculation; NS = *P*>0.05; * = *P*<0.05, ** = *P*<0.01; *** = *P*<0.001.

Conclusions The results show that grain sorghum harvested at a hard dough stage makes excellent silage with or without LAB inoculant, and is comparable in nutritional and fermentation quality to maize silage. The highest content in starch and the lowest fibre content were reached at late dough stage, close to physiological maturity, but silage was more prone to aerobic deterioration.

References

Legarto, J. (2000). L'utilisation en ensilage plante entière des sorghos grains et sucriers: intérêts et limites pour les régions sèches. *Fourrages,* 163, 323-338.

Perennial ryegrasses bred for contrasting sugar contents: manipulating fermentation and aerobic stability using wilting and additives (1) (EU FP V -Project 'SweetGrass')

P. O'Kiely[1], H. Howard[1,2], G. Pahlow[3], R. Merry[4], T. Pauly[5] and F.P. O'Mara[2]

[1]Teagasc, Grange, Dunsany, Co. Meath, Ireland, Email pokiely@grange.teagasc.ie, [2]Department of Animal Science and Production, University College Dublin, Ireland; [3]Fed. Agricultural Research Centre (FAL), D-38116 Braunschweig, Germany; [4]Institute of Grassland and Environmental Research (IGER), Aberystwyth, SY23 3EB, UK; [5]Swedish University of Agricultural Science (SLU), Uppsala, Sweden

Keywords: water-soluble carbohydrate, silage, wilt, additive

Introduction Higher concentrations of water-soluble carbohydrate (WSC) in silage offer ruminant nutrition and environmental attractions. Both successful field wilting and alternative silage additives provide the opportunity to manipulate silage WSC by modifying fermentation and/or improving aerobic stability. This experiment evaluated the fermentation and aerobic stability of silages made from perennial ryegrass cultivars of high or normal WSC genotype that differed in field wilting or additive use.

Materials and methods Aberdart (Ab; bred for high WSC) and Fennema (Fn; normal WSC) perennial ryegrasses were mown on 19 September 2002. Each was precision-chopped and ensiled in laboratory silos (6 kg) after a 0 or 24 h wilt. The additives applied to grass for three silos per treatment were (1) no additive, (2) Add-SafeR (85% ammonium tetraformate salt; Trouw Nutrition UK Ltd.) at 6 ml/kg, (3) *Lactobacillus buchneri, L. plantarum* and *Enterococcus faecium* (Pioneer Hi-Bred) at 3 ml/kg, (4) Powerstart (*L. plantarum* and *Lactococcus lactis;* Genus plc) at 3 ml/kg, (5) and (6) Kofasil Ultra (80 g hexamine, 120 g sodium nitrite, 150 g sodium benzoate, 50 g sodium propionate and 600 g water/kg; Addcon Agrar GmbH) at 2.5 or 5 ml/kg, (7) treatments 4 + 5, and (8) treatments 4 + 6. Silos were filled, sealed and stored (15^0C) for >100 d. Silage composition (n=3/treatment) and aerobic stability (n=2/treatment) measurements were made and the results subjected to 3-way analysis of variance.

Results Mean (s.d.) grass WSC and buffering capacity for unwilted and wilted Ab were 172 (6.1) and 178 (11.0) g/kg dry matter (DM) and 374 (22.8) and 364 (23.1) mEq/kg DM, respectively, with corresponding values for Fn of 158 (11.8) and 186 (5.7) g/kg DM and 379 (7.2) and 386 (7.6) mEq/kg DM. Unwilted and wilted silage DM values were 152 and 199 (s.e. 0.5; $P<0.001$) g/kg, respectively, and cultivar had no significant ($P>0.05$) effect. Wilting increased lactic acid/fermentation products (Table 1). Fn had a more lactic acid fermentation than Ab. Formic acid promoted the dominance of lactic acid in the unwilted silages and restricted fermentation in the wilted silages (reduced fermentation products; $P<0.001$). Except for unwilted Ab, the *L.buchneri* additive reduced ($P<0.001$) lactic acid and increased ($P<0.05$) acetic acid and ethanol. Powerstart increased lactic acid/fermentation products. Kofasil Ultra promoted a more lactic acid fermentation with unwilted Ab but had minor effect with wilted herbage. Little additivity occurred when Powerstart and Kofasil Ultra were co-applied. Unwilted silages were very stable when exposed to air. Powerstart increased susceptibility to aerobic deterioration while Add-SafeR, *L.buchneri* and Kofasil Ultra (high) improved stability with wilted silages.

Table 1 Chemical composition and aerobic stability of silages

Additive (A)		1		2		3		4		5		6		7		8			Significance		
Cultivar (C)		Ab	Fn	Ab	Fn	Ab	Fn	Ab	Fn	Ab	Fn	Ab	Fn	Ab	Fn	Ab	Fn	sem	C	A	CxA
pH	U[4]	4.6	4.2	3.7	3.8	4.5	4.5	3.8	4.0	4.4	4.6	4.1	4.2	3.8	4.1	4.0	4.1	0.05	0.08	***	***
	W[5]	4.2	4.0	4.0	4.1	4.5	4.3	3.8	3.9	4.1	4.1	4.2	4.2	3.9	4.0	4.1	4.2				
Lactic acid	U	243	605	755	760	278	75	762	737	382	374	547	537	819	686	854	837	19.0	**	***	***
(g/kg FP[1])	W	594	771	667	700	291	512	825	825	646	730	559	636	832	829	843	783				
NH$_3$N	U	88	80	100	107	84	143	60	74	94	111	88	96	59	85	67	72	2.8	**	***	***
(g/kg N)	W	109	76	105	99	106	78	51	58	93	78	90	83	60	70	67	70				
Butyric acid	U	0	0	0	0	0	0.5	0.5	1.0	0	0	1.9	10.0	0	1.4	0	0	1.07	**	***	0.06
(g/kg DM)	W	10.7	0.5	1.1	0	11.8	3.6	1.0	0	5.5	0	8.1	0	0.4	0	0	1.1				
Duration to	U	192	192	192	192	192	192	56	135	192	186	192	192	43	43	57	57	2.6	***	***	***
temp. rise[2]	W	96	88	192	192	192	192	55	61	192	109	192	192	66	81	192	192				
ATR to d5[3]	U	3	1	2	1	3	2	30	1	2	2	2	2	33	1	17	1	1.3	***	***	***
	W	12	11	2	1	3	3	46	25	1	6	2	1	26	11	2	1				

[1]FP=fermentation products (lactic+VFA+ethanol); [2]hours; [3]accumulated temp. rise to day 5 (^{0}C); [4]unwilt; [5]wilt; sem = CxA

Conclusions Cultivar had minor effects on ensilability indices, but Fn silages were better preserved. The partial wilt generally promoted a more efficient fermentation but poorer aerobic stability. The most consistent improvement in dominance by lactic acid was from Add-SafeR and Powerstart, but Powerstart silages were prone to aerobic deterioration. Add-SafeR, *L.buchneri* and Kofasil Ultra (high) improved aerobic stability with wilted silages.

Perennial ryegrasses bred for contrasting sugar contents: manipulating fermentation and aerobic stability of unwilted silage using additives (2) (EU-Project 'SweetGrass')

H. Howard[1,2], P. O'Kiely[1], G. Pahlow[3] and F.P. O'Mara[2]

[1]Teagasc, Grange Research Centre, Dunsany, Co. Meath, Ireland; Email: pokiely@grange.teagasc.ie, [2]Dept. of Animal Science and Production, University College Dublin, Belfield, Dublin 4, Ireland; [3]Institute of Crop and Grassland Science, Federal Agric. Research Centre (FAL) Bundesallee 50, D-38116 Braunschweig, Germany

Keywords: water-soluble carbohydrate, silage, additive

Introduction Grass cultivars bred for elevated concentrations of water-soluble carbohydrate (WSC) could have improved silage preservation but possibly disimproved aerobic stability. Additives can be used to manipulate fermentation and thereby increase silage WSC. They can also influence aerobic stability. This experiment evaluated the fermentation and aerobic stability of unwilted silages made from perennial ryegrass cultivars of high or normal WSC genotype that differed in additive use.

Materials and methods Aberdart (Ab; bred for high WSC) and Fennema (Fn; normal WSC) perennial ryegrasses were mown on 17 June, 2003. Each was precision-chopped and ensiled in laboratory silos (6 kg/silo) without wilting. The additives applied to grass, using three silos per treatment, were (1) no additive, (2) and (3) Add-SafeR (85% ammonium tetraformate salt; Trouw Nutrition UK Ltd.) at 3 and 6 ml/kg, (4) Biomax SI (*Lactobacillus plantarum*; Chr. Hansen UK Ltd.) at 5 ml/kg, (5) Biomax SI at 5 ml/kg + potassium sorbate (KSor; 30 g/l) at 5 ml/kg, (6) Biomax SI at 5 ml/kg + sodium benzoate (NaBe; 30 g/l) at 5 ml/kg, and (7), (8) and (9) Bio-Sil (*Lactobacillus plantarum;* Dr. Pieper Technologie- und Produktentwicklung GmbH) at 5 ml/kg alone or with KSor or NaBe at 5 ml/kg, (10) KSor at 5 ml/kg, and (11) NaBe at 5 ml/kg. Silos were filled, sealed and stored (15[0]C) for >100 days. Silage composition and aerobic stability measurements were made on every silage and the results subjected to 2-way analysis of variance.

Results Mean (s.d.) grass dry matter (DM), WSC and buffering capacity for unwilted Ab were 143 (12.6) g/kg, 180 (4.8) g/kg DM and 226 (19.7) mEq/kg DM, respectively, with corresponding values for Fn of 141 (12.8) g/kg, 154 (11.6) g/kg DM and 242 (24.4) mEq/kg DM. Lactic acid bacteria on Ab and Fn at harvesting were 6.1 and 6.2 $\log_{10}$ colony forming units/g, respectively. All silages were well preserved. Ab silages had lower NH_3-N (68 vs. 77 g/kg N) and lactic acid/fermentation products (616 vs. 702 g/kg) values and a higher accumulated temperature rise to day 5 (ATR; 27 vs. 23[0]C) than Fn silages (Table 1). Incremental additions of Add-SafeR restricted (*P<0.05*) fermentation, improved (*P<0.05*) aerobic stability (i.e. duration to temp. rise) and reduced (*P<0.05*) aerobic deterioration (i.e. ATR). Neither of the bacterial inoculants and neither of the salts (KSor or NaBe) altered (*P>0.05*) fermentation, aerobic stability or aerobic deterioration indices.

Table 1 Chemical composition and aerobic stability of unwilted silages

	pH		Lactic acid g/kg FP[1]		NH_3-N g/kgN		Hours to temp. rise		ATR to day 5[2]	
Additive (A)	Ab	Fn	Ab	Fn	Ab	Fn	Ab	Fn	Ab	Fn
No additive	3.73	3.83	636	732	59	66	28	25	34	27
Add-SafeR low	3.80	3.87	633	653	104	105	69	79	24	19
Add-SafeR high	4.20	3.97	379	527	146	140	94	128	14	9
Biomax SI	3.70	3.87	683	686	58	72	51	25	25	27
Biomax SI + KSor	3.97	3.93	491	664	52	69	34	22	29	26
Biomax SI + NaBe	3.80	3.90	594	690	56	64	27	34	31	22
Bio-Sil	3.73	3.80	680	739	53	76	22	38	29	24
Bio-Sil + KSor	3.67	3.73	720	764	62	62	32	32	31	24
Bio-Sil + NaBe	3.67	3.80	702	759	53	61	25	29	30	24
KSor	3.70	3.83	662	740	53	67	24	25	26	28
NaBe	3.77	3.73	597	772	51	65	24	21	27	29
s.e.m. (CxA)	0.072		34.0		8.2		14.7		4.9	
Sig. C (cultivar)	ns		***		*		ns		P=0.076	
A	***		***		***		***		P=0.051	
CxA	ns		ns		ns		ns		ns	

[1]FP=fermentation products (lactic+VFA+ethanol); [2]accumulated temp. rise to day 5

Conclusions The higher WSC and lower buffering capacity for Ab compared to Fn indicate that Ab had better ensilability indices. The higher lactic acid/fermentation products for Fn silage reflects its higher concentration of lactic acid and lower concentration of both acetic acid and ethanol. The formic acid-based additive had the largest impact on fermentation and was the only additive to consistently and significantly improve aerobic stability and reduce aerobic deterioration.

Perennial ryegrasses bred for contrasting sugar contents: manipulating fermentation and aerobic stability of wilted silage using additives (3) (EU-Project 'SweetGrass')

H. Howard[1,2], P. O'Kiely[1], G. Pahlow[3] and F.P. O'Mara[2]
[1]Teagasc, Grange Research Centre, Dunsany, Co. Meath, Ireland; Email: pokiely@grange.teagasc.ie, [2]Dept. of Animal Science and Production, University College Dublin, Belfield, Dublin 4, Ireland; [3]Institute of Crop and Grassland Science, Federal Agric. Research Centre (FAL) Bundesallee 50, D-38116 Braunschweig, Germany

Keywords: water-soluble carbohydrate, silage, wilt, additive

Introduction   Rapid field-drying of grass prior to successful ensilage restricts fermentation and can assist preservation, but can consequently result in silages that are prone to aerobic deterioration at feedout. Additives that directly (e.g. potassium sorbate or sodium benzoate) or indirectly (e.g. formic acid or *Lactobacillus plantarum*, via manipulation of fermentation) alter yeast activity at feedout could modify silage aerobic stability. This experiment evaluated the fermentation and aerobic stability of wilted silages made from perennial ryegrass cultivars of high or normal water soluble carbohydrate (WSC) genotype that differed in additive use.

Materials and methods   Aberdart (Ab; bred for high WSC) and Fennema (Fn; normal WSC) perennial ryegrasses were mown on 17 June, 2003, and field dried for 24 h. Each was then precision-chopped and ensiled in laboratory silos (5 kg/silo). The additives applied to grass, using three silos per treatment, were (1) no additive, (2) and (3) Add-SafeR (85% ammonium tetraformate salt; Trouw Nutrition UK Ltd.) at 3 and 6 ml/kg, (4) Biomax SI (*Lactobacillus plantarum*; Chr. Hansen UK Ltd.) at 5 ml/kg, (5) Biomax SI at 5 ml/kg + potassium sorbate (KSor; 30g/l) at 5 ml/kg, (6) Biomax SI at 5 ml/kg + sodium benzoate (NaBe; 30 g/l) at 5 ml/kg, and (7), (8) and (9) Bio-Sil (*Lactobacillus plantarum;* Dr. Pieper Technologie- und Produktentwicklung Gmbh) at 5 ml/kg alone or with KSor or NaBe at 5 ml/kg, (10) KSor at 5 ml/kg, and (11) NaBe at 5 ml/kg. Silos were filled, sealed and stored (15^{0}C) for >100 days. Silage composition and aerobic stability measurements were made on every silage and the results subjected to 2-way analysis of variance.

Results

Mean (s.d.) grass dry matter (DM), WSC and buffering capacity for wilted Ab were 372 (26.4) g/kg, 165 (4.8) g/kg DM and 208 (11.0) mEq/kg DM, respectively, with corresponding values for Fn of 383 (27.4) g/kg, 144 (4.8) g/kg DM and 235 (16.0) mEq/kg DM. Lactic acid bacteria on Ab and Fn at harvesting were 7.1 and 7.2 $\log_{10}$ colony forming units/g, respectively. All silages were well preserved and were aerobically very stable. Cultivar had relatively little effect on the variables measured (Table 1), although Ab resulted in lower ($P<0.05$) lactic acid/fermentation products (653 vs. 678 g/kg) and duration to temperature rise (136 vs. 158 h). Add-SafeR increased final pH and NH$_3$-N values. Even though it decreased the content of lactic and acetic acids it increased ethanol content. Treatments containing Bio-Sil increased ethanol and reduced acetic acid and the duration to temperature rise. Neither of the bacterial inoculants and neither of the salts altered pH, total fermentation products or the content of lactic acid.

Table 1 Chemical composition and aerobic stability of wilted silages

Additive (A)	pH Ab	pH Fn	Lactic acid g/kg FP[1] Ab	Lactic acid g/kg FP[1] Fn	NH$_3$-N g/kgN Ab	NH$_3$-N g/kgN Fn	Hours to temp. rise Ab	Hours to temp. rise Fn	ATR to day 5[2] Ab	ATR to day 5[2] Fn
No additive	4.00	4.03	669	702	52	61	139	165	5	2
Add-SafeR low	4.10	4.13	613	623	75	88	192	158	2	4
Add-SafeR high	4.23	4.27	459	466	96	80	192	192	3	2
Biomax SI	4.00	4.00	696	716	59	66	110	145	5	4
Biomax SI + KSor	4.03	4.00	672	733	65	89	107	192	5	2
Biomax SI + NaBe	4.00	4.00	702	695	59	58	129	176	5	3
Bio-Sil	4.00	4.00	665	690	50	54	82	109	9	8
Bio-Sil + KSor	4.03	4.00	645	729	52	63	103	159	6	1
Bio-Sil + NaBe	4.00	4.00	673	710	51	58	78	133	12	4
KSor	4.00	4.00	719	705	63	60	171	124	2	8
NaBe	4.00	4.03	666	692	59	56	192	187	2	1
s.e.m. (CxA)	0.019		23.0		8.1		19.8		2.1	
Sig. C (cultivar)	ns		*		ns		*		P=0.089	
A	***		***		***		***		*	
CxA	ns		ns		ns		*		ns	

[1]FP=fermentation products (lactic+VFA+ethanol); [2]accumulated temp. rise to day 5

Conclusions   The higher WSC and lower buffering capacity of Ab at harvesting gave it an apparent ensilability advantage over Fn. However, preservation was quite similar for both cultivars. The high rate of the formic acid-containing additive had the largest effect on fermentation and improving aerobic stability. The rates of inclusion of sorbate or benzoate salts did not improve aerobic stability under the test conditions prevailing.

The effect of additive containing formic acid on quality and aerobic stability of silages made of endophyte-infected green forage

L. Podkówka, J. Mikołajczak, E. Staszak and P. Dorszewski
Department of Animal Nutrition, University of Technology and Agriculture, ul. Mazowiecka 28, 85-084 Bydgoszcz, Poland, E.mail: podkowa@atr.bydgoszcz.pl

Keywords: silage, ensiling additives, meadow fescue, endophytes, *Neotyphodium*

Introduction *Festuca* species grasses are very often infected with endophytic fungi *Neothypodium,* that produce ergotic alkaloids, peramins etc. (Podkówka *et al.,* 2003). Produced green forage can be dangerous to animals. The basic preservation method of such green forages is ensiling, especially with organic acids addition. Organic acids demonstrate destructive action on fungal organisms, as well as influence the quality and aerobic stability of forage. The objective of this study was to determine if the preservation of endophyte-infected green forage by ensiling with formic acid affects quality and aerobic stability of produced fodder.

Material and methods Meadow fescue var. Pasja, *Neotyphodium unicinatum* infected green forage was ensiled in mini silos in two variants: no additive (control E+) and with Kemisile 2000 commercial additive (55% formic acid, 24% 4-ammonium formate, 5% propionic acid, water) (experimental group E+ Kemisile). After 4 months of experiment mini-silos were open and chemical composition and quality were evaluated. Aerobic stability was tested for one week in an air-conditioned room with constant temperature $20^0C \pm 0.5$ according to the method described by Honig (1990). Hourly temperature measurements were recorded using a Squirrel 2000.

Results Nutritive value, some chemical composition and quality of silages are shown in Table 1. Organic matter was significantly higher for control silage. Crude protein level was higher for experimental silage, differences were significant. Nutritive value estimated according to INRA was similar in both groups – 0.8 UFL, 0.72 UFV, 56.9 g PDIN, 58.2 g PDIE. Both variants of silage did not differ significantly based on the scores of the Flieg-Zimmer silage evaluation method – experimental silage 99.5 points, control silage 86.5 points. Butyric acid was found in control silage and pH was similar for both groups. Figure 1 demonstrates average aerobic stability. Control silage was stable for 123 h (±13 h) while experimental one was stable for 152 h (±13 h), difference was significant ($P\leq0.05$).

Table 1 Chemical composition (g/kg DM), nutritive value (in DM) and quality of silages

		DM	OM	CP	UFL	UFV	PDIN	PDIE	pH	points	quality
	Mean	278.4	943.5*	24.0*	0.81	0.72	56.4	59.2	4.19	86.5	Very good
E+	SD	8.1	1.3	0.4	0.005	0.0	6.6	2.1	0.07		
E+	Mean	286.8	939.7*	27.4	0.80	0.72	57.4	57.3	4.06	99.5	Very good
Kemisile	SD	2.3	2.3	1.6	0.008	0.00	3.5	0.6	0.02		
	Mean	282.6	941.6	25.7	0.80	0.72	56.9	58.2	4.18	93.0	Very good
	SD	7.3	2.7	2.1	0.007	0.005	5.3	1.8	0.06		

* differences significant $P\leq0.05$

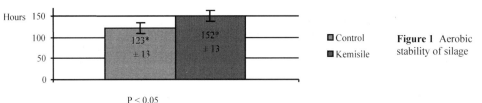

Hours

P < 0.05

■ Control
■ Kemisile

Figure 1 Aerobic stability of silage

Conclusion Quality and nutritive value of silages made with additive was similar to silage with no additives. Chemical additive containing formic acid (Kemisile 2000) improved the aerobic stability of experimental silage comparing to the control one ($P\leq0.05$).

References

Honig H. (1990). Evaluation of aerobic stability, w: S. Lindger and K. Lunden Patterson (ed.) Proceedings of the EUROBAC Conference Grovfoder, Grass and Forage Reports, Uppsala, Sweden 1986, Special issue 3, 76-82.
Podkówka, L., E. Staszak & J. Mikołajczak (2003). Effect of different silage additives on quality and aerobic stability of silages made of Neotyphodium endophyte-infected meadow fescue green forage. Prace Komisji Nauk Rolniczych i Biologicznych, BTN, Seria B, 51, 211-21.

The effect of acetic acid on the aerobic stability of silages and on intake

B. Ruser and J. Kleinmans
Pioneer Hi-Bred Northern Europe GmbH, Apensener Str. 198, D-21614 Buxtehude, Germany,
Email: Barbara.Ruser@Pioneer.com

Keywords: acetic acid, aerobic stability, *L. buchneri*, intake

Introduction Yeasts are the relevant cause for heating in silages. Moulds, bacillus as well as acetobacter are only involved occasionally. Yeasts can be inhibited by short-chain fatty acids like acetic acid. In doing so, undissociated molecules penetrate through passive diffusion into the cells and release an H^- ion, whereby the intracellular pH-value will be quickly lowered and the cells will die. The pH-dependent, minimal inhibition concentration of acetic acid for yeasts and moulds is approximately 94 mmol/l at pH 4.0, this is up to 0.58% in FM (Woolford, 1975). Ensiling trial confirms the inhibiting effect of acetic acid on yeasts and explains the positive correlation between acetic acid concentration and aerobic stability.

Materials and methods Over 4 years different crops were ensiled in 2.8 l mini silos with 4 to 5 replicates per treatment. A total of 80 trials were carried out. In 27 trials the forage was treated with a silage inoculant-based on *Lactobacillus buchneri* (SILA-BAC® Stabiliser/Pioneer brand 11A44) - the *L. buchneri* strain of the silage inoculant converts sugar to lactic acid in the first stage of fermentation. In a second stage a part of the lactic acid is converted to acetic acid and 1.2-propandiol. In another 53 trials a second treatment with 5 l propionic acid/t forage was added. Silos were opened after 7 weeks: aerobic stability and aerobic losses were determined in all 80 trials; fermentation losses, fermentation acids and yeast counts were determined in a subset of 27 trials. Results were statistically analysed using the software JMP 5.1 developed by SAS.

Results and discussion The improvement of aerobic stability through treatment with *L. buchneri* inoculant was demonstrated in the ensiling trials with different types of forages. The significant positive effect of *L. buchneri* in all crops tested can be traced back to the production of acetic acid and 1.2-propandiol. On average, there was a significant increase of the acetic acid content by 0.7 percent units. No 1.2-propandiol was found in the untreated control silage, but in the treated silage 0.5% 1.2-propandiol in fresh matter was detected (Table 1). Occasionally it is discussed as to whether treatment with *L. buchneri* has negative effects on intake because of the increased acetic acid concentration. In several feeding studies with high producing dairy cows the effect of *L. buchneri* treatment was evaluated (Kung *et al.*, 2003; Taylor *et al.*, 2002; Driehuis *et al.*, 1999). Acetic acid concentration in treated silage was increased up to 0.8% FM over untreated silage and in all silages aerobic stability was improved. No effect on feed intake and milk yield was observed.

Table 1 Concentration of acetic acid and 1.2 propandiol in untreated grass and maize silages (average of 27 trials)

	L. buchneri[1]	Untreated
Acetic acid (% FM)	1.1[a]	0.4[b]
1.2 Propandiol (% FM)	0.5[a]	0.0[b]

[1] SILA-BAC® Stabiliser/Pioneer® brand 11A44
[a,b] Within one crop results with different superscripts differ significantly (*P*<0.001)

Conclusions By a significant increase of the acetic acid level and production of 1.2 propandiol the *L. buchneri* treatment (SILA-BAC® Stabiliser/Pioneer® brand 11A44) increased significantly the aerobic stability of grass silage, whole plant maize silage, whole plant cereal silage, maize cob mix and high moisture maize. In all crops, except for maize cob mix and high moisture maize, this effect was even better than with 5 l/t propionic acid.

References

Driehuis, F., S.J.W.H Oude Elferink & P.G. Van Wikselaar (1999). *Lactobacillus buchneri* improves the aerobic stability of laboratory and farm-scale whole crop maize silage but does not affect feed intake and milk production of dairy cows. *Proceedings of the XIIth International Silage Conference*, pp. 264-265.

Kung, L. Jr., C.C. Taylor, M.P. Lynch & J.M. Neylon (2003). The effect of treating alfalfa with *Lactobacillus buchneri* 40788 on silage fermentation, aerobic stability, and nutritive value for lactating dairy cows. *Journal of Dairy Science*, 86, 336–343.

Taylor, C.C.N., J. Ranjit, J.A. Mills, J.M.Neylon & L.Kung Jr (2002). The effect of treating whole-plant barley with *Lactobacillus buchneri* 40788 on silage fermentation, aerobic stability, and nutritive value for dairy cows. *Journal of Dairy Science*, 85, 1793–1800.

Woolford, M.K. (1975). Microbiological Screening of the Straight Chain Fatty Acids (C1-C12) as Potential Silage Additives. *Journal of the Science of Food and Agriculture*, 26, 219-228.

Effectiveness of *Lactobacillus buchneri* to improve aerobic stability and reducing mycotoxin levels in maize silages under field conditions

A. Bach[1,2], C. Iglesias[2], C. Adelantado[3] and M.A. Calvo[3]
[1]ICREA, 08010 Barcelona, Spain, email: alex.bach@irta.es, [2]Unitat de Remugants, IRTA, 08190 Barcelona, Spain, [3]Departament de Sanitat i Anatomia Animal, UAB, 08190, Barcelona, Spain

Keywords: maize silage, mycotoxin, aerobic stability, fungi, yeast

Introduction The aerobic stability of silages may be improved by inhibiting the growth of fungi, moulds and some bacteria after opening the silo. Acetic acid is a potent inhibitor of fungi. Thus, the inoculation of heterolactic bacteria, although energetically less efficient than homolactic bacteria, may result in greater contents of acetic acid. An increased amount of acetic acid in a maize silage may cause greater aerobic stability and reduced mycotoxin levels. *Lactobacillus buchneri* is a heterolactic bacteria able to ferment lactic acid into acetic acid and 1,2-propanediol under anaerobic conditions. However, there are no studies that evaluate the effectiveness of *L. buchneri* on maize silages under field conditions (with greater risk for microbial contamination) and its consequences on mycotoxin content.

Materials and methods A total of 24 different maize silages were used to evaluate the effects of *L. buchneri* inoculation on aerobic stability, fungal growth, and mycotoxin concentrations. During the ensiling process, in each silage, two 7 kg samples of plant material were treated with 200 ml of water (Control) or with 200 ml of water containing 10 g of *L. buchneri* (strain NCIMB 40788) at 4.3×10^5 cfu/g (Treatment) and placed in two permeable nylon bags (320 x 600 mm). These bags were then placed in the bunker silo at the same level and separated about 80 cm from each other in a location that ensured that bags would remain in the silo for at least 3 months. After this period, the materials from each bag were removed and placed in two separate pans and kept at room temperature for 1 wk. Temperature was monitored daily during this time. Also, after retrieving the bags from the silo, a grab sample was analysed for volatile fatty acids, fungal counts, and mycotoxin concentrations. At day four after removal from the silo, another grab sample was taken from the pans to determine fungal counts and mycotoxin concentrations.

Results and Discussion Inoculation of *L. buchneri* tended to result in lower temperatures compared with the Control silages after removing the bags from the silo, especially right after opening the silage (Table 1). Acetic acid concentrations were greater in Treated than in Control silages. Fungal counts (cfu/g) were numerically lower in Treated than in Control silages on day one and day four after opening the silages. Aflatoxin concentrations tended to be lower in Treated than in Control silages, but concentrations of deoxynivalenol and zearalenone did not differ between treatments.

Table 1 Fermentation characteristics of corn silages treated or untreated with *Lactobacillus buchneri*

Item	Control	Treated	SE	*P*-value
Temperature at opening the silage (°C)	19.07	18.10	1-3	0.065
pH	4.00	3.88	0.069	<0.05
Acetate (m*M*)	248.7	279.4	19.12	<0.05
Propionate	14.98	14.5	3.95	0.68
Fungal counts on day one (cfu/g)	538.5	7846.2	5444.6	0.36
Fungal counts on day four (cfu/g)	2130.4	25934.8	15418.7	0.29
Alfatoxin (ppm)	1.09	0.28	0.352	0.08
Deoxynivalenol (ppm)	0.069	0.057	0.026	0.82
Zearalenone (ppm)	465.2	342.4	91.22	0.35

Conclusion Inoculation of *L. buchneri* increases acetic concentrations in maize silages, improves aerobic stability, and may reduce aflatoxin concentrations.

The effect of Lalsil Dry inoculant on the aerobic stability of lucerne silage

J.P. Szucs and Z. Avasi

University of Szeged College of Agriculture, 6800 Hódmezővásárhely, Andrássy u. 15. HUNGARY, Email: szucsne@mfk.u-szeged.hu

Keywords: aerobic stability, lucerne silage, inoculant

Introduction Preventing aerobic deterioration during feedout is an essential goal in every silage programme. Silage deterioration on exposure to air is inevitable and usually results in high losses of DM and important nutritional components through the oxidation of lactic acid and water soluble carbohydrates (WSC). The use of biological silage additives has produced variable results for aerobic stability according to the experiences of some researchers. Recently, heterolactic LAB have been used as silage inoculants with the purpose of increasing aerobic stability. *Lactobacillus buchneri* has the ability to ferment lactic acid to acetic acid and 1,2 propandiol, and might produce other yet unidentified metabolites with antifungal activity (Driehuis *et al.*, 1996).

Material and methods The lucerne originated from third cut in early flowering maturity. The medium wilted lucerne was applied with Lalsil Dry inoculant containing *Pediococcus acidilactici* (CNCM MA 18/5M), *L. buchneri* (NCIMB 40788) and enzyme. The treated and untreated control forages were round-baled and wrapped with plastic film 200 pcs/each. The aerobic stability was determined 4 different times after ensiling using the standard of Honig system (1986). The bales for observation were selected randomly on days 45, 60, 67 and 122 and the samples were exposed to air during days 7, 9, or 14 at 22-24.5°C room temperature. The registration of the temperature of the samples was on the basis of one hour by computer.

Results The considerable deterioration of silage starts when the temperature of the sample is at least 2°C higher than the ambient temperature (first observation). After 45 d of ensiling, the control samples were undamaged until fifth day of exposure to air and then heated and spoiled on day 7. The temperature of Lalsil Dry-treated samples did not change, which means that 45 d fermentation resulted in a stable silage. The undisturbed 60 d fermentation resulted in stable silages made with Lalsil Dry or without any treatment in the second observation. In the third observation, the first phase of determination confirmed the conclusion of the second observation, all samples remained undamaged for 1 wk. The spoilage of control silages started on the second week, and culminated on days 9-11. While the Lalsil Dry-treated samples remained undeteriorated during 14 d exposure to air (fourth observation). After 122 d of ensiling, the stability of Lalsil Dry-treated silages remained, there was no deterioration during 9 d exposure to air. On the contrary, the control silages started to deteriorate rapidly: first sample from second day, second sample from fourth day while the third sample after 1 wk exposure to air (Figures 1 and 2).

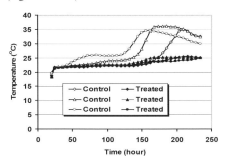

Figure 1 Aerobic stability of lucerne silages   **Figure 2** Ammonia and pH of lucerne silages

Conclusion The aerobic stability of Lalsil Dry-treated silages is strong. There was no deterioration at either 7 or 14 d exposure to air while the untreated samples spoiled after 5-7 d and were totally moulded by day 14.

Reference

Driehuis, F., S.F. Spoilstra, C.C.J. Cole & R. Morgan (1996). Improving aerobic stability by inoculation with Lactobacillus buchneri. The XIth International Silage Conference, Aberystwyth, Wales pp. 106-107.

Honig, H. (1986). Evaluation of aerobic stability. Proceedings of the Eurobac Conference, Uppsala, Sweden pp.76-81.

Acknowledgement: Thanks for the contribution to the experiments of Lallemand SAS.

Section 3

Developments in ensiling techniques

C. Nutritive value

Ensiling characteristics and ruminal degradation of Italian ryegrass with or without wilting and added cell wall degrading enzymes

Y. Zhu[1], H. Jianguo[1], Z. He[1], X. Qingfang[1], B. Chunsheng[1] and N. Nishino[2]
[1]Institute of Grassland Science, China Agricultural University, No 2 Yuan mingyuan Xilu,Haidian District Beijing 100094,China, Email:yuzhu3@sohu.com; [2]Department of Anima and Technology, Faculty of Agriculture, Okayama University, Japan

Keywords: wilting, enzymes, ryegrass, silage, degradation

Introduction The previous experiment (Zhu *et al*., 1999) has shown that the efficacy of added enzymes varied greatly according to the DM content of the material crop. The silage DM did not alter the effects of enzymes on the *in vitro* digestion of NDF (Zhu *et al*., 1999, Zhu *et al*., 2000). The aim of this experiment was to study the effect of wilting and enzymes on fermentation quality, chemical composition and *in situ* digestion of Italian ryegrass (*Lolium multiflorum* Lam.) silage.

Materials and methods Primary growth of Italian ryegrass was harvested at the late heading stage. They were chopped into approximately 25 mm length and ensiled in laboratory silos (1 L) directly or after being wilted for 2 h with or without added cell wall degrading enzymes. The enzymes (1:2 mixture of Acremonium and Trichoderma cellulase based on avicelase activity) were added at 0.1 g/kg just before ensiling. Triplicate silos for each treatment were stored for 45 d at room temperature, then sampling for the analysis of fermentation quality (Zhu *et al*., 1999). Three castrated mature goats about 19 kg body weight were used. Nylon bag incubation was conducted using the silages samples and the disappearance of DM and NDF after 0, 3, 6, 12, 24, 48 and 72 h was determined. The parameters explaining the degradation were estimated by non-linear regression analysis using Syatat (Ver 5.2 for Macintosh) and subjected to two-way analysis of variance with wilting and addition of enzymes as main factors.

Results Wilting increased the contents of DM, CP, NDF, ADF, ADL, and the buffering capacity, decreased the WSC content, and had little effects on fructose, glucose and sucrose contents. This difference suggested a significant amount of fructosan. The addition of enzymes increased (*P*<0.01) the DM, and decreased the NDF and ADF contents, but did not affect the CP and ADL contents of silage. Higher contents(*P*<0.01) of NDF, ADF and ADL were recorded in wilted rather than direct-cut silage. There remained more WSC (*P*<0.01) in enzyme-treated silage, while the difference appeared less when treated with wilted crops. The main fermentation quality and degradation characteristics of Italian ryegrass silage are show in Table 1.

Table 1 Fermentation quality and degradation characteristics of silage

Item	Direct cut -E	+E	Wilted -E	+E	Pooled SE	ANOVA E	W	E×W
pH	4.11	3.98	4.88	4.62	0.08	*	**	NS
Lactic acid (g/kg DM)	59.7	67.3	21.3	27.7	7.61	NS	**	NS
Butyric acid (g/kg DM)	17.9	6.43	35.8	32.2	6.05	NS	**	NS
NH3-N (g/kg N)	90.2	93.5	108	97.3	5.43	NS	NS	NS
Dry matter								
Potential degradation (g/kg)	791	801	750	790	17.7	NS	NS	NS
Rate of degradation	0.047	0.045	0.047	0.034	0.006	NS	NS	NS
Neutral detergent fibre								
Potential degradation (g/kg)	702	676	673	693	36.8	NS	NS	NS
Rate of degradation	0.045	0.043	0.043	0.034	0.006	NS	NS	NS

NS; not significant.*:*P*<0.05,**:*P*<0.01, –E; no enzyme, +E; with enzyme, W; wilting

The *in situ* degradation of DM was reduced (*P*<0.01) by wilting. The addition of enzymes increased (*P*<0.01) the degradation at 3, 6, and 12 h of incubation, while the effect on the DM degradation was diminished with wilted crops. The degradation of NDF was also enhanced by enzymes (*P*<0.05) at the initial incubation time. However, at 6 and 12 h of incubation, the effect of enzymes was found to be opposite in the silage which was wilted prior to the treatment.

Conclusions Added enzymes may have a potential of enhancing the digestibility of silage, although the benefits would be hindered by wilting. It appeared necessary to consider, in addition to the species, the DM content of the forage, when cell wall degrading enzymes were used to improve the silage utilisation.

References

Zhu, Y. & N. Nishino (1999). Ensiling characteristics and ruminal degradation of Italian ryegrass and lucerne silages treated with cell-wall degrading enzymes. *Journal of the Science of Food and Agriculture*, 11, 111-117.
Zhu, Y. & N. Nishino (2000). Fermentation of rhodesgrass and guineagrass silages with or without wilting and added cell wall degrading enzymes. *Grassland Science*, 4, 235-239.

Quality and nutritive value of grass-legume ensiled with inoculant Lactisil 300

J. Jatkauskas and V. Vrotniakiene
Lithuanian Institute of Animal Science, R. Žebenkos 12, LT-5125 Baisogala, Radviliškis distr., Lithuania, Email: lgi_pts@siauliai.omnitel.net

Keywords: big bale, inoculant, fermentation, aerobic stability

Introduction Technical progress in forage harvesting, the development of effective additives and application systems the main reasons for the rapid increasing of silage production (Wilkins, 2000). Because of an increasing interest in environment control in Europe and safety of handling and corrosion of harvesting machinery use of inoculants became more popular over organic acids. The aim of this experiment was to investigate the effect of inoculant *Lactisil 300* on grass-legume mixture silage fermentation, aerobic stability, dry matter losses and energy value.

Materials and methods The first cut 8 to 10 h wilted grass-legume sward (20% *Festuca pratense*, 30% *Trifolium pratense*, 50% *Lolium perenne*) was baled in round bales (1.2 m in diameter and 1.2 m high) (C) and either left untreated or treated with inoculant *Lactisil 300* (Medipharm, Sweden) containing a mixture of *Lactobacillus plantarum* Milab 393, *Pediococcus acidilactici* P6 and P11, *Enterococcus faecium* M74, and *Lactococcus lactis* SR 3.54 (L). The rate of inoculant was 5×10^5 cfu/g fresh matter and the additive was applied by spraying on the swath. Herbage samples were taken directly from swath and silage samples were taken for every ten bales. Five big bale silages from each treatment were weighed after wrapping and again after 82 d storage for measurement of DM loss. Aerobic stability was measured by changes in silage temperature following exposure to air for 10 d. The data were analysed by one-way ANOVA, and a mean comparison by Fisher'PLSD.

Results Fermentation quality of both silages was good, although dry matter content of inoculated silage was lower because of rainfall during the application of the additive (Table1). Silage inoculated with *Lactisil 300* showed a lower final pH, significantly higher ($P<0.05$) concentrations of total acids and lactic acid and numerically lower concentrations of butyric acid and ammonia-N than control silage. A higher fermentation quality of the inoculated silage decreased significantly ($P<0.01$) fermentation (DM) losses and significantly increased ($P<0.01$) energy concentration in silage DM. In this experiment inoculated silage was more aerobically stable than untreated, however there were no significant differences in temperature rise between treatments (Figure 1).

Table1 Chemical composition of herbage and silages and fermentation quality of silages

	Herbage	Control	Lactisil	LSD 0.01
Dry matter (DM) (g/kg)	348	337	328	20.7
Crude protein (g/kg DM)	133	138	140	34.2
WSC (g/kg DM)	115	52	37*	19.8
LA (g/kg DM)		28	46*	20.5
AA (g/kg DM)		12	14	14.1
BA (g/kg DM)		4.7	2.2	8.06
Ammonia N (g/kg total N)		46	35	27.3
pH		4.58	4.17	0.70
ME (MJ/kg DM)		9.91	10.6**	0.36
DM losses (g/kg DM)		105	86**	4.8

* and ** denotes significant at level 0.05 and 0.01 respectively; WSC-water-soluble carbohydrate; LA-lactic acid; AA-acetic acid; BA-butyric acid; ME- metabolisable energy.

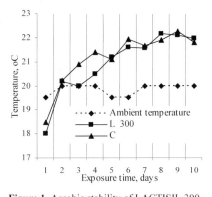

Figure 1 Aerobic stability of LACTISIL 300 treated and untreated silages

Conclusions The addition of inoculant improved silage quality by increasing lactic acid content and by reducing proteolysis of protein and dry matter losses. Inoculated silage was more resistance to aerobic deterioration.

References

Wilkins, R.J. (2000). Grassland in the twentieth century. *IGER Innovations 2000*, Aberystwyth, 26-33.

Effect of additives in grass silage on rumen parameters in Rusitec

A. Potkański, A. Cieślak, K. Raczkowska-Werwińska, M. Szumacher-Strabel and A. Gubała
August Cieszkowski Agricultural University, Department of Animal Nutrition and Feed Management Wołyńska 33, 60-637 Poznań, Poland Email: potkansk@jay.au.poznan.pl

Keywords: grass silage, Rusitec, fermentation, inoculant

Introduction Silage effluent is a major agricultural pollutant, because of its high content of fermentation acids and soluble carbohydrates. Grass silage is important forage in rations of dairy cattle. The quality of silage depends on the quality of the crop at ensiling, type of fermentation, rate on pH decrease, moisture content of the crop and maintenance of anaerobic conditions. The use of a lactic acid bacterial inoculant and chemical additives may improve fermentation of silage as well as animal production. The aim of this study was to determine the effect of biological (VTT; *Lactobacillus plantarum* E-78076) and chemical additives (KemiSile 2000 composed of formic acid, ammonia formate, propionic acid and benzoic acid/ethyl benzoate) on silage quality and rumen fermentation *in vitro* (RUSITEC; RUmen SImulation TEChnic).

Materials and methods In this experiment silages were prepared from grass-clover mixture into three barrels. They were ensiled during 60 days with chemical additive KemiSile 2000 (barrel 1) and biological additive VTT (barrel 2). After this period samples of silage were taken and analysed. The *in vitro* study was carried out using, RUSITEC, system simulating rumen fermentation with simultaneous flow of substrate and outflow of fermentation products (Czerkawski & Breckenridge, 1977). RUSITEC consisted of 4 vessels, each with a volume of 1 l. On days 5-10, every vessel was sampled for analysis of pH, ammonia nitrogen (NH_3-N), volatile fatty acid and micro organism count (Table 1).

Results Ammonia N decreased when both additives were used whereas pH was about 7.0 in all groups. No differences in dry matter (DM) digestibility (DMD) were detected in silage (Table1), VTT caused decrease in bacteria count. The content of DM and pH in silages with KemiSile 2000 and VTT were on the same level. The silages with the additives had higher lactic acid amounts and lower acetic acid content than in control group.

Table 1 Effect of silage additives on basic rumen parameters *in vitro*

Estimate Parameters	Forage	SE	Silage Control	SE	Silage KemiSile 2000	SE	Silage VTT	SE
PH	6.91	0.02	7.02	0.07	7.01	0.05	7.07	0.04
NH_3–N (mmol/l)	3.72	0.22	6.30	0.36	4.21	0.27	5.40	0.38
DMD (%)	79.94	2.80	75.31	2.26	75.14	2.42	76.81	3.54
Bacteria x 10^8	1.60×10^8	0.38	3.36×10^8	0.55	3.92×10^8	0.75	2.44×10^8	0.36

Conclusions Silage additives improved the quality of silages and influenced rumen metabolism e.g. decreased ammonia in both experimental groups and bacteria count when VTT was used.

References

Czerkawski, J.W. & G. Beckenridge (1977). Design and development of long-term rumen simulation technique (RUSITEC). *British Journal of Nutrition* 38, 271-384.

The quality and nutritive value of big bale silage harvested from bog meadows

H. Żurek[1], B. Wróbel[2] and J. Zastawny[2]
[1]Experimental Farm of Institute for Land Reclamation and Grassland Farming in Biebrza, 19-200 Grajewo, Poland; [2]Institute for Land Reclamation and Grassland Farming at Falenty, 05-090 Raszyn, Poland, Email: j.zastawny@imuz.edu.pl

Keywords: bog grasslands, silage, hay, nutritive value, gains of heifers

Introduction In Poland bog meadows are mown exclusively for hay which is used mostly for horse feeding. Because of the high silica content and structural fibre, cattle are reluctant to eat this hay (Denisiuk, 1980). This study was aimed at finding methods for improving the palatability and nutritive values of feeds harvested from bog meadows and ensiled in big bales.

Materials and methods During the years 2002-2003 studies on big bale silage made from sedge herbage composed of different *Carex* species were carried out. Treatments comprised: control sedge silage (I), sedge silage with Polmazym (lactic acid bacteria + enzymes at a rate of 1 l/t of herbage) (II), hay from sedge herbage (III) and grass silage from cultivated meadow sward (IV). The four forages were fed over 40 d in 2002 and 63.5 d in 2003 to 32 heifers (age of 18-23 months and weight over 300 kg) divided into four groups. Animals were fed the forages *ad libitum* together with 1.5 kg of concentrate supplement. The live weights and daily feed intake were recorded. Dry matter, pH, organic acids and nutritive components were analysed in the feeds. To verify the zero hypothesis on body gains of animals the Fisher-Snedecor F test was used.

Results The quality of sedge silages and grass silage according to Flieg–Zimmer scale was good and very good. The content of lactic acid in the sedge silage, both with and without Polmazym, was similar to that in grass silage. The silage prepared from sedges contained considerably more crude protein and less crude fibre than hay prepared from similar herbage. Sedge silages (Treatment I and II) had lower contents of crude protein and crude fat than grass silage. In all feeds from sedge herbage, both hay and silage, the phosphorus content was lower than in the grass silage. Ensiling of sedge herbage improved the nutritive value of feeds and animal performance. Animals receiving sedge silages had significantly higher liveweight gains than animals fed hay and similar to animals fed grass silage (Table 1). Heifers fed silage prepared with Polmazym addition obtained higher body gains than control silage but they were not significantly different. The enzymes in Polmazym caused the decay of cell wall membranes and liberation of cell contents (McDonald *et al.*, 1991).

Table 1 Chemical composition and nutritive value of tested feeds (2002-2003)

Year	2002				2003			
Feeding groups	I	II	III	IV	I	II	III	IV
DM (g/kg)	553.7	523.7	-	661.9	439.6	496.5	-	511.8
pH	5.00	4.90	-	5.60	4.90	4.70	-	5.10
Lactic acid (g/kg FM)	2.70	2.67	-	2.79	2.03	2.62	-	2.93
Acetic acid (g.kg FM)	0.24	0.33	-	0.22	0.73	0.46	-	0.35
Butyric acid (g/kg FM)	0.00	0.00	-	0.00	0.11	0.06	-	0.00
Sum of acids (g/kg FM)	2.94	3.00	-	3.01	2.88	3.14	-	3.28
Evaluation acc. to Flieg - Zimmer scale	very good	very good	-	very good	good	good	-	very good
Crude protein (g/kg DM)	151.0	186.3	134.8	190.7	150.1	146.8	131.4	199.7
Crude fibre (g/kg DM)	229.1	257.3	323.3	286.6	343.0	332.0	399.1	321.9
Crude fat (g/kg DM)	37.5	37.6	35.6	37.9	-	-	-	-
Phosphorus (g/kg DM)	2.4	2.7	2.3	3.4	3.6	4.4	2.5	6.5
Daily gains of animals (g/head)	575ab	695a	384b	717a	690a	748a	367b	813a
Daily intake of silage (kg/head)	7.47	8.30	5.94	7.81	7.66	8.82	6.21	9.58

Conclusions Sedge silage had better quality, higher nutritive value and gave better animal performance than hay made from sedge. The quality and nutritive value of sedge silage was similar to grass silage and was considered suitable for feeding to farm animals. Feeds made from sedge herbage had lower phosphorus than grass silage.

References

Denisiuk Z. (1980) Wartość gospodarcza łąk turzycowych oraz możliwości ich rolniczego wykorzystania. W: Łąki turzycowe Wielkopolski. PWN. Warszawa- Kraków.
McDonald P., A.R. Henderson & S.J.E. Heron (1991) The biochemistry of silage, Chalcombe Publications, Second Edition.

The aerobic stability and nutritive value of grass silage ensilaged with bacterial additives

B. Wróbel and J. Zastawny
Institute for Land Reclamation and Grassland Farming at Falenty, Department of Meadows and Pastures, 05-090 Raszyn, Poland, Email: b.wrobel@imuz.edu.pl

Keywords: aerobic stability, bacterial inoculant, lactic acid bacteria, grass silage, nutritive value

Introduction The use of biological inoculants usually improves silage quality and increases feed intake and animal performance (Andrieu & Demarquilly, 1996). The lack of aerobic stability of silage with biological additives is their main weakness. However, recently obtained bacterial additives, thanks to the suitable selection of LAB, improve aerobic stability of silage (Driehuis *et al.*, 2000). The aim of this study was to investigate the influence of bacterial additives containing homofermentative LAB on the quality and nutritive value of meadow silage.

Materials and methods During the years 1999-2002 studies on big bale silage made from a pre-wilted meadow sward with no additives or the addition of two bacterial inoculants: K1 (*Enterococcus faecium, Lactobacillus plantarum, Lactobacillus casei* and *Pediococcus* spp.) and K2 (*Lactobacillus plantarum* K) were conducted. Every year 30 big bales per treatment were prepared. During the 100 d feed experiment, silage was fed to three groups of 10 heifers (200 kg) *ad libitum*. The live weights and daily feed intake were recorded. Dry matter, pH, organic acids and nutritive components were analysed in the silage. The changes of temperature in silage samples placed in boxes in aerobic conditions were about 21°C.

Results The mean DM content and pH values in silages were similar with tendency of higher pH values in the control silage. The concentration of organic acids in FM depended on the treatment. In all silage samples lactic acid dominated among other acids. The quality of K2 silage was very good while the quality of the control silage was only satisfactory. The mean stability of the tested silages was 6.6 days (Control), 8.6 days (K1) and 9.6 days (K2). The mean concentration of crude fibre in all silage samples was about 270 g/kg and crude protein about 135 g/kg. Heifers daily consumed 5.10 to 5.33 kg of dry matter of silage. The highest average weight gains were obtained with K2 treated silage (0.67) and the lowest gains with heifers fed the K1 treated silage (0.62 kg; Table 1).

Table 1 The quality and nutritive value of grass silage made with the addition of LAB (1999-2002)

	Control	K1	K2	Additive (1)	Year (2)	Interaction (1) x (2)
Dry matter (g/kg)	429.8	388.2	402.6	NS	NS	NS
Ph	5.14	4.80	4.67	**	*	NS
Lactic acid (g/kg FM)	13.2	18.6	22.5	**	**	**
Acetic acid (g/kg FM)	4.0	3.5	3.5	NS	**	**
Butyric acid (g/kg FM)	1.1	0.8	0.3	NS	*	NS
Quality in Flieg-Zimmer scale	satisfactory	good	very good	-	-	-
Stability (days)	6.6	8.6	9.6	NS	NS	NS
Crude protein (g/kg DM)	131.9	136.0	138.4	NS	**	NS
Crude fibre (g/kg DM)	269.8	271.1	275.3	NS	NS	NS
Crude fat (g/kg DM)	34.8	35.7	33.8	NS	*	NS
Intake of DM (kg/d)	5.33	5.10	5.16	NS	NS	NS
Daily gains (kg)	0.64	0.62	0.67	NS	NS	-

NS- not significant, ** $P > 0.01$, * $P > 0.05$

Conclusions Addition of bacterial inoculants improved the quality and aerobic stability of grass silage made in big bales. The most effective appeared to be the inoculant K2 (*Lactobacillus plantarum* K). Tested bacterial inoculants had no influence on nutritive value of tested feeds.

References

Andrieu J-P. & C. Demarquilly (1996). Efficiency of biological preservatives for improving the fermentation quality of grass silage: results of approval tests carried out in France. *Proceedings of the XIth International Silage Conference,* Aberystwyth, Wales, pp. 136-137.

Driehuis F., S.J.W.H. Oude Elferink & P.G. Van Wikselaar (2000). Fermentation characteristics and aerobic stability of grass silage inoculated with *Lactobacillus buchneri* alone and in mixture with *Pediococcus pentosaceus* and *Lactobacillus plantarum. Grassland Science in Europe,* 5, 41-43.

Section 3

Developments in ensiling techniques

D. Big bale silage production

Factors affecting bag silo densities and losses

R.E. Muck[1] and B.J. Holmes[2]
[1]USDA, Agricultural Research Service, US Dairy Forage Research Center, Madison, Wisconsin 53706 USA, Email: remuck@wisc.edu, [2]Biological Systems Engineering Dept., University of Wisconsin-Madison, Madison, Wisconsin 53706 USA

Keywords: bag silo, density, loss, lucerne, maize

Introduction Bag silos (polyethylene tubes, 30 to 90 m length, 2.4 to 3.7 m diameter, 0.22 mm thick) are used on approximately one-third of the dairy farms in the U.S.A. for making silage, and the level of adoption is increasing rapidly. Unfortunately, almost no research data have been published on these types of silos. Our objective was to measure densities and losses in bag silos at three farms, looking for causes of variation in both.

Materials and methods Bag silos made on three research farms over the course of two years were monitored at filling and emptying. These consisted largely of lucerne and whole-crop maize silages. All loads of forage entering the bags were weighed and sampled. Average density was calculated based on bag length and nominal bag diameter. At emptying, the weight of all silage removed from a bag was recorded. Any spoiled silage not fed was weighed, sampled and specifically identified on the emptying log. A grab sample from the face of each silo was taken periodically, one per filling load. Factors influencing density and losses were determined through data analysis using a combination of the CORR, GLM, and STEPWISE procedures in SAS®.

Results Over two years, 47 bag silos were made at the three farms, 23 of lucerne, 23 of whole-crop maize and 1 of red clover. Density ranged from 160 to 280 kg dry matter (DM)/m^3. Density increased as DM content increased (Figure 1). The operator and how the bagging machine was set were important factors affecting density. The same bagging machine (Kelly-Ryan, KR) was used at the Arlington (Arl) and West Madison (WM) farms, and Arl consistently got higher densities. The Prairie du Sac (PDS) farm had higher densities the second year after training from a manufacturer's representative. Density declined with longer particle size (Figure 2). Kernel processing in maize reduced density at PDS where there was a planned comparison.

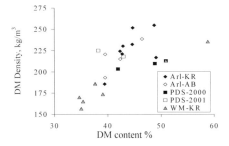

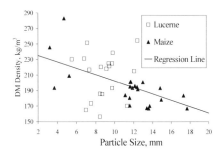

Figure 1 Dry matter density in hay crop bag silos as affected by dry matter content, farm (Arl, PDS, WM), and bagging machine (Ag-Bag, AB;

Figure 2 Dry matter density in bag silos as affected by average particle size at ensiling

Dry matter losses were measured on 39 of the bag silos and ranged from 0 to 40%. Average DM losses were 9.2% invisible plus uncollected losses and 5.4% spoilage losses for a total loss of 14.6%. Six silos had excessive spoilage losses (>15%) due to damaged plastic or overly dry silage (>40% DM) being fed out in warm weather. In contrast, 11 silos had no spoiled silage, and 15 bags had less than 5% spoilage loss, representing bags with spoilage largely at the ends. Invisible losses were reduced in high porosity silages (where spoilage losses were exacerbated), greater in warm weather, and affected by emptying procedures (reduced at WM where bag silos were emptied in 2 to 3 one-day periods as opposed to daily removal for cattle at PDS and Arl). Spoilage losses in bags without damaged plastic were greater in dry, porous silages, from emptying silos in warm weather, and at lower feed out rates. Both invisible and spoilage losses were not affected by crop or bagging machine.

Conclusions These results indicate that low DM losses (<10%) are regularly achievable in bag silos. However, deviations from good management (harvesting between 30 and 40% DM, operating the bagging machine to get a smooth bag of high density, monitoring routinely for and patching holes, and feeding out at a minimum of 300 mm/d) can result in substantial (>25%) losses. Because higher losses occur during warm weather, silage from the best preserved bags should be reserved for summer use.

Transport of wrapped silage bales

Å.T. Randby[1] and T. Fyhri[2]
[1] Norwegian University of Life Sciences, P .O. Box 5003, NO-1432 Ås, Norway, Email: ashild.randby@umb.no,
[2] The Royal Society for Rural Development, Hellerud, P. O. Box 115, NO-2026 Skjetten, Norway

Keywords roundbales, plastic wrapping, transport equipment, silage, mould

Introduction Round bales of grass silage have traditionally been transported unwrapped from the field for subsequent wrapping at the storage place. Today's commonly used combined baler and wrappers necessitate transport of plastic-wrapped bales. Such transport may damage the plastic cover and cause fungal growth.

Materials and methods Grass silage bales were produced in 2003 using a Vicon RF 130 Balepack on three different timothy/meadow fescue swards: an immature wet crop baled at 166 g/kg DM on June 4 (crop 1), a mature wet sward baled at 245 g/kg DM on June 18 (crop 2), and a one-day wilted mature sward baled at 450 g/kg DM on June 18 (crop 3). Bales produced from crops 1, 2 and 3 weighed 917, 775 and 650 kg, and contained 90, 112 and 173 kg DM/m^3, respectively. Wrapped bales were transported from the field to the storage place (approximately 300 m) using tractor with 4 different equipments: (1) Dalen silage bale grip 1591, (2) Kverneland Silagrip 7700 lifting tubes fitted with two rollers, (3) Trailer, loaded and unloaded by the same rollers as used in treatment 2 and (4) Bale fork with 3 tines (the holes in the plastic, caused by the tines were immediately taped with bale-tape). Additionally, 4 bales from each sward were transported unwrapped from the field and were wrapped at the storage place using a Kverneland Silawrap UN 7558. All bales were wrapped with 6 layers of 750 mm x 0.025 mm white plastic film and stored on their round sides in a single layer. Half of the bales were transported from the field immediately after baling while the other half were left on the field for 4-6 days.

Results Transport of bales prior to wrapping produced bales with least mould and wasted silage (Table 1). Transport with rollers, or with trailer loaded by rollers, gave the most gentle handling of wrapped bales. Grip damaged the plastic film more than rollers, but use of bale fork was clearly the worst method. The tape used to repair the tine holes in the plastic film was torn off by rain and wind and did not protect the silage against air over time. This was especially evident with the wet bales that lost their circular shape. Such bales (crops 1 and 2) moulded least when transported to the storage place immediately after wrapping. In contrast, the wilted and more stable shaped bales moulded least when transported from the field 4-6 days after wrapping. If rollers (alone, not with trailer) were used, however, wet bales also moulded least when transported 4-6 days after wrapping.

Table 1 Fungal growth, wasted silage and aerobic stability of bales transported with different equipment

Crop	Transport equipment[1]	No of bales	Silage weight (kg)	Wasted silage (%)	Surface mould (% of area covered)			Aerobic stability (h)
					ends	sides	total	
1	Unwrapped	4	872	0^a	0^a	0^a	0^a	76
	Grip	4	864	0.9^a	0^a	1.5^a	1.0^a	82
	Rollers	4	835	<0.1^a	0^a	<0.1^a	<0.1^a	64
	Trailer[2]	3	879	5.2^b	0^a	11.7^b	7.5^b	68
	Fork[3]	4	840	57.4^c	52.5^b	31.3^c	38.9^c	44
2	Unwrapped	4	761	0^a	0^a	0.3^a	0.2^a	47^a
	Grip	4	726	0.4^a	0.5^b	3.5^b	2.4^b	44^a
	Rollers	4	755	0.3^a	0^a	1.0^a	0.6^a	45^a
	Trailer	4	725	0.1^a	0.1^a	0.3^a	0.2^a	36^b
	Fork	4	741	7.9^b	31.0^c	5.3^c	14.5^c	31^b
3	Unwrapped	4	646^b	0.1	0.3^a	0.3	0.3	98
	Grip	4	604^a	1.6	0.4^a	2.4	1.7	82
	Rollers	4	659^b	0.9	0.5^a	2.3	1.6	71
	Trailer	4	645^b	0.8	0.4^a	1.8	1.3	104
	Fork	4	645^b	1.1	4.3^b	1.5	2.5	76

[1] Due to heavily moulding of bales transported by fork, this treatment was omitted in the statistical evaluation of differences in moulding among the other treatments. [2] Two of the three bales were loaded with grip instead of rollers. [3] Two of the four bales had to be wasted in full, and were therefore not assessed for aerobic stability.

Conclusions Transport of bales to the storage place prior to wrapping gave least mould. Wrapped bales should preferably be transported from the field using rollers some days after wrapping. Bale fork should not be used.

Wrapping rectangular bales with plastic to preserve wet hay or make haylage

D. Undersander[1], T. Wood[1] and W. Foster[2]
[1]University of Wisconsin, Department of Agronomy, 1575 Linden Drive, Madison WI 53706, USA Email: djunders@facstaff.wisc.edu and 2Consultant, 8755 Co Rd A, Bloomington, WI 53804 USA

Keywords: baleage, haylage, Medicago sativa, silage

Introduction Medium square bales (350 to 450 kg) are increasingly being used across the northern dairy regions of the US to reduce the labour associated with hay and haylage making. It is often difficult to get hay dried to 16 percent or less moisture for safe baling so it is frequently necessary to use a preservative or to wrap bales with plastic to avoid spoilage.

Materials and Methods Studies on square bale wrapping at the University of Wisconsin Lancaster Research Station have examined effect of moisture at baling, thickness of plastic wrapping, and time of wrapping variables. All studies used 0.86 by 0.81 by 1.5 m bales that were lucerne or lucerne/grass mixtures. In one study, the thickness of plastic necessary to preserve either wet hay or to make haylage was examined. Bales were wrapped with 2 to 10 wrappings of either 2.2 μ or 3.3 μ thick plastic. Bales averaged either 30% or 56% moisture at baling. In a second study, a lucerne/grass mixture was baled at either 36% or 63% moisture and then the bales wrapped either immediately (0), 12, 24, 36, 48, 72, or 96 hours after baling. All bales were wrapped with 13 μ of plastic. All studies were repeated to get different hay making conditions.

Results Internal bale temperatures over time from study one are shown in Figure 1. Unwrapped bales rose in temperature, reaching a maximum of about 57°C. Bales wrapped with 4.4 μ of plastic remained between 40 and 43°C, indicating that oxygen was leaking through the plastic to support continued microbial activity. Increases in neutral detergent fibre were noted in these bales. They also had significant mould throughout when opened for feeding. Bales wrapped with 13 μ or more of plastic, began to decline in temperature immediately after wrapping and fell to ambient temperature in eight to nine days. These bales had only a little white mould on the exterior of the bales.

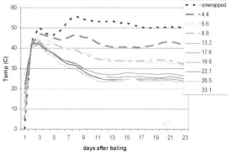

Figure 1 Effect of plastic wrap thickness on internal temperature of bale over time, Lancaster WI

Results of the second study for internal temperature of bales made for haylage at 63% moisture are shown in Figure 2 with 'Day 1' being the time of wrapping, not baling. In all cases, the internal bale temperature began to fall within 48 hours of wrapping. Unwrapped bales reached 66°C, which is near to the temperature that spontaneous combustion could start. Bales wrapped more than 24 hours after baling had significantly higher neutral detergent fibre, indicating loss of non fibrous carbohydrate due to respiration (Nai and Wittenburg, 2000). Bales wrapped within 24 hours after baling had the highest quality forage.

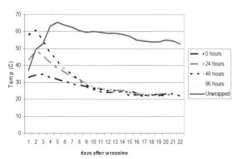

Figure 2 Effect of timing of bale wrapping on internal temperature of bale over time, Lancaster WI

Conclusion: Square bales can be wrapped over a range of moisture contents (from 23 to 63% in our study) and preserved. Huhnke et al. (1997) reported success in preserving wrapped bales between 25% and 65% moisture. Success depends on sufficiently thick plastic to maintain anaerobic conditions within the bale and wrapping within 24 hours after baling.

References
Huhnke, R.L., R.E. Muck & M.E. Payton (1997). Round bale silage storage losses of ryegrass and legume-grass forages. *Applied Engineering in Agriculture,* 13, 451-457.
Nia, S.A.M. & K.M. Wittenberg (2000). Effect of delayed wrapping on preservation and quality of whole crop barley forage ensiled as large bales. *Canadian Journal of Animal Science,* 80,145-151.

Bacteria and yeast in round bale silage on a sample of farms in County Meath, Ireland

J. McEniry[1,2], P. O'Kiely[1], N.J.W. Clipson[2], P.D. Forristal[3] and E.M. Doyle[2]
[1]Teagasc, Grange Research Centre, Dunsany, Co. Meath, Ireland, Email: jmceniry@grange.teagasc.ie,
[2]Department of Industrial Microbiology, University College Dublin, Belfield, Dublin 4, Ireland; [3]Teagasc, Crops Research Centre, Oak Park, Co. Carlow, Ireland

Keywords: baled silage, microorganisms, bacteria, yeast

Introduction Baled silage is made on almost three-quarters of all Irish farms. As in conventional clamp silage, a dominant lactic acid fermentation is required to reduce pH and minimise both quantitative and qualitative losses caused by undesirable microorganisms (e.g. yeast, *Clostridia*, *Bacilli* and *Enterobacteria*). However, the higher pH achieved in baled silage compared with conventional silage is less inhibitory to these microorganisms. The thickness of the plastic barrier used to seal ensiled forage from air is thinner with conventionally wrapped bales (70 μm for 4 layers) than clamp silos (250 μm for 2 layers). Furthermore, with baled silage, ~0.5 of the silage is within 12 cm of the plastic wrap compared to less than 0.1 with conventional clamp silage. This may create a more aerobic environment at the bale surface reducing the hygienic quality of the silage. The microbial composition within baled silage in Ireland has not been quantified. This study enumerated the predominant bacteria and yeast in the outer and inner layers of baled silage at feedout.

Materials and methods Two bales of silage were sampled on each of 10 farms from February to April 2004, using an electrically powered aseptic cylindrical core bit (length, 65 cm; internal diameter, 3.5 cm) at eight points around the bale. Subsampling points were at 1500, 1800, 2100 and 0000 h clock positions on the bale barrel, *ca* 40 cm from each end. There was no visible evidence of aerobic deterioration at these points. At each point subsamples were taken from both the outer 20 cm and through to the centre of the bale. The outer and inner core subsamples were each composited per bale, mixed and stored at 4°C for microbiological analysis. MRS nystatin agar, malt extract agar (MEA) containing streptomycin (pH 3.5), violet red bile glucose (VRBG) agar, nutrient agar (NA) and reinforced clostridial agar (RCA) with neutral red were used to enumerate for lactic acid bacteria (LAB), yeast, *Enterobacteria*, *Bacillus* spores and clostridial spores, respectively (Seale *et al.*, 1990). VRBG was incubated at 37°C for 2 d. RCA was incubated anaerobically in a GasPak 150 jar for 7 d at 37°C while all other media were incubated for 3 d at 30°C. The number of colony forming units (cfu) on each plate was enumerated and the number of microorganisms/g silage calculated.

Results The bales sampled had a mean dry matter of 354 (s.d. 118.5) g/kg and pH of 4.5 (s.d. 0.50). These are typical of values found on bales throughout the country, and indicate that the preservation of the wilted silage was satisfactory (Table 1). Type of microorganism had a major impact on the mean counts, with LAB > yeast > *Clostridia* > *Bacilli* > *Enterobacteria*. In contrast, the standard errors suggest the range in counts was in the order yeast > *Enterobacteria* > *Bacilli* > *Clostridia* > LAB. Yeast, LAB and *Enterobacteria* numbers were higher in the outer layer. There was no significant difference (*P*>0.1) in pH, DM, *Bacillus* and *Clostridia* numbers between the outer and inner bale layers.

Table 1 Composition of outer layer compared with inner layer of baled silage

	Outer	Inner	s.e.m.	Sig.
Dry matter (g/kg)	352	355	8.5	NS
PH	4.51	4.53	0.034	NS
LAB ($\log_{10}$ cfu/g silage)	5.83	5.53	0.093	*
Yeast ($\log_{10}$ cfu/g silage)	5.24	4.37	0.197	**
Enterobacteria ($\log_{10}$ cfu/g silage)	1.41	0.78	0.186	*
Bacilli ($\log_{10}$ cfu/g silage)	2.49	2.33	0.138	NS
Clostridia ($\log_{10}$ cfu/g silage)	3.75	3.87	0.096	NS

* = *P*<0.1, ** = *P*<0.01, NS = not significant

Conclusions The numbers of yeast, LAB and *Enterobacteria* were higher in the outer layer than the inner layer, but the preservation appeared similar in both parts of the bales. Higher yeast and *Enterobacteria* numbers may reflect more aerobic conditions at the bale surface. This may be due to an imperfect seal by the plastic wrap. The data provides a benchmark for future research on the fermentation kinetics of baled silage.

References

Seale, D.R., G. Pahlow, S.F. Spoelstra, S. Lindgren, F. Dellaglio & J.F. Lowe (1990) Methods for the microbiological analysis of silage. *Proceedings of the Eurobac Conference 1986, Uppsala, Sweden. Grass and Forage Reports*, 3, 147-164.

Schizophyllum on baled grass silage in Ireland: national farm survey 2004

M. O'Brien[1,3], P. O'Kiely[1], P.D. Forristal[2] and H. Fuller[3]

[1]*Teagasc, Grange Research Centre, Dunsany, Co. Meath, Ireland, Email mobrien@grange.teagasc.ie, [2]Crops Research Centre, Oak Park, Co. Carlow, Ireland, [3]Department of Botany, University College Dublin, Belfield, Dublin 4, Ireland*

Keywords: baled silage, fungus, *Schizophyllum commune*

Introduction   Since the early 1990's, a mushroom-type growth has been appearing with increasing frequency protruding through the plastic film on baled grass silage in Ireland. The fungus was identified as *Schizophyllum commune,* a gilled bracket fungus that is known primarily as a white rot fungus, found worldwide on fallen branches in woodlands (Brady *et al.*, 2005). A countrywide survey conducted in 1999 recorded the presence of *Schizophyllum* on 0.53 farms in Ireland (Fuller *et al.*, 2000). The authors of that survey concluded that *Schizophyllum* was widely distributed throughout Ireland and had the potential to cause considerable loss of feedstuff on farms. The objective of this survey was to assess the current prevalence of *Schizophyllum* on baled silage.

Materials and methods   In February 2004, collections of baled silage were inspected along six routes (Fuller *et al.*, 2000). These routes represented different geographical locations and farm enterprises in Ireland. Thirty farms were visited along each route and bale collections were examined for the presence and extent of the visible protrusion of *Schizophyllum* basidiomes through the plastic stretch film, in addition to other bale parameters. Silage samples were collected from two bales on each farm and silage dry matter (DM) and pH were determined.

Results   Bales had a mean DM concentration of 309 (standard deviation (SD) 111.4) g/kg and a pH of 4.5 (0.43). Bales examined along Route 6 had lower silage DM concentrations relative to the other five routes. *Schizophyllum* was visible on 106 out of 180 (0.58) farms surveyed, ranging from 0.13 (Route 6) to 0.76 (Route 1) (Table 1). Proportionally 0.36 farms had less than 0.1 bales affected by this fungus and 0.2 farms had between 0.1 and 0.5 bales of their bales affected.

Table 1 Occurrence and extent of *Schizophyllum commune* on Irish farms (n=30 farms/route, 180 in total) in 2004

Occurrence and incidence of *Schizophyllum*	Number (and proportion) of farms on which *Schizophyllum commune* was observed on bales						
	Route 1 (Midlands)	Route 2 (West midlands)	Route 3 (South-west)	Route 4 (South-east)	Route 5 (North midlands)	Route 6 (North-west)	Overall
Present	23 (0.76)	21 (0.7)	21 (0.7)	20 (0.66)	17 (0.57)	4 (0.13)	106 (0.58)
<0.1 bales affected	10 (0.33)	10 (0.33)	14 (0.47)	15 (0.5)	12 (0.40)	4 (0.13)	65 (0.36)
0.1 – 0.5 bales affected	12 (0.40)	11 (0.37)	4 (0.13)	4 (0.13)	5 (0.17)	0 (0)	36 (0.20)
>0.5 bales affected	1 (0.03)	0 (0)	3 (0.10)	1 (0.03)	0 (0)	0 (0)	5 (0.02)
DM (SD) (g/kg)	335 (118.7)	334 (105.4)	312 (113.6)	346 (114.5)	296 (94.2)	236 (83.8)	309 (111.4)
pH (SD)	4.4 (0.32)	4.5 (0.35)	4.6 (0.57)	4.6 (0.49)	4.4 (0.31)	4.5 (0.45)	4.5 (0.43)

Conclusions   *Schizophyllum* is widespread on baled grass silage in Ireland and on a slightly higher proportion of farms (0.58) than recorded five years previously by Fuller *et al.* (2000). In that study, a lower incidence of *Schizophyllum* was also recorded on farms in the north-west region (Route 6). A low grass DM at ensiling, as recorded for bales on Route 6, may be a factor that prevents bales being successfully colonised with *Schizophyllum.*

References

Brady, K.C., P. O'Kiely, P.D. Forristal & H. Fuller (2005). *Schizophyllum commune* on big-bale grass silage in Ireland. *Mycologist,* 19(1), 30-35.

Fuller, H., K. McNamara, K. Brady, P. O'Kiely, P.D. Forristal & J.J. Lenehan (2000). *Schizophyllum* on big-bale silage: results of farm survey January 1999. *Proceedings of the Agricultural Research Forum, Dublin, Ireland,* 51-52.

Bagged silage: Mechanical treatment applied by packing rotor improves fermentation

M. Sundberg[1] and T. Pauly[2]

[1]JTI – Swedish Institute of Agricultural and Environmental Engineering, P.O. Box 7033, SE-750 07 Uppsala, Sweden, Email: martin.sundberg@jti.slu.se, [2]Swedish University of Agricultural Sciences, Department of Animal Nutrition and Management, SE-753 23 Uppsala, Sweden, Email: thomas.pauly@huv.slu.se

Keywords: bagged silage, ensilability, mechanical treatment, packing rotor

Introduction   For the filling of silage bags, a considerable input of mechanical energy is required for the packing rotor to compress forage into the plastic tube. This means that forage is subjected to an intensive mechanical treatment that also affects the structure of the plant material. Many studies have shown that treatments which damage the plant structure and release plant juice facilitate and speed up the ensiling process, which may improve silage quality. If the mechanical treatment applied by a bagging machine would result in a faster and more substantial decrease in pH, the growth of undesirable micro-organisms, most notably clostridia, could be considerably restricted. The aim of this project was to test the hypothesis that the mechanical treatment applied by the packing rotor on the fresh forage could produce a favourable effect on the ensiling process.

Materials and methods   During the ensiling with a bagging machine (Winlin 5400) precision-chopped grass was sampled either before or after passing through the packing rotor of the bagger. For investigation of fermentation pattern, forages from these two treatments were ensiled in small laboratory-scale silos (1.7 l), 12 replicates per treatment. After 2-3, 6, 18 and 100 days of storage, three silos per treatment were opened and the silage analysed with respect to dry matter (DM), pH, fatty acids, ethanol, ammonia-N and for 100-d silages number of clostridial spores. Samples were also taken from the fresh crop for chemical and microbial analyses. The experiment was repeated on two occasions with different grass-dominated crops.

Results   DM content of the wilted grass at filling was approximately 28% in the first experiment and 32% in the second. In the first experiment, the number of lactic acid bacteria in the fresh forage increased after passage through the rotor by almost a factor of 100, from 3.5 to 5.4 log cfu/g FM. In the second experiment, extremely high numbers of lactic acid bacteria were found in the windrows in the field (5.7 log cfu). After chopping, the number of lactic acid bacteria increased to 10^7, with no further increase after passing the rotor. All silages fermented well, which resulted in good silage quality with negligible clostridial activity (butyric acid <0.05% of DM, clostridial spores <100). The drop in pH was faster and more pronounced for the forage that had passed through the rotor ($P<0.001$ for whole storage period). The two experiments gave similar results (Table 1 and Table 2).

Table 1 Experiment 1: Fermentation parameters (means from 3 replicates). Treatment A: Grass ensiled prior to passage through the rotor, B: after passage through the rotor

	Day 3		Day 6		Day 18		Day 100		
	A	B	A	B	A	B	A	B	LSD $_{0.05}$
pH	4.85	4.41	4.52	4.26	4.25	4.07	4.15	4.02	0.03
Ammonia-N (% of N)	3.3	3.2	4.4	3.9	5.9	5.7	6.3	5.6	0.18
Lactic acid (% DM)	3.64	4.46	5.41	5.47	7.25	7.42	8.57	8.27	0.27
Acetic acid (% DM)	0.95	1.33	1.38	1.53	1.75	1.87	1.89	2.09	0.08
Ethanol (% DM)	0.71	0.51	0.68	0.51	0.73	0.59	0.73	0.79	0.05

Table 2 Experiment 2: Fermentation parameters (means from 3 replicates). Treatment A: Grass ensiled prior to passage through the rotor, B: after passage through the rotor

	Day 2		Day 6		Day 18		Day 100		
	A	B	A	B	A	B	A	B	LSD $_{0.05}$
PH	4.87	4.65	4.70	4.38	4.43	4.20	4.30	4.16	0.02
Ammonia-N (% of N)	6.2	6.2	7.7	7.1	9.5	8.8	12.8	12.4	0.32
Lactic acid (% DM)	3.56	4.73	4.43	6.69	6.14	8.53	7.40	9.15	0.10
Acetic acid (% DM)	1.24	1.51	1.48	1.82	1.69	1.99	2.12	2.34	0.04
Ethanol (% DM)	0.34	0.25	0.34	0.27	0.34	0.29	0.38	0.33	0.02

Conclusions   The results from this study confirmed the hypothesis that the mechanical treatment of forage by the packing rotor had a beneficial effect on the fermentation process, resulting in faster acidification. This could improve the possibilities of producing high quality silages, especially for crops that are more difficult to ensile.

Carbon dioxide permeation properties of polyethylene films used to wrap baled silage

C. Laffin[1], G.M. McNally[1], P.D. Forristal[2], P. O'Kiely[3] and C.M. Small[1]
[1]Polymer Processing Research Centre, Queen's University Belfast, Belfast, BT9 5AH, Northern Ireland, Email: c.laffin@qub.ac.uk, 2Teagasc, Crops Research Centre, Oak Park, Carlow, Ireland; 3Teagasc, Grange Research Centre, Dunsany, Co. Meath, Ireland

Keywords: baled silage, polyethylene film, gas transmission rate, permeation coefficient

Introduction Polyethylene film wrapping for baled silage was introduced in the mid 1980s. This system is now used on 73% of farms in Ireland, accounting for one-third of the total national silage tonnage, or in excess of 9 million bales annually (O'Kiely et al., 2000). The film must possess good gas barrier and mechanical properties to ensure satisfactory levels of anaerobosis are achieved and maintained in the bale. This investigation determined the CO_2 barrier properties of a number of commercially available films.

Materials and methods Five commercial films (designated A to E) were investigated. Film C was reported to have been pre-stretched during manufacture. In each case a single layer of black film was tested, with 5 samples being obtained from the top, middle and bottom sections of a roll of film. A Davenport gas permeability measuring apparatus was used to determine the gas transmission rate (GTR) and permeation coefficient (PC) of film samples to CO_2 (99.8% purity). The GTR was measured according to B.S.2782, method 821A, ASTM D. 1434 and indicates the volume of gas transmission, with PC indicating the permeation per unit thickness. Youngs modulus was measured from tensile analysis of the films in both the machine and transverse directions. The co-monomer type of the films was predicted from differential scanning calorimetry (DSC) analysis. Each variable was subjected to two-way analysis of variance appropriate for a 5 x 3 factorial arrangement of treatments.

Results and discussion Table 1 shows that the gas transmission rates (GTR) for films A, B, D and E were in good agreement with previous reports for 25 μm polyethylene film (Briston, 1983). The GTR is influenced by film thickness, with Film C (thickness 12 μm) showing the highest GTR (49×10^3 ml/m^2/day/atm). Film C also exhibited the lowest permeation coefficient (24×10^6 ml at standard temperature and pressure (STP) cm^2 versus $31 - 36 \times 10^6$ ml (STP) cm^2). The higher GTR of film C at least partially reflected its lower thickness (12 μm), but the lower permeation coefficient appeared to be as a result of pre-stretching during manufacture leading to greater crystallinity and molecular orientation, which would improve the barrier properties of the film (Laffin et al., 2004). Mechanical analysis also showed significant variation ($P<0.001$) in the Youngs modulus across the width of the roll, with the largest variation being recorded for film D (108.3 – 146.5 MPa in the machine direction). Film C exhibited the lowest Youngs modulus of all five films given its different co-monomer type. All films showed significant variation ($P<0.001$) in mechanical and gas permeation properties across the width of the roll, indicating non-perfect manufacture.

Conclusions All films had similar permeation coefficient and Youngs modulus values, except for Film C which was pre-stretch and has a different co-monomer. The considerable variation in permeation and mechanical properties across the width of the roll for all films indicates an aspect of manufacture that needs improvement.

Table 1 CO_2 permeation and mechanical properties of five commercial films tested

Film	Thickness (μm)		Location in roll of film (L)		
			Top	Middle	Bottom
A	25	GTR[#]	27,7	27,2	38,5
		PC[+]	27,7	28,6	37,6
		YMm[a]	103,6	125,0	116,6
		YMt[b]	129,3	135,5	140,5
B	25	GTR[#]	35,3	33,6	45,1
		PC[+]	32,5	33,0	42,9
		YMm[a]	133,2	117,7	131,8
		YMt[b]	151,2	150,7	132,5
C	12	GTR[#]	49,8	47,4	49,1
		PC[+]	25,8	23,5	21,7
		YMm[a]	98,9	111,1	115,1
		YMt[b]	109,6	128,4	136,5
D	25	GTR[#]	34,3	37,6	41,3
		PC[+]	33,4	34,0	40,5
		YMm[a]	108,3	120,3	146,5
		YMt[b]	135,5	140,2	172,8
E	25	GTR[#]	34,9	31,6	34,5
		PC[+]	38,4	33,4	35,6
		YMm[a]	115,9	117,8	124,8
		YMt[b]	117,8	127,3	125,3

#GTR = CO2 gas transmission rate ((ml/m2/d/atm)(103))
+PC = CO2 permeation coefficient (ml(STP)cm2(10-6))
aYMm = Youngs modulus machine direction (MPa)
modulus transverse direction (MPa)
GTR SEM = 1.5, PC SEM = 1.1, YMm SEM = 2.5, YMt = 4.0. bYMt = Youngs Film (F), Location (L) and FxL were each significant at $P<0.001$

References
Briston, J.H. (1983). Plastics Films, New York: The Plastics and Rubber Institute, 2nd Edition: pp 362.
Laffin, C., G.M. McNally, P.D. Forristal, P. O'Kiely & C.M. Small (2004). "The Effect of Extrusion Conditions and Material Properties on the Gas Permeation Properties of LDPE/LLDPE silage wrap Films" ANTEC, pp. 230–234.
O'Kiely, P., K. McNamara, D. Forristal & J.J. Lenehan. (2000). "Characteristics of baled silage on Irish farms" Farm and Food, 10 (3), 33-38.

National survey to establish the extent of visible mould on baled grass silage in Ireland and the identity of the predominant fungal species

M. O'Brien[1, 3], P. O'Kiely[1], P.D. Forristal[2] and H. Fuller[3]
[1]Teagasc, Grange Research Centre, Dunsany, Co. Meath, Ireland, Email: mobrien@grange.teagasc.ie, [2]Crops Research Centre, Oak Park, Co. Carlow, Ireland, [3]Department of Botany, University College Dublin, Belfield, Dublin 4, Ireland

Keywords: baled silage, fungi, mould, *Penicillium roqueforti*, yeast

Introduction On an annual basis, some 12 million bales of grass silage are made on Irish farms. A pilot survey of 35 farms in March 2003 recorded 90% of bales with some fungal growth present. The predominant fungus identified was *Penicillium roqueforti*, a species that can produce harmful mycotoxins under certain conditions (O'Brien *et al.*, 2004). Evidence is increasing that mycotoxins are formed regularly under ensiling conditions (Fink-Gremmels, 2003). The objectives were to determine (i) the proportion of bales with visible fungal growth on farms in Ireland, (ii) the extent of fungal contamination on the surface of bales and (iii) the identity of the largest fungal colony present on bales.

Materials and methods In February 2004, 180 farms were visited on six separate routes throughout Ireland. Each route was sub-divided into five sections with six farms per quintile visited. All routes were 100-150 km in length and were representative of geographical locations and farm enterprises. Two bales in readiness for feeding were examined on each farm (total of 360 bales). Protocols were as described by O'Brien *et al.* (2004) with the exception that in this survey only the largest fungal colony on each bale was sampled.

Results The majority of bales examined were harvested between June and August 2003. Bales had a dry matter (DM) concentration of 309 (standard deviation (SD) 111.4) g/kg and a pH of 4.5 (SD 0.43). Fungal growth was visibly present on 334/360 (0.92) bales examined. A total of 334 mould and yeast colonies were sampled and 344 fungi were isolated and identified as being the contaminant visible fungal growth on the bales. The extent of fungal growth on bales ranged from 0 to 0.82 surface coverage, with mean proportion coverage of 0.06 (Table 1). The fungus affecting the largest surface area on bales was *P. roqueforti* (Table 2).

Table 1 Prevalence of visible fungal growth on baled grass silage on 180 farms in Ireland (60 bales/route)

Route location	Proportion of bales contaminated	Proportion of bale surface area affected Mean ±(SD)
North west	0.77	0.09 (0.12)
North midlands	0.98	0.04 (0.03)
West midlands	0.98	0.09 (0.11)
Midlands	0.90	0.08 (0.08)
South west	0.98	0.07 (0.08)
South east	0.91	0.02 (0.04)
Overall mean	0.92	0.06 (0.06)

Table 2 Predominant fungi on baled silage in Ireland

Fungal genera/species	No. of bales on which predominant	Proportion of bales
Penicillium roqueforti	146	0.42
Schizophyllum commune	71	0.20
Yeasts	45	0.13
Penicillium paneum	15	0.04
Geotrichum	12	0.03
Fusarium	4	0.01
Trichoderma	3	0.01
Mixed mycobiota[a]	35	0.10
Unidentified moulds	13	0.03

[a]>1 fungus was isolated from the predominant fungal

Conclusions The results of this study reflect the findings made in the previous pilot survey. Visible fungal growth on baled silage is extensive throughout Ireland, with some differences in the extent of contamination on bales between different regions. As a consequence, some baled silage will have an inferior feeding value, which in turn will negatively influence farm profits. Although a relatively small number of fungal species was responsible for most of the contamination, at least two of these fungi (*P. roqueforti* and *P. paneum*) could potentially cause health problems in livestock by their known ability to produce harmful mycotoxins.

References
Fink-Gremmels, J. (2003). Moulds and mycotoxins in silage. The 2nd World Mycotoxin Forum, 17–18 February, Noordwijk aan Zee, The Netherlands, 49-50.
O'Brien, M., P. O'Kiely, P.D. Forristal & H. Fuller (2004). Pilot survey to establish the identity and extent of occurrence of visible fungi on baled grass silage. *Proceedings of the Agricultural Research Forum, Tullamore, Ireland*, 50.

Section 4

Ensilage of tropical forages

Effect of ensiling temperature, delayed sealing, and simulated rainfall on the fermentation and aerobic stability of maize silage grown in a sub-tropical climate

A.T. Adesogan and S.C. Kim
Department of Animal Sciences, IFAS, University of Florida, PO Box 110910, Gainesville, FL 32611, USA, Email: adesogan@animal.ufl.edu

Keywords: maize silage, temperature, moisture

Introduction Dairy farmers in many parts of the world rely on maize silage as a source of digestible fibre and readily fermentable energy for their cattle. In Florida and many tropical countries, such farmers face several climatic challenges that are perceived to complicate maize silage production such as greater disease pressure due to a conducive climate for proliferation of pathogens and inclement weather at harvest. Delayed bunker sealing also occurs because of long distances between contract growers' fields and dairy farms. Each of these factors in isolation adversely affects the fermentation and/or aerobic stability of the silages, but little is known about their combined effect on maize silage fermentation. Therefore, this determined how simulated rainfall, delayed sealing, and ensiling temperature affect the fermentation and aerobic stability of maize silage.

Materials and methods A maize hybrid (31R87RR Pioneer, Des Moines, Iowa) was grown on four replicated 6 x 1.5 m plots. Half of each plot was harvested at 350 g DM/kg and chopped with a two-row forage harvester (Dry). The other half was harvested and chopped (Wet) after it was sprayed with sufficient water from a tanker until 4 mm were collected in a rain gauge in the centre of the plots. Wet and dry forage samples from each plot were either ensiled (2 kg) immediately (Quick) or after a 3 h wilt (Delay) in plastic bags in quadruplicate. Half of the bags from each moisture by sealing time treatment combination were placed in a 40°C incubator (Hot) and the other half in a 20°C, air-conditioned room (Cool). After 82 d of ensiling the bag silos were opened and the silages were chemically characterised and analysed for aerobic stability and yeasts and mould counts. A 2 (moisture treatments) x 2 (sealing times) x 2 (ensiling temperatures) factorial design was used for the study.

Results and discussion There were no three-way interactions ($P<0.05$) between the treatments for the measurements except for aerobic stability. Wetting the maize silage increased ($P<0.05$) concentrations of acetate and NH_3N and reduced the DM concentration (Table 1). Ensiling the maize at the higher temperature increased the pH and NH_3N concentration, and reduced lactate, acetate and propionate concentrations. Delaying sealing for 3 h, increased the DM concentration and reduced NH_3N ($P<0.05$) concentration. Yeast counts were reduced ($P<0.05$) by wetting the forage, ensiling at the higher temperature and delaying sealing. Mould counts and lactate to acetate ratio were unaffected ($P>0.05$) by treatment. Aerobic stability was greater ($P<0.05$) in the maize silage that was wetted, ensiled at 40°C and sealed after a 3 h delay than in the other treatments.

Table 1 Effect of moisture, temperature and delayed sealing on the composition (g/kg DM) of maize silage

	Dry				Wet				Contrasts		
	Cool		Hot		Cool		Hot		Moisture	Temp	Seal
	Delay	Quick	Delay	Quick	Delay	Quick	Delay	Quick			
DM (g/kg)	36.2	34.1	39.4	33.8	33.1	30.4	36.6	31.6	***	**	***
PH	3.76	3.82	4.11	4.30	3.72	3.73	4.25	4.27	NS	***	NS
NH_3-N	0.35	0.46	0.53	0.70	0.49	0.55	0.57	0.84	*	***	***
Lactate	54.2	67.8	39.5	39.7	73.1	75.3	32.2	45.3	NS	***	NS
Acetate	18.9	23.5	16.6	14.4	26.8	26.3	18.6	19.0	*	*	NS
Propionate	6.0	8.2	0.0	0.0	5.1	7.9	0.0	0.0	NS	***	NS
Yeasts[1]	6.70	7.70	7.00	6.43	6.83	7.30	4.43	6.07	**	***	*
Aerobic stability (h)	15.33	13.00	12.00	12.50	14.00	10.67	45.66	16.00	*	*	*

[1] Log cfu/g; NS Not significant $P>0.05$; * $P<0.05$; ** $P<0.01$; *** $P<0.001$. Temp = temperature.

Conclusions This study suggests that high temperatures during ensiling can hinder the fermentation of maize silage. The simulated rainfall had conflicting effects on fermentation since it increased proteolysis, but resulted in higher lactate concentrations. Delaying sealing for 3 h reduced proteolysis.

Effect of different densities on tropical grass silages

T.F. Bernardes, R.C. Amaral, G.R. Siqueira and R.A. Reis
Universidade Estadual Paulista, FCAV, Jaboticabal, SP, Brasi,14.884-900, Email: tfbernardes@yahoo.com

Keywords: compaction, aerobic stability, fermentation, losses

Introduction Grasses in tropical areas, in spite of their low quality (55-60% NDT), are justified in their utilisation by their high dry matter production (30-40 t DM/ha). National machines used for harvesting these kinds of forages, cut particles with large sizes which prejudice mass compaction during silo filling. High densities promote anaerobic conditions, reduce storage cost by amortization of structure and reduce losses caused by deterioration (Muck *et al.*, 2003). This work aimed to evaluate Marandu-grass (*Brachiaria brizantha*) silages ensiled at different densities.

Material and methods Forage with 35% dry matter (DM) was harvested using a tractor-drawn machine with rotating knives obtaining particles in which 72% were bigger than 31.7 mm. The forage was ensiled in 8 l capacity laboratory silos. Forages were submitted to four different compaction pressures to give four different densities: 97.42 (D1), 118.63 (D2), 139.21 (D3) and 164.02 kg DM/m^3 (D4). After 65 d of fermentation, each silo was opened and its content removed, homogenised and divided into two samples for evaluation of aerobic stability. Temperature variation over 120 h was measured in one sample. In the other sample, pH after three and six d of aeration was measured. A completely randomised design with four treatments and four repetitions was used. Data were analysed by SAS proceedings.

Results Treatments D2, D3 and D4 presented the highest DM recovery (DMR), average 94.7% (Table 1). In the treatment with the lowest density (D1) the DMR was 83.1%. The *in vitro* digestibility of DM (IVDM) was highest (63.1%) in the silage with the highest density, followed by D3, D1 and D2 treatments. The pH values were considered high, but the lowest value was found in treatment D4 (4.8). The aerobic deterioration of the silages was concomitant to the increase of the pH (Table 1) and temperature. Silages with lower density (D1, D2 and D3) reached the highest temperature between 81 and 84 h of air exposition. The D4 treatment showed the highest temperature after the longest air exposition period (Figure 1).

Table 1 Silage characteristics

Item	Dl	D2	D3	D4	SE
DM (%)	27.8b	28.4b	32.7a	31.1ab	0.80
DMR (%)	83.1b	96.4a	94.4a	93.5a	2.25
Gas (% DM)	7.8ab	6.0b	5.3b	12.8a	1.40
Effluent (kg/t)	9.75a	2.2a	3.8a	7.6a	3.17
IVDM (%)	55.7ab	52.8b	58.6ab	63.1a	1.82
Ammonia/N	9.5a	10.4a	11.4a	9.3a	1.72
pH (0)[1]	6.7ab	6.8a	4.9ab	4.8b	0.47
pH (3)[1]	7.8a	8.8a	5.4b	5.4b	0.55
pH (6)[1]	8.5ab	9.1a	6.7b	6.8b	0.46

Means followed by different letters in row are different by Tukey test ($P<0.05$)
[1]pH values at zero, three and six d of air exposure

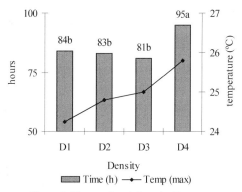

Figure 1 Temporal variations in silages on exposure to air

Conclusions Increasing the density of grass at ensiling helped the desirable fermentative processes and increased silage DM recovery. Increasing compaction pressure might help ensilage when forage is harvested with machines producing bigger particle sizes.

References

Muck, R.E., L.E. Moser & R.E. Pitt (2003). Postharvest factors affecting ensiling. In: D.R. Buxton, R.E. Muck, J.H. Harrison (eds). Silage Science and Technology. *American Society of Agronomy*, 251-304.

Sugarcane silage compared with traditional roughage sources on performance of dairy cows

O.C.M. Queiroz, L.G. Nussio, M.C. Santos, J.L. Ribeiro, P. Schmidt, M. Zopollatto, M.C. Junqueira, M.S. Camargo, S.G.T. Filho, L.G. Vieira, M.O. Trivelin, L.J. Mari and D.P. Souza
Department of Animal Science, University of São Paulo, ESALQ, Av. Pádua Dias 11, Piracicaba, SP, 13418-900, Brazil, Email: nussio@esalq.usp.br

Keywords: sugarcane silage, maize silage, green chopped sugarcane, milk yield, DM intake

Introduction The utilisation of green chopped sugarcane as a forage resource in Brazil has increased significantly due to the agronomic potential of biomass yield and better accuracy on diet formulations including this forage. The adoption of this system is, however, limited mainly for larger herds because of the logistic restrictions created by the daily harvesting. By producing sugarcane silages dairy farmers might improve the overall operational efficiency of the farm and overcome domestic struggling. Based on results of previous studies (Pedroso *et al.*, 2002), the present trial aimed to evaluate if sugarcane silage properly fermented may support intermediary lactation performances when compared with traditional forage sources.

Material and methods Forty-eight Holstein cows were randomly assigned in 4x4 multiple Latin square design during a 84-d trial. Cows were fed during the adaptation (14 d) followed by sampling periods (7 d) resulting in 21-d periods. Treatments consisted of rations formulated for a standard cow (90 d in milk) with an initial milk yield of 30 kg/d, according to NRC-Dairy Cattle (2001) (Table1) considering the inclusion of green chopped (GC) sugarcane (T1), sugarcane silage (T2), maize silage (T3) and GC sugarcane (50%) + corn silage (50%) (T4). After chopping, sugarcane forage was inoculated with *Lactobacillus buchneri* NCIMB 40788 $5x10^4$ cfu/g (Lalsil cana - Lallemand) during ensiling. Both maize silage and sugarcane silage were stored for more than 2 months before feeding. GC sugarcane was chopped twice daily and TMR (Table1) were offered both in the morning (6 am) and evening (6 pm) in all diets. Forage DM contents were measured weekly to adjust ration composition and to calculate the DM collective intake. During the sampling period rations and feed samples were taken daily and milk yield and milk samples twice daily, every other day. Milk composition was measured by NIRS equipment. Data were analysed using GLM procedure (SAS, 1996).

Results Milk yields and 4% FCM were not changed across treatments and both, T2 and T4 diets resulted in higher DMI (8.8%), however, feed efficiency (4% FCM/DMI) and milk fat (g/kg) were higher for T3 diet. Sugarcane silage properly inoculated resulted in similar performance as supported by GC, which turned out to be a very positive result considering the traditional trend for sugarcane silages. Furthermore, both GC and sugarcane silage diets were not different ($P<0.05$) in milk yield and 4% FCM from a traditional maize silage diet, even considering the high nutritive value of this forage (65% TDN). The DMI and milk yield observed in T4 diets supports the benefits from the theory of forage or probably carbohydrate complementarity.

Table 1 Ingredient composition of the diets

Ingredients (% DM)	T1	T2	T3	T4
GC sugarcane	40.0	-	-	-
Sugarcane silage	-	40.0	-	22.51
Corn silage	-	-	50.0	22.51
Grounded corn	7.99	7.99	-	-
Citrus pulp	12.9	12.9	14.8	17.52
Whole cottonseed	10	10	10	10
Cotton meal	12	12	10	10
Soybean meal	14.0	14.0	12.1	14.39
Na bicarbonate	0.70	0.70	0.70	0.70
Mineral	2.38	2.38	2.38	2.38

Table 2 Performance of dairy cows and milk composition

Parameters	T1	T2	T3	T4
DMI (kg/d)	22.3^b	23.5^a	21.3^c	23.5^a
Milk yield (kg/d)	24.6	24.4	25.5	25.2
4% FCM (kg/d)	22.0	22.1	24.0	23.0
4% FCM/DMI (kg/kg)	0.99^b	0.95^b	1.13^a	0.99^b
Milk fat (g/kg)	33.4^b	33.8^b	36.1^a	34.8^{ab}
Milk protein (g/kg)	32.4	31.7	31.7	32.2
Milk total solids (g/kg)	116	116	118	118

[a,b] ($P<0.05$)

Conclusions The comparisons of this present trial supported the potential for sugarcane resources even for high producing dairy cows, suggesting the needs for accuracy on diets formulations. The performance achieved by cows fed sugarcane silage diets stimulates strongly further research on this particular forage resource.

References

NRC (2001). *Nutrient Requirements for Dairy Cattle*. National Academy Press, Washington, D.C.
Pedroso, A.F. L.G. Nussio, S.F. Paziani, D.R.S. Loures, M.S. Igarasi, L.J. Mari, R.M. Coelho, J.L. Ribeiro, M. Zopollatto & J. Horii (2002). Bacterial inoculants and chemical additives to improve fermentation in sugar cane (*Saccharum spp*) silage. In: *International Silage Conference*, Auchincruive, Scotland, p. 66-67.
SAS (1996). SAS Users Guide, Statistics (1996, version 5). SAS Inst. Incorporated Cary, North Carolina, USA.
Financial support provided by CNPQ, SP (Brazil)

Moisture control, inoculant and particle size in tropical grass silages

S.F. Paziani, L.G. Nussio, D.R.S. Loures, L.J. Mari, J.L. Ribeiro, P. Schmidt, M. Zopollatto, M.C. Junqueira and A.F. Pedroso
Department of Animal Science, University of São Paulo, ESALQ, Av. Pádua Dias 11, Piracicaba, SP, 13418-900, Brazil, Email: nussio@esalq.usp.br

Keywords: grass silage, *Panicum maximum*, silo losses, moisture, wilting

Introduction Decreased fermentation and spoilage losses with improved aerobic stability during feed out can be accomplished by several strategies, such as wilting, addition of microbial additives and moisture absorbents. Particle size reduction may increase bulk density and improve the fermentation. The objective of this trial was to evaluate the effects of particle size, moisture content and a microbial additive on chemical-physical parameters and losses in silages made from Tanzania grass.

Material and methods The trial was carried out during the summer on a 90 d vegetative regrowth cut of Tanzania grass (*Panicum maximum*) which was harvested and ensiled with the following treatments: T1 - fresh forage, large particle size, no microbial additive; T2 - fresh forage, small particle size, no microbial additive; T3 - wilted forage, large particle size, no microbial additive; T4 - fresh forage, large particle size, no microbial additive + ground pearl millet grain (GM); T5 - fresh forage, small particle size, microbial additive (Ecosyl®, UK). Pressed bag silos (40 t each) with 2.7 m diameter were packed under pressure (80 pounds/inches^2) and opened after 90 d storage. A core sample (30x30x30 cm) was taken weekly for analysis. Spoilage losses were measured daily as a % of the silage unloading rate. Chemical analyses were carried out according to AOAC (1980), mean particle size following Lammers *et al.* (1996) and porosity according to Williams (1994). Repeated measurements were taken in a complete randomised design during eight weeks and analysed using a mixed procedure (SAS, 1996).

Results Wilting and pearl millet grain addition increased the dry matter (DM) content (Table 1). The small particle size in the forage did not increase wet or DM silage bulk densities (Table 2), even though the addition of pearl millet grain showed a trend for higher DM density-DMD (156 kg/m^3) compared to the other treatments. Forage wilting tended to lower the wet density of the silage (460 kg/m^3), but DM density was not affected due to the compensatory effect of the higher DM content. Reducing the particle size in the forage (T2 and T5) did not reduce the porosity, in contrast to the expected results. This may have arisen because fewer and larger pores with longer forage were compensated by many smaller pores. The wilted forage (T3) showed higher losses when compared to the addition of pearl millet (29% vs 18%). Particle size reduction did not change the spoilage losses (*P*=0.60) but the addition of bacterial inoculant showed a trend (*P*=0.09) for increased losses.

Table 1 Chemical parameters of tropical grass silages

Parameters	T1	T2	T3	T4	T5
DM (%)	24.8	24.0	27.7	28.5	24.0
CP (% DM)	9.2	10.2	9.6	11.0	8.5
NDF (% DM)	67.8	69.4	69.0	49.8	69.3
ADF (% DM)	45.0	45.4	46.4	33.7	45.4
ASH (% DM)	10.9	10.5	11.2	8.3	10.8
WSC (% DM)	1.8	1.8	2.4	1.4	1.2
N-NH$_3$ (% total N)	8.2	5.8	4.6	2.4	10.1
pH	4.9	4.9	4.8	4.8	4.7

Table 2 Physical parameters of tropical grass silages

Parameters	T1	T2	T3	T4	T5
Mean particle size (cm)	2.4	2.2	3.4	2.2	2.0
Sieve retention (%)	47.4	53.1	67.4	54.0	36.9
Bulk density (kg/m^3)	535[a]	523[a]	460[b]	505[ab]	487[ab]
DMD (kg/m^3)	142[ab]	131[b]	135[ab]	156[a]	122[b]
Porosity (%)	45[b]	52[a]	50[ab]	48[ab]	55[a]
Spoilage losses (%)	17[ab]	14[b]	29[a]	18[ab]	23[ab]

[a,b] *(P<0.05)*

Conclusions High spoilage losses suggested that wilting may not be a suitable strategy for ensiling tropical grasses when harvested with larger particles and stored in pressed bag silos. The bacterial inoculant also increased spoilage losses during feed out.

Acknowledgement Financial support provided by FAPESP, SP (Brazil)

References
AOAC (1980). Official Methods of Analysis.13th.ed. Washington, 1015 pp.
Lammers, B.P., D.R. Buckmaster & A.J. Heinrichs (1996). A simple method for the analysis of particle sizes of forage and total mixed rations. *Journal of Dairy Science*, 79, 922-928
Williams, A.G.(1994). The permeability and porosity of grass silage as affected by dry matter. *Journal of Agricultural Engineering Science*, 59, 133-140.
SAS (1996). SAS Users Guide Statistics (Version 5), SAS Institute Incorporated, Cary North Carolina, USA.

Stability of silage wrapped round bales in Réunion Island

P. Grimaud[1], V. Barbet-Massin[2], P. Thomas[2] and D. Verrier[1]
[1]Cirad-Elevage, 7 Chemin de l'IRAT, F97410 Saint Pierre, Email:grimaud@cirad.fr, [2]UAFP, PK 23, F97418, Plaine des Cafres

Keywords: silage, tropical and temperate grass, dairy cows

Introduction Réunion Island is a French tropical island. Pastures consist of both temperate and tropical grasses depending on the altitude. Due to seasonal variations in pasture availability, farmers preserve their pastures in many different ways in order to meet the nutritional needs of livestock throughout the year (Grimaud and Thomas, 2002). Many of them preserve their forage as silage, through wrapped round bales made from April to June to profit by the high grass growth at the end of the rainy season (Paillat, 1995). This study was designed to investigate the variability of weight, DM and pH of silage preserved in this form, and consequently advise farmers.

Materials and methods Thirty silage bales were weighed and analysed for DM and pH on eleven dairy farms. Fifty percent of the bales consisted of tropical grass (*Pennisetum clandestinum*, *Chloris gayana*), and the other half consisted of temperate *Lolium perenne* and *Dactylis glomerata*. A t-test was used to compare tropical and temperate grasses and the relationships between DM content and pH were also compared with the standard (pH of stability = 0.04 DM + 3.2), defined by Paillat (1995).

Results The nature of grass did not affect the gross weight and the pH of silage bales (Table 1). However, the DM contents of tropical grass silages were significantly lower than those made from temperate grasses. Consequently, dry weight of tropical bales was significantly lower. Results further indicate silage instability for bales with DM contents less than 38% (intersection between the curves, Figure 1).

Table 1 Gross weight (GW), Dry matter contents (DM) Dry weight (DW) and pH of silage bales

	Temperate grass Avg (*stdev*) Min-Max	Tropical grass Avg (*stdev*) Min-Max
GW (kg)	605 (*108*) 394-756	593 (*94*) 426-720
DM (%)	40.6[a] (*12.2*) 23.3-70.6	32.6[b] (*12.4*) 16.8-56.0
DW (kg)	237[a] (*46*) 162-325	190[b] (*73*) 113-347
pH	4.9 (*0.5*) 4.2-5.8	4.9 (*0.4*) 4.1-5.6

Values on the same row with different superscripts are significantly different (*P*<0.05)

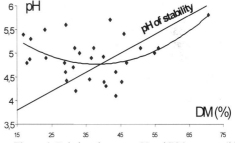

Figure 1 Relations between pH and DM contents (%)

Conclusions Results of the present study suggest variability of silage parameters depending on the type of grass. The survey results indicate instability of silage wrapped round bales with DM contents lower than 38%, which is greater than the value of 27% reported by Paillat (1995) in standard conditions. This weakness is higher for bales made of tropical grasses; thus, to ensure a good conservation of silage, mechanisms that improve the quality of silage preserved in this form must be applied, with particular attention paid to tropical grasses: For example, longer period of drying in the fields, larger incorporation of molasses. This study also shows that, from very simple criteria, farmers could be informed about the stability of their bales, and thus on the economic gain of the method they use to preserve their forage.

References

Grimaud, P. & P. Thomas (2002). Diversité des rations à base de graminées et gestion des prairies en élevage bovin sur l'île de la Réunion, *Fourrages*, 169, 65-78.
Paillat, J.M. (1995). Etude de l'ensilage en balles enrubannées sous climat tropical d'altitude: cas des fourrages tempérés et tropicaux récoltés à l'île de la Réunion. PhD thesis, INA Paris Grignon, France, 197 pp.

Ensilage of tropical grasses and legumes using a small-scale technique

M. Delacollette[1], S. Adjolohoun[2], R. Agneesens[3] and A. Buldgen[1]
[1]Faculté universitaire des Sciences agronomiques de Gembloux, Unité de Zootechnie, 2 Passage des Déportés, 5030, Gembloux (Belgium), Email address: delacollettem@yahoo.fr, [2]Faculté des Sciences agronomiques, Université d'Abomey - Calavi, 01-BP 526 Cotonou (Benin), [3]Centre Wallon de Recherches Agricoles, Section systèmes agricoles, rue du Serpont 100, 6800 Libramont(Belgium), Email address: agneesens@crawallonie.be

Keywords: silage, small-scale technique, grass, legumes

Introduction In many developing tropical countries, storage methods for forage supplies during the dry season must be considered for ruminants. Silage allows the forage to be harvested at an optimal stage and avoid, or highly decrease, field-drying periods under wet climates (Ashbell *et al.*, 2001). For small-holders with limited production means, the development of a small-scale technique is however necessary.

Materials and methods Pure silages of 3 tropical grasses (*Andropogon gayanus, Panicum maximum, Pennisetum purpureum*) and 4 tropical legumes (*Aeschynomene histrix, Stylosanthes humilis, Mucuna pruriens, Cajanus cajan*) were tested. The forages were first chopped (5-10 cm) and wilted during several hours to reach a dry matter (DM) content of 30% or more. Plastic bags were filled with 5 kg of each forage, pressed to expel air and closed manually (node). The bags were put in a second bag hermetically closed as well and stored in a dark room at ambient temperature. Six replicates were taken per silage. All the bags were opened after a 4-week conservation time. Laboratory analyses were carried out for the main fermentation characteristics: pH, ammonia/total nitrogen (NH_3/N), volatile fatty acids (VFA) and lactic acid. Near infrared reflectance spectroscopy was used to determine the chemical composition of the forages.

Results During ensiling the bags were difficult to seal hermetically, but the different forages always presented good structures, colours and smells after ensilage. Butyric and propionic acid concentrations were close to 0 g/kg DM (results are not presented in Table 1). Except for *Mucuna pruriens*, pH values were over 5, which is representative of an unstable fermentation and characteristic of ensiled forages with high DM contents (González and Rodríguez, 2003). Lactic acid concentrations were low, but proteolysis was very limited, even with *Mucuna pruriens* which had the highest CP content.

Table 1 Chemical composition and fermentation characteristics of the studied forages (n = 6, mean ± standard deviation)

	Silage composition (% DM)			Silage fermentation characteristics			
	DM	Crude protein	Crude fibre	pH	NH_3/N (%)	Lactic acid (g/kg DM)	Acetic acid (g/kg DM)
Grasses							
A. gayanus	44.66[a1]	5.62[b]	37.99[b]	5.56 ± 0.16[a]	2.36 ± 0.46[a]	1.08 ± 1.28[b]	1.48 ± 0.41[b]
P. maximum	40.50[a]	2.93[c]	39.86[a]	5.87 ± 0.09[b]	5.53 ± 0.51[b]	1.84 ± 0.53[b]	7.43 ± 2.57[b]
P. purpureum	47.88[a]	8.22[a]	35.48[c]	5.77 ± 0.16[ab]	4.92 ± 1.44[b]	6.04 ± 0.63[a]	38.36 ± 12.42[a]
Legumes				±			
A. histrix	38.71[c]	11.71[a]	36.24[a]	5.24 ± 0.06[b]	5.56 ± 1.05[b]	4.18 ± 0.41[a]	10.68 ± 1.93[a]
M. pruriens	29.00[d]	16.56[b]	36.79[a]	4.84 ± 0.04[c]	5.85 ± 0.62[b]	14.20 ± 1.4[b]	11.36 ± 1.22[a]
C. cajan	55.91[a]	15.92[b]	38.56[a]	6.30 ± 0.08[a]	7.64 ± 0.93[a]	2.75 ± 0.99[a]	10.17 ± 0.72[a]
S. humilis	40.65[b]	10.90[a]	36.88[a]	5.26 ± 0.06[b]	3.42 ± 0.28[c]	3.34 ± 0.73[a]	9.06 ± 1.02[a]

[1] For a same family of forage, values followed by different letters in a column differ significantly (*P*<0.05)

Conclusions The studied grasses and legumes may be conserved using the silage small-scale technique if they are chopped and, if necessary, slightly wilted. This conservation method is promising for smallholders, because it is feasible manually, with low inputs and without mechanisation.

References
Ashbell, G., T. Kipnis, M. Titterton, Y. Hen, A. Azrieli & Z.G. Weinberg (2001). Examination of a technology for silage making in plastic bags. *Animal Feed Science and Technology*, 91, 213-222.
González, G. & A.A. Rodriguez (2003). Effect of storage method on fermentation characteristics, aerobic stability, and forage intake of tropical grasses ensiled in round bales. *Journal of Dairy Science*, 86, 926-933.

The use of *Lactobacillus buchneri* inoculation to decrease ethanol and 2,3-butanediol production in whole crop rice silage

N. Nishino and H. Hattori
Department of Animal Science, Okayama University, Okayama 700-8530, Japan,
Email: jloufeed@cc.okayama-u.ac.jp

Keywords: 2,3-butanediol, ethanol, *Lactobacillus buchneri*, silage, whole crop rice

Introduction There is increasing interest in Japan in the utilisation of whole crop rice as a feed for ruminants. More than 400,000 ha of paddy field are unused due to production adjustment, while about 25% of the domestic roughage demand is supplied by imports mainly from China, Australia and North America. Evidence has shown that whole crop rice is usually low in sugars and lactic acid bacteria, and produces silages rich in ethanol rather than lactic and volatile fatty acids. As moulds are found in silage when stored until the warm season, there is interest in the use of additives to inhibit fungal growth even after a long-term storage. Yeasts and moulds can be suppressed when silage is inoculated with *Lactobacillus buchneri*; therefore, laboratory experiments were carried out to investigate the influence of *L. buchneri* on the fermentation and microbial composition of whole crop rice silage when stored up to 12 months.

Materials and methods Whole crop rice (*Oryza sativa* L., cv. Hamasari) was harvested manually at yellow-ripe stage (DM 507 g/kg) and ensiled in plastic pouches after being chopped into 13 mm length. *L. buchneri* was inoculated at 10^5 cfu/g, while sterile saline was sprayed onto the control. After the air was removed by a vacuum pump, the silos were stored at 25°C for 1, 3, 6 and 12 months. Microbial counts, fermentation products and aerobic stability were determined as previously described (Nishino *et al.*, 2004).

Results Ethanol and 2,3-butanediol prevailed over the fermentation in untreated whole crop rice silage; the lactic and acetic acid concentrations were low at 1 month after the ensiling but gradually increased as the storage was extended. Ethanol was lower at 12 compared with 6 months in both treatments whereas 2,3-butanediol was almost unchanged after reaching the maximum at 1 month of the storage. Addition of *L. buchneri* increased the lactic acid and decreased the ethanol and 2,3-butanediol. Acetic acid increased as the ensiling was prolonged and at 12 months took over the fermentation in the treated silage. The inhibitory effect of *L. buchneri* on fungal growth was unclear because no yeast was detected in silage with or without inoculation; however, the reduction of alcohols may imply some favourable effects on ensiling in the prevention of fungal activity.

Conclusions Enhanced ethanol and 2,3-butanediol production may occur when whole crop rice is ensiled without any additives. Inoculation of *L. buchneri* can decrease the alcohols, facilitating energy recovery and long-term storage.

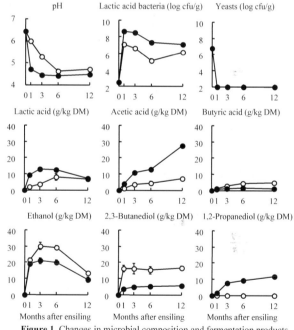

Figure 1 Changes in microbial composition and fermentation products during ensiling of whole crop rice inoculated with (●) or without *L. buchneri* (○)

References

Nishino, N., H. Wada, M. Yoshida & H. Shiota (2004). Microbial counts, fermentation products and aerobic stability of whole crop corn and a total mixed ration ensiled with and without inoculation with *Lactobacillus casei* or *Lactobacillus buchneri*. *Journal of Dairy Science*, 87, 2563-2570.

Microorganism occurrence in Tanzânia (*Panicum maximum* Jacq. cv. Tanzânia) grass silage exposed to the environment

R.M. Coan, R.A. Reis, G.R. Garcia, R.P. Schocken-Iturrino and E.D. Contato

Departamento de Zootecnia da Universidade Estadual Paulista "Julio de Mesquita Filho" Campus de Jaboticabal, Jaboticabal - SP. CEP: 14.870-000, Email: rogeriocoan@netsite.com.br

Keywords: aerobic stability, enterobacterium, *listeria*, pH

Introduction Silage stability depends on the aerobic fermentation (post-fermentation), which occurs after silo opening. Kung Jr. & Ranjit (1998) reported that in silages exposed to environmental air opportunist microorganisms, as those from *Enterobacterium* and *Bacillus* groups, start their metabolic activity producing heat, consuming nutrients and provoking losses around 15%. In addition, the growth of some pathogenic microorganisms should be propitiated, as an example of the *Listeria monocytogenes,* which causes listeriosis.

Material and methods Tanzânia (*Panicum maximum* cv. Tanzânia) grass harvested at 64 days of growth and mixed with 0, 5% or 10% of pellet citric pulp (PCP) (wet basis). Experimental silos were metal pipes with 200 litres of volume. Density adopted was around 550 kg/m^{-3}. After 210 days, silos were opened to evaluate the activity of the *Enterobacteria, Bacillus, Listeria* spp., *Listeria monocytogenes,* pH values and dry matter content measured at 0, 2, 4 and 6 days after silo opening.

Results In Figure 1 is illustrated the development of *Enterobacteria* and *Bacillus,* as well as the dry matter content in Tanzânia grass silage mixed with 0, 5% or 10% of PCP. *Enterobacteria* were not found in Tanzânia grass silage, probably due to the pH values (Figure 2) responsible by an inhibitory effect on these microorganism growths (McDonald *et al.* 1991). In relation to *Bacillus* occurrence, with the increased exposure period there was an increase of these microorganisms growth in silage with 0% of PCP. In silages with 5% and 10% of PCP, population of *Bacillus* decreased with the increase of exposure period. Dry matter content was very variable during the period of air exposure. During all periods of air exposure, *Listeria* spp. populations were found in silages mixed with 5 or 10% of PCP; and in silages with 0% of PCP were found *Listeria monocytogenes* (Figure 2).

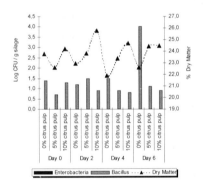

Figure 1 *Enterobacteria, Bacillus* and DM values

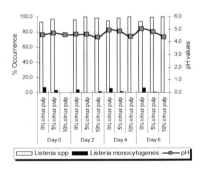

Figure 2 *Listeria* and pH values

Conclusions Pellet citric pulp addition did not produce pH conditions enough for reducing *Listeria* spp. and *Listeria monocytogenes* growths, which evidenced the sanitary risks related to the use of this kind of silage in ruminant nutrition.

References

Kung Jr., L. & N.K. Ranjit (1998). The effect of *Lactobacillus buchneri* and other additives on the fermentation and aerobic stability of barley silage. *Journal Dairy Science,* 84, 1149-1155.

McDonald, P., A.R. Herderson & S.J.E. Heron (1991). Biochemistry of silage. 2. ed. Marlow: Chalcombe Publication, 340 p.

Forage variety and maturity on fermentative losses of sugarcane silages added with urea

P. Schmidt, L.G. Nussio, C.M.B. Nussio, A.A. Rodrigues, P.M. Santos, J.L. Ribeiro, L.J. Mari, M. Zopollatto, M.C. Santos, O.C.M. Queiroz and D.P Souza
Department of Animal Science, University of São Paulo, ESALQ, Av. Pádua Dias 11, Piracicaba, SP, 13418-900, Brazil, Email: nussio@esalq.usp.br

Keywords: sugarcane silage, urea, fermentation losses, forage variety

Introduction   Sugarcane is one of the most important forage resources in Brazil. However, it has great limitations for conservation as silage, due to the yeast activity which yields high amounts of ethanol (7-15% of dry matter) and reduces the nutritive value of silage. Fermentation losses of dry matter (DM) can be greater than 30%. Former studies (Alvarez & Preston, 1976) showed the positive effects of urea addition inhibiting ethanol production and enhancing silage quality. The objective of this trial was to evaluate the effects of urea addition, forage maturity and variety on fermentative losses in sugarcane silages.

Material and methods   Sugarcane was harvested during the summer (December), autumn (March) and in the winter (June), when forage reached 12, 15 and 18 months of maturity, respectively. Treatments were assigned to a randomised factorial design (3x2x2x2) as, 3 forage maturities, 2 varieties (IAC87-3184 and IAC86-2480), 2 urea levels (0.0 and 0.5% - wet basis) and 2 storage periods (60 and 120 days), with four replications each. Experimental silos (20 litre plastic buckets) contained dry sand at the bottom to absorb effluent drained and a Bunsen type valve to allow gases to escape. Samples were taken at ensilage and silo opening, to determine DM content as described by AOAC (1980). Total DM losses, gases losses and effluent yield were determined considering initial and final silo and silage weights. Data were analysed by using PROC GLM of SAS (1996).

Results   Conservation losses were not changed across storage periods in sugarcane silages ($P<0.05$). There were significant effects of forage maturity and variety interactions (Table 1). The 3184 variety showed linear decrease in DM content as forage maturity increased, however, the 2480 variety did not follow this same pattern. The 3184 variety had higher effluent yield, which is in contrast with the lowest gases and DM losses, mainly at 12 and 15 months of maturity. Those results are not consistent with the current literature data. An interaction between sugarcane variety and urea level is shown in Table 2. Increased DM content in the silage was explained by urea addition only for the 2480 variety but not for 3184. Even though the 2480 variety showed lower DM content, those were the silages with less effluent. Urea addition was effective to decrease gases and total DM losses only within the 2480 variety which creates the opportunity for further research on this interaction.

Table 1 Forage maturity and variety effects on sugarcane silages losses

Variables	12 months 2480	12 months 3184	15 months 2480	15 months 3184	18 months 2480	18 months 3184	SEM
DM (%)	22.7^c	30.0^a	22.5^c	28.0^b	23.0^c	23.1^c	0.49
Effluent (l/t)	46.6^c	57.7^b	47.6^c	69.2^a	35.7^d	38.7^d	1.70
DM losses (%)	31.3^a	20.3^b	14.1^c	10.8^d	31.0^a	30.8^a	0.97
Gases losses (% DM)	27.7^a	15.2^b	9.7^c	4.0^d	28.3^a	27.9^a	1.01

^{a,b,c,d} ($P<0.05$)

Table 2 Forage variety and urea level effects on sugarcane silages losses

Variables	3184 0%	3184 0,5%	2480 0%	2480 0,5%	SEM
DM (%)	27.5^a	26.5^a	21.1^c	24.3^b	0.40
Effluent (l/t)	55.0^a	55.4^a	43.9^b	42.7^b	1.39
DM losses (%)	20.7^b	20.6^b	28.7^a	22.2^b	0.79
Gases losses (% DM)	15.8^c	15.6^c	25.3^a	18.6^b	0.82

^{a,b,c} ($P<0.05$)

Conclusions   The DM losses observed in sugarcane silages are extremely high and the control of fermentation is required to assure technical efficiency on the conservation process. Forage variety and maturity influenced silage DM losses. Urea added at 0.5% showed lower efficiency to decrease silage losses.

References
Alvarez, F.J. & T.R. Preston(1976). Ammonia/molasses and urea/molasses as additives for ensiled sugarcane. Tropical Animal Production, v.1, p. 98-104.
AOAC (1980). Official methods of analysis.13.ed. Washington, 1015 pp.
SAS (1996). SAS Institute Inc. SAS/STAT User's Guide. Volume 2, GLM-VARCOMP. Version 6. Fourth Edition. Cary.

Financial support provided by EMBRAPA (Brazil).

Effect of moisture on the fermentation and the utilisation by cattle of silages made from tropical grasses

M. Niimi, O. Kawamura, K. Fukuyama and S. Sei
Faculty of Agriculture, University of Miyazaki, Gakuen- kibanadai- nishi- 1- 1 ,Miyazaki, 889- 2192, Japan, Email: mitsu-n@cc.miyazaki-u.ac.jp

Keywords: silage fermentation, voluntary intake, daily gain, tropical grass

Introduction In tropical and subtropical regions, tropical grasses have been used for hay and silage making as well as grazing and soiling. Ensiling tropical grasses often results in different silages from temperate grass silages, especially more acetic acid (Catchpool & Henzell, 1971; Sujahta *et al.,* 1986; Niimi & Kawamura, 1998). However, there are few data on the fermentation and utilisation by cattle of silage made from tropical grasses. In this study, silages were made from guineagrass and rhodesgrass at different levels of moisture.

Materials and methods Guineagrass and rhodesgrass grown in the experimental field of Miyazaki University were harvested at heading stage on 5 August, 2004. They were wilted and rolled into round bale silages, wrapped with white elastic film. About two months later, the silages were unloaded and observed for fermentation characteristics and voluntary intake by cattle. The animal experiment was carried out in a 3x3 Latin square design, using 3 Japanese Black cattle.

Table 1 Silage fermentation and utilisation by cattle

		Moist (%FM)	pH	LA (%DM)	AA (%DM)	BA (%DM)	VBN/TN (%)	FMI (kg/day)	DMI/MLW (g/day/kgBW$^{0.75}$)	DG (g)	DG/DMI
	MA	76.6	5.68	0.02	0.03	0.00	2.5				
	70%	72.1	4.79	0.22	1.50	1.68	15.0	28.7	83.1	238	30
Guineagrass	60%	61.5	5.21	0.75	0.74	1.12	16.1	19.6	78.0	1048	139
	50%	49.7	5.14	1.39	0.29	0.41	12.8				
	40%	40.5	5.34	1.09	0.39	0.03	9.5	13.8	84.7	3048	370
	MA	77.7	6.19	0.02	0.12	0.00	2.1				
	70%	75.6	5.01	0.12	2.44	3.37	29.3	31.3	76.6	143	18
Rhodegrass	60%	63.4	5.67	0.94	0.97	1.84	24.6	22.0	80.1	1333	166
	50%	47.3	5.72	2.09	0.51	0.80	16.3				
	40%	37.0	5.96	0.99	0.30	0.01	8.4	18.2	112.7	3667	322

Moist:Moisture, FM:Fresh Matter, DM:Dry Matter, LA:Lactic Acid, AA:Acetic Acid, BA:Butyric Acid, VBN:Volatile Basic Nitrogen, TN:Total Nitrogen. FMI:Fresh Matter Intake, DMI:Dry Matter Intake, MLW:Metabolic Live Weight, DG: Daily Gain, MA:Material.

Results In most silages, pH values were over 5. The values for low moisture silages were relatively high, compared with the high moisture silages. The concentration of lactic acid was highest at 50% moisture level. The acetic acid, butyric acid and VBN concentrations were higher at the high moisture level than at the low moisture level. In the voluntary DM intake of guineagrass silage, there was little difference among the moisture levels. On the other hand, in rhodesgrass, the low moisture silage resulted in higher intakes than the high moisture silage. The gain of cattle fed the low moisture silage was higher than that of the high moisture silage.

Conclusions Wilting to 50% moisture level increased lactic acid concentration but more wilting did not. The production of acetic acid, butyric acid and VBN were depressed with decreasing moisture content. Effects of moisture on DM intake by cattle was different between guineagrass and rhodesgrass, but daily gain was consistently higher in cattle fed low moisture silages.

References

Catchpool, V.R. & E.F. Henzell (1971). Silage and silage-making from tropical herbages species. *Herbage Abstracts*, 41, 213-221.
Niimi, M. & O. Kawamura (1998). Degradation of cell wall constituents of guineagrass (*Panicum maximum* Jacq.) during ensiling. *Grassland Science*, 43, 413-417.
Sujahta, P., V.G. Allen, J.P. Fontenot & M.N. Jayasuriya (1986). Ensiling characteristics of tropical grasses chopping as influenced by stage of growth, additives and chopping length. *Journal of Animal Science*, 63, 197-207.

Section 5

Chemical and biological characterisation of silages

Estimation of legume silage digestibility with various laboratory methods

A. Olt[1], M. Rinne[2], J. Nousiainen[3], M. Tuori[4], C. Paul[5], M.D. Fraser[6] and P. Huhtanen[2]
[1]Estonian Agricultural University, Department of Animal Nutrition, Kreutzwaldi 1, 51014 Tartu, Estonia, [2]MTT Agrifood Research Finland, Animal Nutrition, FIN-31600 Jokioinen, Finland, marketta.rinne@mtt.fi, [3]Valio Ltd., Farm Services, P.O. Box 10, FIN-00039 Valio, Finland, [4]University of Helsinki, Dept. Animal Science, P.O. Box 28, FIN-00014 University of Helsinki, Finland, [5]Federal Agricultural Research Centre (FAL), Bundesallee 50, D-38116 Braunschweig, Germany, [6]Institute of Grassland and Environmental Research, Plas Gogerddan, Aberystwyth, Ceredigion SY23 3EB, UK

Keywords: *in vitro*, *in situ*, *in vivo*, indigestible fibre, red clover, lucerne, galega

Introduction   The current feed evaluation and ration formulation systems require accurate and precise measurements of digestibility. The objective of the present study was to compare the potential of different laboratory methods in estimating digestibility of silages made from leguminous forage species as pure stands.

Materials and methods   Legume silage samples with measured *in vivo* organic matter (OM) digestibility (OMD; measured with sheep using total faecal collection) were obtained from 7 trials conducted in Finland, Germany and UK. The dataset comprised 22 red clover (*Trifolium pratense*), 7 lucerne (*Medicago sativa*) and 4 galega (*Galega orientalis*) samples. Ash, crude protein (CP), neutral detergent fibre (NDF), acid detergent fibre (ADF), lignin, organic matter pepsin-cellulase solubility (OMS) and indigestible NDF (INDF; measured *in situ* by a 12-day rumen incubation in a nylon bag with a pore size of 17 µm) of the samples was determined in the laboratory of MTT. *In vitro* OM digestibility was assessed according to the Tilley and Terry method ($OMD_{T\&T}$) at the University of Helsinki. The relationships between OMD and the various laboratory methods were estimated using overall fixed or mixed (trial as a random factor) regression models using SAS MIXED procedure.

Table 1 Description of samples (g/kg)

	Mean	Std. Dev.
Ash	107	18.7
Crude protein	208	33.3
NDF	384	71.4
ADF	270	67.8
Lignin	52	22.6
INDF	132	64.1
OMS	736	66.8
$OMD_{T\&T}$	621	62.7
In vivo OMD	692	61.8

Results   The average chemical composition and digestibility of the samples is presented in Table 1. The digestibility of legume samples could be estimated most accurately with INDF when the fixed model was used (Table 2). The results are consistent with earlier results from silages prepared from gramineous forage species (Nousiainen, 2004). Of the two *in vitro* digestibility measurements, commercial fungal cellulase method (OMS) was more accurate than the method based on rumen liquor ($OMD_{T\&T}$). Accuracy of different cell wall fractions (ADF, NDF and lignin) was quite similar, but CP was clearly poorer in estimating OMD. The potential of the chemical components was not markedly improved even if used in a bivariate regression analysis (results not shown). The ranking order of the various methods was altered if a mixed regression model with trial as a random factor was used (results not shown). OMS was superior (RMSE 11.3 g/kg, adj. R^2 0.965), $OMD_{T\&T}$ second (RMSE 14.9 g/kg, R^2 0.940) and INDF third (RMSE 15.7 g/kg, R^2 0.930). The small difference in RMSE between fixed and mixed regression models for INDF reflects a more universal relationship with OMD compared with the other laboratory methods.

Table 2 Prediction of *in vivo* OMD with various laboratory methods (g/kg; overall fixed regression model)

Model	Intercept	SE[1]	*P*-value	Slope	SE	*P*-value	RMSE[2]	R^2 adj.[3]
INDF	811	9.1	<0.001	-0.90	0.062	<0.001	16.8	0.925
OMS	42	36.5	0.262	0.88	0.049	<0.001	18.7	0.909
$OMD_{T\&T}$	122	40.7	0.005	0.92	0.065	<0.001	23.1	0.860
ADF	916	18.9	<0.001	-0.83	0.068	<0.001	26.1	0.822
NDF	994	25.3	<0.001	-0.79	0.065	<0.001	26.2	0.821
Lignin	819	12.0	<0.001	-2.46	0.214	<0.001	27.3	0.805
Crude protein	479	58.5	<0.001	1.03	0.278	<0.001	52.4	0.282

[1] Standard error, [2] Residual mean square error, [3] Adjusted for degrees of freedom

Conclusions   The OMD of legume silages could be estimated with high accuracy by INDF and *in vitro* OMD methods, but not from the chemical composition. The results suggest that INDF could replace traditional *in vitro* OMD measurements in feed evaluation, but the need for surgically modified animals may limit its usefulness.

References

Nousiainen, J. (2004). Development of tools for the nutritional management of dairy cows on silage-based diets. Available at *http://ethesis.helsinki.fi/julkaisut/maa/kotie/vk/nousiainen/*

Evaluation of prediction equations for metabolisable energy concentration in grass silage used in different energy feeding systems

T. Yan and R.E. Agnew
Agricultural Research Institute of Northern Ireland, Hillsborough, Co Down BT26 6DR, UK,
Email: tianhai.yan@dardni.gov.uk

Keywords: grass silage, sheep, metabolisable energy, prediction equation, evaluation

Introduction The metabolisable energy (ME) concentration in grass silage is used in energy feeding systems across the world and is normally estimated from digestible nutrient concentrations. The objective of the present study was to evaluate these recommended equations using grass silage data obtained at this Institute.

Materials and methods The data used in the present evaluation were derived from grass silages (n = 136) offered to sheep (4 animals/silage) as the sole diet at maintenance feeding level. Parameters measured included silage chemical composition, nutrient digestibility and urine energy output. The sheep were male Greyface and approximately two years old with live weights between 45 and 50 kg; and were housed in individual pens for three weeks before a 6-day total collection of faeces and urine in metabolism crates. The silages encompassed primary growth and first and second regrowth perennial ryegrass. The grass was either unwilted or wilted prior to ensiling and ensiled with or without application of silage additives. Silage DM concentration was determined on a toluene basis and ME concentration estimated using predicted methane energy output. A number of prediction equations were evaluated, with silage ME concentration predicted from digestible OM (DOM) alone (AFRC, 1993; SCA, 1990; Van Es, 1978 (Equation 1)), DOM and digestible crude protein (Van Es, 1978, Equation 2), or digestible energy concentration which is estimated from nutrient digestibilities (NRC, 2001; INRA, 1989).

Results The silages had a wide range in quality and a relatively even distribution over the range. Silage DM ranged from 0.155 to 0.413 kg/kg, ME from 6.8 to 13.3 MJ/kg DM, DOM from 0.528 to 0.769 kg/kg DM, pH from 3.5 to 5.5 and ammonia-N/total-N from 0.037 to 0.385 g/g. The evaluation results are presented in Table 1. The mean ME concentration predicted from NRC (2001) and INRA (1989) was similar to the mean actual value, but AFRC (1993) over-predicted ME concentration by proportionately 0.02 and SCA (1989) and Van Es (1978, Equations 1 and 2) under-predicted by 0.02, 0.04 and 0.03 respectively. In comparison with other equations, INRA (1989) prediction produced a higher R^2 in the relationship between actual and predicted ME concentration, a lower standard error and a lower mean-square prediction error (MSPE). Similarly, INRA (1989) had a lower standard deviation and a smaller range in the residual ME concentration (actual minus predicted).

Table 1 Evaluation of a range of prediction equations for metabolisable energy concentration in grass silage

	ME (MJ/kg DM)					Actual – Predicted ME (MJ/kg DM)			
	Actual	Predicted	s.e.	R^2	MSPE	Mean	s.d.	Min	Max
AFRC (1993)	10.2	10.5	0.50	0.72	0.40	-0.22	0.59	-1.71	1.68
SCA (1989)		10.0	0.56	0.72	0.43	0.27	0.60	-1.24	2.03
Van Es (Equation 1)		9.8	0.47	0.72	0.51	0.39	0.60	-1.14	2.37
Van Es (Equation 2)		9.9	0.50	0.72	0.44	0.30	0.59	-1.06	2.11
INRA (1989)		10.2	0.32	0.91	0.12	0.04	0.34	-0.71	1.42
NRC (2001)		10.2	0.66	0.70	0.44	0.03	0.67	-1.69	2.00

Conclusion Use of the INRA (1989) equation to estimate grass silage ME concentration produced a more accurate prediction than other equations evaluated in the present study.

References
AFRC (1993). Energy and protein requirements of ruminants. CAB International, Wallingford, Oxon, UK.
INRA (1989). Ruminant nutrition - Recommended allowances and feed tables. John Libbey Eurotext, Paris, France.
NRC (2001). Nutrient requirements of dairy cattle. 7th revised edition, National Academic Science, Washington, D.C., USA.
SCA (1990). Feeding standards for Australian livestock - Ruminants. CSIRO, Australia.
Van Es, A.J.H. (1978). Feed evaluation for ruminants. 1. The systems in use from May 1977 onwards in the Netherlands. *Livestock Production Science*, 5, 331-345.

The effect of fermentation quality on voluntary intake of grass silage by growing steers

S.J. Krizsan and Å.T. Randby
Department of Animal and Aquacultural Sciences, Norwegian University of Life Sciences, P.O. Box 5003, NO-1432 Ås, Norway, Email: sophie.krizsan@umb.no

Keywords: fermentation quality, silage, voluntary intake

Introduction The variation in fermentation quality in low dry matter (DM) grass silage is large and affects the voluntary intake. There is no definitive agreement of which indices of fermentation quality that should be included in intake prediction equations (Rook & Gill, 1990; Steen *et al.*, 1998; Huhtanen *et al.*, 2002). The objective of the present study was to describe differences in feed intake by fermentation quality alone. Among several analysed parameters, organic acids, N-fractions and biogenic amines are included in the present study.

Materials and methods On 3 to 5 June 2002, 24 different silage qualities were produced from the same sward. The grass was harvested with a flail-harvester, a precision chopper, a self-loading wagon, or a roundbaler. Different additives and application rates were allocated the experimental silages. The grass was ensiled in 6 m^3 tower silos, a stack silo or in round bales. The silages were offered 30 Norwegian Red steers (initially 137 kg live weight, s.e. 16.4) as sole feed in a partially balanced changeover experiment, with pre-experimental periods where standard silage was fed. DM intake was recorded daily. Silage samples were analysed for DM, crude protein (CP), neutral detergent fibre (NDF) and quality parameters as presented below. Statistical analysis was carried out using MIXED procedures of SAS. The model was fitted to comprise silage effect with intake data from the pre-experimental periods as covariate. Linear and quadratic relationships between individual silage parameters or groups of parameters and corrected silage intake were examined. Multiple regression relationships between silage components and intake have been produced in SAS with forward and stepwise selection.

Results Daily corrected silage DM intake (SDMI) varied between 1.79 and 2.65, with mean 2.38 kg/100 kg live weight. Ranges (mean value) for the chemical components of the silages were; DM 166–237 (213) g/kg, CP 163–193 (174), NDF 476–601 (544), water soluble carbohydrates 16.3–70.9 (33.0), acetate (AA) 11.5–64.7 (28.6), propionate (PA) 0-5.2 (1.0), butyrate (BA) 0-25.1 (6.0), lactate (LA) 2.2-102 (49.3), total volatile fatty acids (TVFA) 11.5-85.9 (35.5), total acids (TA) 48.3-141 (84.9), LA/TA 0.025-0.827 (0.584), 2-phenyl-ethylamin 0-0.257 (0.100), histamine (HIS) 0-1.43 (0.347), tryptamine (TRP) 0-0.643 (0.085), tyramine 0.294-2.68 (1.49), putrescine 0.174-3.73 (1.44), cadaverine (CAD) 0.122-5.41 (1.36) total amines (TAM) 0.975-13.8 (4.82) g/kg DM, ammonium-N (NH$_3$-N) 89.3-255 (153), acid detergent insoluble-N (ADIN) 12.2-28.9 (18.9), true soluble protein 4.00-75.2 (37.0) and nonprotein-N 461-719 (605) g/kg total N. Best significant (P-value <0.05) simple regressions on intake are presented in Table 1. All chemical components and their squared values if significant in the quadratic relationship were included in the multiple selection methods. The best multiple model was: SDMI = 2.58 – 0.082 PA + 0.0272 BA – 0.00168 BA^2 – 0.0000378 LA^2 (R^2$_{adj}$ = 0.838 R^2$_{pred}$= 0.729).

Table 1 Best significant linear (Y = A + BX) and quadratic (Y = A + BX + CX^2) relationships.

	A	B	C	P-value	R^2$_{adj}$
DM	0.8937	0.0070		0.0010	0.367
AA	2.6262	-0.0086		0.0007	0.384
PA	2.4674	-0.0896		<0.0001	0.527
BA	2.3943	0.0290	-0.00184	0.0069	0.318
LA	1.9700	0.0193	-0.00018	0.0018	0.399
TVFA	2.5811	-0.0057		0.0008	0.380
TA	2.8369	-0.0054		0.0050	0.276
LA/TA	1.8955	1.8344	-1.48839	0.0062	0.325
ADIN	2.7058	-0.0172		0.0355	0.149
NH$_3$-N[1]	1.2966	0.0150	-0.00005	0.0060	0.328
HIS	2.4707	-0.2600		0.0081	0.245
TRP	2.4542	-0.8721		0.0003	0.432
CAD	2.4744	-0.0692		0.0189	0.191
TAM	2.5078	-0.0264		0.0337	0.152

[1] Not corrected for additive-derived N.

Conclusions In this study, where an early harvested grass crop was used, neither the N compounds nor WSC, were of importance in predicting SDMI. The best multiple model included only individual organic acids.

References
Huhtanen, P., H. Khalili, J.I. Nousiainen, M. Rinne, S. Jaakkola, T. Heikkilä & J. Nousiainen (2002). Prediction of the relative intake potential of grass silage by dairy cows. *Livestock Production Science*, 73, 111-130.
Rook, A.J. & M. Gill (1990). Prediction of the voluntary intake of grass silages by beef cattle. 1. Linear regression analyses. *Animal Production*, 50, 425-438.
Steen, R.W.J., F.J. Gordon, L.E.R. Dawson, R.S. Park, C.S. Mayne, R.E. Agnew, D.J. Kilpatrick & M.G. Porter (1998). Factors affecting the intake of grass silage by cattle and prediction of silage intake. *Animal Science*, 66, 115-127.

Determination of toxic activity of mould-damaged silage with an *in vitro* method

A. Solyakov[1] and T. Pauly[2]
[1]*Department of Animal Feed, National Veterinary Institute (SVA), SE-751 89 Uppsala, Sweden, Email: Alexey.Solyakov@sva.se, 2Department of Animal Nutrition, Swedish University of Agricultural Sciences (SLU), SE-753 23 Uppsala, Sweden*

Keywords: silage, moulds, mycotoxins, toxicity, *in vitro*

Introduction   Mould growth in animal feeds can cause the production of mycotoxins (the toxic secondary metabolites of moulds). Intoxication by mycotoxins can cause suffering in animals, economic losses and, possibly, the transfer of toxins to humans via foodstuffs. Therefore, there is a need to be able to identify feeds of questionable quality that may pose hazards to animals and/or humans. The objective of the study was to evaluate the possibility of using a cellular *in vitro* technique as a screening method for monitoring toxic components in mould-damaged silage.

Materials and methods   A grass crop was harvested on two occasions, in June and August 2003, wilted to approximately 45% and 55% DM, respectively, and ensiled in laboratory silos (volume 25 litres) in three different ways (6 silos/method):
1. In laboratory silos with a hole (d = 2.7 mm) in the lid, no spore suspension applied.
2. A spore suspension containing *P. roqueforti* and *A. fumigatus* (the most common moulds in silage in Scandinavia) was applied to the fresh forage (approx. 10^3 cfu/g FM) before it was ensiled in silos as above.
3. In laboratory silos with oil-filled fermentation traps on the lids (anaerobic storage, referred to below as control method).

Silage was stored at ambient temperature (range 10-23°C) and samples were taken from 3 silos/method after 45 and 90 days of ensiling. The area of visible mycelium was estimated using a 3×3 cm grid. Samples represented the homogenised upper part (2 kg of total 9.5 kg) of the silage. The samples, including the fresh forage, were extracted using acetonitrile/water (86:14) (for both non- and polar compounds), and purified using a solid-phase extraction/purification method. The purified extracts were applied to human neuroblastoma SH-SY5Y cells, and the general cytotoxicity was determined as described by Wenehed *et al.* (2003). Positive controls, spiked with the mycotoxin patulin, were included in the study.

Results and discussion   No mould growth was visible in any of the silage samples stored anaerobically (method 3). After 45 days, between 20% and 55% of the silage surface was mould infested in the silos open to air (methods 1 and 2). Equivalent values after 90 days were 45% and 85% surface coverage by moulds in these silos. Visible mould occurrence was estimated to be slightly more widespread in forage treated with the spore suspension (method 2) than in the untreated silage (method 1). After 45 days of ensiling, mould-damaged silage (methods 1 and 2) was more cytotoxic than control silage (method 3), showing up to 67% inhibition, compared with up to 45% inhibition in the control method. After 90 days ensiling the samples showed a similar trend, up to 88% inhibition for the mouldy samples from methods 1 and 2. Interestingly, samples ensiled using method 1 (no spore suspension) were more toxic than samples from method 2 (spore suspension applied). One explanation of this may be that *P. roqueforti* and *A. fumigatus* (method 2) inhibited the growth of other moulds, while in the silage treated according to method 1 different moulds could grow unrestrained. These (other) moulds may have produced secondary metabolites, which had toxic effects on the SH-SY5Y cells. The search for toxic secondary metabolites with standard chemical methods (e.g. HPLC, LCMS) is possible but very time and labour consuming, as well as expensive. Thus the reported cell-based *in vitro* method can provide a practical and more realistic method of evaluating general toxicity in silage and other feed matrices. Further studies are ongoing with the aim of evaluating the usefulness of this method for other animal feed matrices and moulds. Other investigations will, however, be necessary to determine toxic substances other than mycotoxins. The results of this study are important in the light of the requirements of EU legislation concerning the safety of animal feeds.

Conclusions   The results show that this *in vitro* technique based on SH-SY5Y cells can be useful for screening toxic mould metabolites in mould-damaged silage.

Acknowledgment   The financial support of the Swedish Farmers' Foundation for Agricultural Research (Stiftelsen Lantbruksforskning, SLF) is gratefully acknowledged.

References
Wenehed, V., A. Solyakov, I. Thylin, P. Häggblom & A. Forsby (2003). Cytotoxic response of *Aspergillus fumigatus*-produced mycotoxins on growth medium, maize and commercial animal feed substrates. *Food Chemistry and Toxicology*, 41, 395-403.

Butyric acid bacteria spores in whole crop maize silage

F. Driehuis and M.C. te Giffel
NIZO food research, PO Box 20, 6710 BA Ede, The Netherlands, Email: Frank.Driehuis@nizo.nl

Keywords: *Clostridium tyrobutyricum*, butyric acid bacteria, spores, maize silage, aerobic deterioration

Introduction Growth of *Clostridium tyrobutyricum* is usually associated with anaerobically unstable silages. Bacteria of this species, also called butyric acid bacteria (BAB), are capable of converting lactic acid into butyric acid under acidic conditions. BAB spores from silage can be transmitted to milk via the cow's faeces. The spores survive the pasteurisation of milk and cause off-flavours and texture defects in various cheese types. Dry matter, sugar and nitrate content of the crop are key factors for clostridial growth in silage. Generally, whole crop maize silage is not considered a hazard of BAB spores, since the ensilability of the crop is high (high acidification rate, final pH usually <4). However, increases in BAB spore count may occur after opening the silo, when pH rises due to aerobic deterioration, as has been shown for grass silage (Jonsson, 1991) and in laboratory studies with maize silage too (Driehuis, unpublished results). The objective of this study was to investigate whether maize silage is a significant source of BAB spores at dairy farms in The Netherlands.

Materials and methods BAB spores were detected with a most probable number method (Van Beynum and Pette, 1935). Samples analysed: core samples of 184 maize silages from Dutch farms (sampled before opening of the clamps); samples from the open front of maize and grass silages from an experimental dairy farm, sampled every 2 weeks during one year; samples from the open front of maize and grass silages, mixed roughage ration and bulk tank milk from six dairy farms suspected to produce milk with a high BAB spore concentration.

Results Low levels of BAB spores were detected in core samples of maize silage: 77% contained <10^3 spores/g and 23% contained between 10^3 and 10^4 spores/g. In contrast, during a 1-year survey of an experimental dairy farm, BAB spores in samples from the open front of maize silage exceeded 10^4/g in 6 out of 24 instances and exceeded 10^6/g in 2 instances (Figure 1). BAB spores in grass silage samples of the same farm exceeded 10^4/g in 1 instance only. BAB spores exceeded 10^4/g in mixed roughage ration of all 6 'suspected' dairy farms (Table 1). For 5 of these farms maize silage appeared the major source of BAB spores. High spore counts were detected in surface layers and/or moulded spots, whereas low counts were detected in the central area of the silage front.

Table 1 BAB spores in silage, mixed roughage ration (log MPN/g) and milk (MPN/ml) of dairy farms

Farm	Maize			Grass		Mixed ration	Milk
	cen[1]	sur	mou	cen	sur		
A	2.6	4.6	6.4	2.6	5.0	4.4	0.2
B	3.4	4.9	6.0	2.4	3.0	4.4	0.4
C	4.0	6.0	6.0	<1.6	<1.6	5.2	0.4
D	4.4	7.4	7.5	3.2	4.2	5.1	1.5
E	4.2	6.2	6.0	<1.6	2.2	5.7	2.3
F	2.6	4.6	4.9	3.6	3.4	5.4	4.6

[1]cen, centre; sur, surface; mou, moulded spot

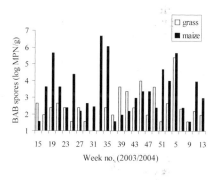

Figure 1 BAB spores in silage front samples

Conclusions Maize silage is an important source of BAB spores at dairy farms in The Netherlands. The results indicate that BAB spores in maize silage accumulate in surface layers and moulded spots. This suggests a relationship with aerobic instability problems. We hypothesise that growth of BAB in maize silage occurs in anaerobic niches of aerobically deteriorating layers.

References
Jonsson, A. (1991). Growth of Clostridium tyrobutyricum during fermentation and aerobic deterioration of grass silage. Journal of the Science of Food and Agriculture, 54, 557-568.
Van Beynum, J. & J.W. Pette (1935). Zuckervergärende und laktatvergärende Buttersäurebakterien. Zentralblatt für Bakteriologie, Parasitenkunde, Infektionskrankheiten und Hygiene (Abt. II), 93, 198-212.

Effects of the stage of growth and inoculation on proteolysis in field pea silage

L. Cavallarin[1], G. Borreani[2], S. Antoniazzi[1] and E. Tabacco[2]
[1]ISPA CNR, Via L.da Vinci 44, 10095 Grugliasco, Italy, Email: laura.cavallarin@ispa.cnr.it, [2] Dip. Agronomia, Selvicoltura e Gestione del Territorio, Università degli Studi di Torino, Italy

Keywords: legume, field pea, lactic acid bacteria, stage of development, proteolysis

Introduction Ensiling legumes is a good way of providing home-grown protein in dairy farms but severe protein degradation can occur when conserving legumes. Peas (*Pisum sativum* L.) are legumes with a high crude protein and starch content, that provide a high forage yield in a short growing period. Very little information is available on the protein value of field pea silage. The aim of the study was to investigate the effect of stage of maturity and inoculant application on proteolysis in field pea silage in the Po Valley, NW Italy.

Material and methods Stands of semi-leafless field peas were sown on 21 March 2001. The herbage was harvested at 4 progressive morphological stages (end of flowering, I; beginning of pod filling, II; advanced pod filling, III; beginning of ripening, IV) between 1 and 21 June. The herbage was chopped and ensiled in 2-l glass silos with (I) and without (C) an inoculant (*Lactobacillus plantarum* strain for legume crops, CSL, Italy*).* The silages were analysed after 60 d for the nitrogen fractions and the amino acid content. Free amino acids were determined according to Winters *et al.* (2002) and total amino acids as reported in Cavallarin *et al.* (2005).

Results The stage of growth significantly affected all the nitrogen fraction concentrations. Inoculation treatment lowered the ammonia and total amino acid concentrations in all the silages, except in the first stage. Extensive proteolysis occurred in the I, II and III stages, as shown by the NPN and free amino acid values, while it was significantly reduced in the IV stage. The amino acid composition of silages (d 60) made at the beginning of ripening (IV) was close to that of fresh herbage (d 0), with minimal losses of nutritionally essential amino acids for ruminants. Major changes occurred between 0 and 60 d in silages made beginning of pod filling (II). Differences were also evident, to a lesser extent, in the silages from the I and III stages (data not shown).

Table 1 Nitrogen fractions of control (C) and inoculated (I) pea silages at four stages of growth

S[A]		DM[B]	TN	NH₃-N	NPN	FAA	TAA
I	C	143	39	137	782	28.7	29.8
	I	148	39	134	880	29.4	30.1
II	C	163	35	116	728	29.2	28.9
	I	158	36	52	708	23.1	22.3
III	C	188	32	131	708	25.4	32.3
	I	198	32	68	673	24.1	28.6
IV	C	212	33	132	517	17.7	32.2
	I	209	32	105	503	19.0	31.8
	S	**[C]	***	***	***	***	***
	I	NS	NS	***	NS	NS	*
	SxI	NS	NS	*	NS	**	**

[A] S = stage of growth, C = Control, I = inoculation;
[B] DM = dry matter (g/kg);
TN = total nitrogen (g/kg DM); NH₃-N and NPN (g/kg TN);
FAA and TAA = free and total amino acids (mol/kg TN);
[C] NS = $P>0.05$; * = $P<0.05$; ** = $P<0.01$; *** = $P<0.001$

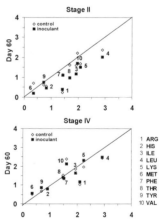

Figure 1 Concentration of essential amino acids (mol/kg TN) in fresh herbage (day 0) plotted against values for 60 d silages

Conclusions In silages at the beginning of ripening proteolysis was reduced in comparison to the earlier stages and the concentration of nutritionally essential amino acids for ruminants was close to that of herbage at cutting.

References
Cavallarin L., S. Antoniazzi, G. Borreani & E. Tabacco (2005). Effects of wilting and mechanical conditioning on proteolysis in sainfoin (*Onobrychis viciifolia* Scop.) wilted herbage and silage. *Journal of the Science of Food and Agriculture* (in press).

Winters A.L., J. Lloyd, R. Jones & R. Merry (2002). Evaluation of a rapid method for estimating free amino acids in silage. *Animal Feed Science and Technology*, 99, 177-187.

Ruminal proteolysis in forages with distinct endopeptidases activities

G. Pichard, C. Tapia and R. Larraín
Department of Animal Sciences, School of Agriculture, Pontificia Universidad Católica de Chile Email: gpichard@puc.cl

Keywords: forages, proteolysis, rumen, plant peptidases

Introduction Improving livestock efficiency in utilisation of nitrogen resources continues to be a major environmental and economic objective. Zhu *et al.* (1999) have shown that plant endopeptidases are activated as a response to cutting stress. Previous work in our laboratory explored over 300 entries of forage genotypes and found a broad diversity in enzymatic activity by means of hydrolysis in gelatine and direct autolysis assays in forage tissues. The objective of this work was to assess if the species previously identified as having high or low endopeptidase activity, would behave consistently when exposed to ruminal microbial proteolysis.

Materials and methods Two groups of forages were selected according with their level of peptidase activity (Table 1). They were grown in a greenhouse, fresh leaf samples were collected in vegetative stages, submitted to a molar-like pressing device and further chopped to 1 cm. The rumen fluid inocula was subjected to 3 hours pre-incubation with sugar for depletion of free N and further 2 hours with hydrazine-cloramphenicol inhibitors (Broderick, 1987). Fresh forage samples (2 g) were incubated *in vitro* during 6 hours (T_0 to T_6) with inhibited rumen fluid (IRF). The residue of neutral detergent insoluble nitrogen (NDIN) (Licitra *et al.*, 1996) and non protein soluble nitrogen (NPSN) were determined.

Table 1 Extent and rates of solubility of N compounds incubated in inhibited rumen fluid

Species		EPA*	Total N (g/kg)	NDIN (% of TN)			NPSN (% of TN)		
				T_0	T_6	k_{0-6}	T_0	T_6	k_{0-6}
Avena strigosa cv. Negra		High	32	46	20	4.4	19	32	2.
Festuca arundinacea cv. Conway		High	23	62	17	7.5	22	31	1.5
Medicago sativa cv. Innovator		High	35	56	18	6.3	21	35	2.4
Trifolium repens cv. Blanca		High	37	51	19	5.4	22	34	1.9
	Mean			54	19	5.9	22	33	1.9
Bromus unioloides cv. M. Fierro		Low	25	62	31	5.3	20	27	1.2
Lolium hybridum cv. Galaxy		Low	25	54	29	4.3	20	28	1.4
Trifolium pratense cv. Resistenta		Low	37	48	32	2.7	19	27	1.3
Trifolium repens cv. Kopu		Low	33	50	38	1.9	17	25	1.2
	Mean			53	32	3.6	19	27	1.3
Effect of endopeptidases (*P* value)				NS	0.003	0.06	NS	0.012	0.03

* Endopeptidases activity

Results and discussion The extents and rates of NDIN disappearance and NPSN accumulation were statistically different between the groups of high and low enzymatic activity (Table 1). The extent of nitrogen solubilised from the ND residue was not accounted for by the fraction of NPSN, thus suggesting that it remains as a soluble true protein. Also, within the group with high endopeptidase activity, no major differences were observed between legumes and grasses.

Conclusion The activity of plant endopeptidases varies among different germplasms and affects ruminal proteolysis. Such plant diversity supports the idea that one way of improving nitrogen utilisation in pasture-based animal production systems is the genetic improvement of germplasms that have the potential for delayed protein hydrolysis. The laboratory enzymatic assays with gelatine showed to be consistent with the *in vitro* rumen microbial proteases. The large fraction of soluble proteins released in rumen soon after eating suggests that the potential for rumen by pass with the liquid phase may be high and should be reassessed.

Acknowledgements Funded by Fondecyt Grant 1030918

References
Broderick, G.A. (1987). Determination of protein degradation rates using a rumen *in vitro* system containing inhibitors of microbial nitrogen metabolism. *British Journal of Nutrition*, 58, 463-475.
Licitra, G., T.M. Hernández & P.J. Van Soest (1996). Standardization of procedures for nitrogen fractionation of ruminant feeds. *Animal Feed Science and Technology*, 57, 347-358.
Zhu W-Y., A.H. Kingston-Smith, D. Troncoso, R.J. Merry, D.R. Davies, G. Pichard, H. Thomas & M.K. Theodorou (1999). Evidence of a role for plant proteases in the degradation of herbage proteins in the rumen of grazing cattle. *Journal of Dairy Science*, 82, 2651-2658.

Effects of particle size in forage samples for protein breakdown studies

G. Pichard and C. Tapia

Department of Animal Sciences, School of Agriculture, Pontificia Universidad Católica de Chile, Email: gpichard@puc.cl

Keywords: forages, particle size, proteolysis, rumen, nitrogen solubility

Introduction Coupling ruminal processes of hydrolysis and synthesis continues to be a research issue where more progress is needed. This requires the development of good protein assessment methods, particularly when representing the breakdown processes that occur in fresh pastures eaten by herbivores. Laboratory analyses need to deal with small and homogeneous samples, but the mechanical reduction of particle size may not reflect the actual digestion kinetics occurring when the original fresh forage is consumed. Such physical traits may alter the release of non-structural compounds and the penetration of microbial enzymes (Boudon *et al.*, 2002). The objective of this work was to assess in fresh samples the effect of reducing particle size upon the *in vitro* breakdown of proteins during the early rumen fermentation period.

Materials and methods Eight fresh forage samples with contrasting endopeptidase activities were subjected to different strategies for particle size reduction. Protein hydrolysis was assessed by measuring the residual neutral detergent insoluble nitrogen (NDIN) (Licitra *et al.*, 1996) and the accumulation of non-protein soluble nitrogen (NPSN) after 6 h *in vitro* rumen fermentation (IIV, Broderick, 1987). In fresh samples mastication-like damage was obtained with a device in which forage samples were pressed between two stony surfaces that simulated the animal molar surfaces. During the development of this method, microscopic observation was used in order to obtain a similar damage to that observed in samples obtained from the cardias of a fistulated adult cow fed the same type of fresh long forage. Three chopping sizes and two macerations were tested. Chopping was preceded by laboratory-mastication and further cutting to 3 cm, 1 cm or 0.25 cm; maceration was thoroughly done in a mortar with dry ice (CO_2) or liquid nitrogen. Treatments means were compared by Tukey-Kramer test at $P<0.05$.

Results and discussion Sample size significantly affected ($P<0.05$) the fractions of NDIN and NPSN (Figure 1), but the two macerates were essentially identical. As expected, the smaller the chopping the greater the solubility, with this effect being more pronounced in cultivars with lower endopeptidase activity. Mechanical particle comminution may facilitate access of external enzymes and activate the endogenous enzymatic system.

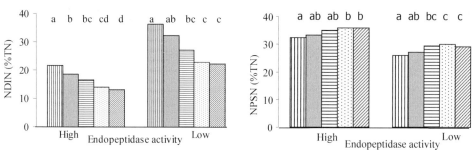

Figure 1 Effect of forage particle size on protein solubilisation during 6 h *in vitro* rumen fermentation
▥ : 3cm.; ■ : 1cm.; ☰ : 0.25cm.; ⠿ : CO_2; ▨ : N $_{Liq}$

Conclusion Our results show that particle size is a major source of variation when studying kinetics of protein breakdown. The mechanical damage affects the release of fermentable substrates as well as the accessibility of bacteria or their enzymes into the plant cells and the activity of plant endopeptidases. Larger particles may be more representative of actual animal behaviour, but they present practical problems for analytical purposes.

Acknowledgement This research was funded by Fondecyt Grant No. 1030918.

References
Boudon, A., S. Mayne, J-L Peyraud & A.S. Laidlaw (2002). Effects of stage of maturity and chop length on the release of cell contents of fresh ryegrass (*Lolium perennne* L.) during ingestive mastication in steers fed indoors. *Animal* Research, 5, 349-365.
Broderick, G.A. (1987). Determination of protein degradation rates using a rumen *in vitro* system containing inhibitors of microbial nitrogen metabolism. *British Journal of Nutrition*, 58, 463-475.
Licitra, G., T.M. Hernández, & P.J. Van Soest (1996). Standardisation of procedures for nitrogen fractionation of ruminant feeds. *Animal Feed Science and Technology*, 57,347-358.

A new system for the evaluation of the fermentation quality of silages

E. Kaiser and K. Weiß
*The Institute of Animal Science, Humboldt-University of Berlin, Invalidenstraße 42, D-10115 Berlin, Germany,
Email: ehrengard.kaiser@agrar.hu-berlin.de*

Keywords: silage quality, fermentation quality, fermentation process

Introduction Depending on the content of nitrate in green forage, the pattern of fermentation products in silages differ significantly (Weiß & Kaiser, 2001). The systems, which are now common in practice for evaluating the quality of silage fermentation, characterise fermentation quality incorrectly because the evaluation is influenced by the chemical composition of green forage. The aim of this work was to derive an evaluation system for fermentation quality, which is independent from the chemical composition of green forage.

Materials and methods Under laboratory conditions, 570 silages were produced from different green forages of known chemical composition. Fermentation quality parameters were selected which were suitable to characterise all stages of fermentation quality independent of the chemical composition of the green forage. An evaluation system was developed on the basis of the relations between the parameters of undesirable decomposition (butyric acid (BA), acetic acid (AA), ammonia (NH_3)) considering recent information of metabolism in silages during the fermentation process. It was applied to 3503 silages from green forages with unknown chemical composition obtained from farms of different regions in Germany.

Results and discussion The results confirmed that all stages of fermentation quality (anaerobic stability, "turn over" of fermentation products and increased spoilage) can be evaluated by BA and AA concentration exclusively (Kaiser *et al.*, 1999, 2000). The parameters pH-value and ammonia content in silages are inappropriate for evaluation, because they are influenced by variation in the chemical composition of green forage (see also Kaiser *et al.*, 2000). The suggested new estimation system is presented in Table 1. The content of 3.0% AA in DM as an upper limit for anaerobically stable silages is derived from its relationship with BA and ammonia. If the content of BA is low, the classes are very narrow because the evaluation of the fermentation quality is strongly influenced by the production of BA in anaerobically unstable silages from green forage low in nitrate.

Table 1 Evaluation system for the fermentation quality of silages from contents of butyric acid and acetic acid

Butyric acid (% DM)	Points	Acetic acid (% DM)	Points (Discount)	Evaluation Score	Evaluation Mark
0 - 0.3	100	≤ 3.0	0	90 to 100	1
> 0.3 - 0.4	90	> 3.0 - 3.5	-10	72 to 89	2
> 0.4 - 0.7	80	> 3.5 - 4.5	-20	52 to 71	3
> 0.7 - 1.0	70	> 4.5 - 5.5	-30	30 to 51	4
> 1.0 - 1.3	60	> 5.5 - 6.5	-40	<30	5
> 1.3 - 1.6	50	> 6.5 - 7.5	-50		
> 1.6 - 1.9	40	> 7.5 - 8.5	-60		
> 1.9 - 2.6	30	> 8.5	-70		
> 2.6 - 3.6	20				
> 3.6 - 5.0	10				
> 5.0	0				

Conclusions From the evaluation of 3503 silages made under practical conditions, this new system, based only on the content of BA and AA, was able to characterise the fermentation quality of all green forage silages, including maize, more correctly than previous systems.

References

Kaiser, E., K. Weiß & R. Krause (1999). Vorschlag zur Beurteilung der Gärqualität von Grassilagen. Proceedings 111. VDLUFA-Kongreß, Halle, 385-388.
Kaiser, E., K. Weiß & R. Krause (2000). Beurteilungskriterien für die Gärqualität von Grassilagen. Proceedings of the Society for Nutritional Physiology, 9, 94.
Weiß, K & E. Kaiser (2001). Fermentation patterns in silage depending on chemical composition of herbage. Grassland Science in Europe, 6, 150-153.

Prediction of indigestible NDF content of grass and legume silages by NIRS

L. Nyholm[1], M. Rinne[2], M. Hellämäki[1], P. Huhtanen[2] and J. Nousiainen[1]
[1]Valio Ltd. Farm Services, P.O. Box 10, FIN-00039 Valio, Finland, Email: juha.nousiainen@valio.fi, [2]MTT Agrifood Research Finland, Animal Production Research, FIN-31600 Jokioinen, Finland

Keywords: NIRS, indigestible NDF, grass silage, legume silage

Introduction   The future feed evaluation systems based on mechanistic digestion models require reliable estimates of forage digestible and indigestible NDF content (DNDF and INDF). The objective of this study was to examine the potential of near infrared reflectance spectroscopy (NIRS) in predicting INDF content of grass and legume silages.

Materials and methods   The INDF content of silages was determined by 12 d ruminal incubation in nylon bags (NB; pore size 6 or 17 µm) with two dairy cows fed grass silage-based diets. After incubations, the NBs were washed with a household washing machine and the residues were analysed for NDF. INDF (g/kg DM) was expressed as ash-free NDF of NB-residues. Dried and milled samples were scanned with a NIRSystems 6500 spectrometer using a small ring cup. After treatments (math treatments 1,4,4,1 and standard normal variate and de-trending) the spectra (1100-2498 nm) were used to develop MPLS equations for INDF with WinISI II 1.50a software. The basic calibration included 52 experimental and 42 farm grass silages (Nousiainen *et al.*, 2004), which was extended with 50 farm grass silages and 36 legume silages (Table 1). Both basic and extended calibrations were used to predict the INDF content of 50 grass and 36 legume silages (Figure 1). The majority of the samples were of Finnish origin, but some legume samples were provided by FAL (Germany), IGER (UK) and Estonian Agricultural University.

Results   The performance of calibration (all samples included) was satisfactory (SD/SECV >2.5). The prediction error (SEP) of grass silages was noticeably higher than the SECV of the basic calibration (10.0 vs. 18.0 g/kg DM, (see Nousiainen *et al.*,

Table 1 NIRS calibration statistics in predicting indigestible NDF content (g/kg DM) of grass and legume samples

| Outliers | n | Calibration | | | | Cross-validation | | |
		Mean	SD	SEC[a]	R^2	SECV[b]	R^2	SD/SECV
Included	180	90.0	44.69	14.88	0.889	17.56	0.847	2.5
Excluded	176	89.6	44.29	13.87	0.902	15.24	0.883	2.9

[a]SE of calibration, [b]Pooled SE over cross validation sub-sets

2004), but it decreased to 15.6 g/kg DM after extending the calibration with both grass and legume samples (Figure 1a). The prediction performance of legume samples was essentially improved when the basic calibration (only grass silages) was extended with legume samples (Figure 1b: SEP 49.7 vs. 17.3 g/kg DM).

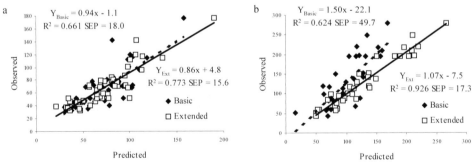

a
$Y_{Basic} = 0.94x - 1.1$
$R^2 = 0.661$ SEP = 18.0
$Y_{Ext} = 0.86x + 4.8$
$R^2 = 0.773$ SEP = 15.6
◆ Basic
□ Extended

b
$Y_{Basic} = 1.50x - 22.1$
$R^2 = 0.624$ SEP = 49.7
$Y_{Ext} = 1.07x - 7.5$
$R^2 = 0.926$ SEP = 17.3
◆ Basic
□ Extended

Figure 1 Relationship between observed and predicted INDF content of (a) grass (n = 50) and (b) legume (n = 36) silages

Conclusions   The results imply that NIRS has a great potential in predicting INDF content of silages. A general calibration both for grasses and legumes may be developed. Further work is needed to extend the INDF calibration, and to examine whether NIRS can be used to predict the potential rate of silage DNDF digestion.

References

Nousiainen, J., S. Ahvenjärvi, M. Rinne, M. Hellämäki & P. Huhtanen (2004). Prediction of indigestible cell wall fraction of grass silage by near infrared reflectance spectroscopy. *Animal Feed Science and Technology*, 115, 295-311.

Analysis of silage fermentation characteristics using transflectance measurements by near infrared spectroscopy

A. Martínez, A. Soldado, R. García, D. Sánchez and B. de la Roza-Delgado
Servicio Regional de Investigación y Desarrollo Agroalimentario (SERIDA), PO Box 13, 33300 Villaviciosa (Asturias), Spain Email:admartinez@serida.org

Keywords: silage juice, fermentation parameters, FT-NIR

Introduction The fermentation end products as lactic acid, volatile fatty acids and ammonia-nitrogen, are important indicators of the efficiency of silage fermentation and are closely related to nutritive value of them (Jaster, 1995). Drying is problematic in the case of silage as many fermentation products are volatile and may get lost during the drying process. This may be a reason why NIR technology is being little used for the evaluation of silage fermentation characteristics. The feasibility of using near infrared transflectance spectroscopy to evaluate the content in fermentation end products of grass and maize fresh silage was investigated in this study.

Materials and methods One hundred and forty-seven representative grass and sixty maize silage samples were collected over a 24 month period. Each of the silage samples were pressed immediately to obtain the extract. After centrifugation (8500 rpm, 10 min) to clarify the extract, the spectra were taken directly on extract. Samples were scanned in transflectance mode over the range 1000 to 2500 nm in a FT-NIR Spectrum One from Perkin Elmer using IdentiCheck Reflectance Accessory (ICRA) liquids sampling accessory. To develop calibration and prediction models the Quant + software was employed. Reference data were determined on the extracts for lactic acid and total volatile fatty acids by HPLC using a WATERS Alliance 2690 instrument and ammonia-N by SPECTROQUANT ammonium test, (Merck).

Results Different mathematical pre-treatment and calibration models were developed using partial least squares (PLS) regression with internal full cross-validation. The optimum equation for each fermentation parameter was selected according to a low standard error of cross validation (SECV) and a large correlation coefficient in the calibration. R^2 values were acceptable on all fermentation parameters with $R^2 > 0.70$, except for acetic acid in grass silage (Table 1). Therefore the results indicate a reliable relationship between chemical analysis and NIR predicted values determined directly on extract. According to ASTM (American Society for Testing and Materials) guidelines for the judgement of a model's validity for NIR calibration (ASTM, 1994) and Williams and Sobering, (1996), this study obtained acceptable values of RDP (ratio of the standard deviation of the reference data to the SECV) for screening purposes, except for ammonium-N in maize silage.

Table 1 Range and statistical parameters for prediction of fermentative characteristics on grass and maize silage by FT-NIR

Grass silage				Maize silage			
Parameter (g/dl)	Range	R^2	SECV	Parameter (g/dl)	Range	R^2	SECV
Lactic acid	0-8.08	0.71	0.506	Lactic acid	0.841-5.75	0.80	0.268
Acetic acid	0-1.27	0.61	0.138	Acetic acid	0-2.30	0.85	0.148
Butyric acid	0-4.15	0.78	0.379	Butyric acid	-	-	-
Ammonium-N	0-0.261	0.82	$299 *10^{-3}$	Ammonium-N	0-0.0012	0.71	$153 *10^{-3}$

Conclusions The results of this investigation indicated that silage fermentation characteristics can be predicted with a fair amount of accuracy by FT-NIR analysis using ICRA as liquids sampling accessory directly on silage extract, in order to avoid the volatilisation of fermentation end products. These results may be of great practical significance when silage is used as a roughage component in dairy rations.

References

ASTM Designation E 1655-94 (1994). American Society for Testing and Materials, 100 Barr Harbor Dr., West Conshocken, PA 19428, Annual Book of ASTM Standards.
Jaster, E.H. (1995). Legume and grass silage preservation. In: K.J. Moore and M.A. Peterson (eds). Post-harvest physiology and preservation of forages. CSSA Special Publication 22, Madison, USA, 91-115.
Williams, P. & D. Sobering (1996). How do we do it? A brief summary of the methods we use in developing near infrared calibrations. In: A.M.C. Davies and P. Williams (eds). *Near Infrared Spectroscopy: The future waves.* NIR Publications, Chichester, UK, 185-188.

This work was supported by Spanish INIA infrastructure project

A simple method for the correction of fermentation losses measured in laboratory silos

F. Weissbach

Goesselweg 12, D-18107 Elmenhorst, Germany, Email: Prof.F.Weissbach@t-online.de

Keywords: silage, fermentation losses, laboratory experiments

Introduction Dry matter (DM) losses caused by formation of gaseous fermentation products can be measured by different methods. The most common method (A) is measuring the difference between the DM input and output of a silo. Other methods are based on the measurement of the fermentation gases which spontaneously leave the silo, either directly by collecting them (B) in a special absorbent like KOH or, much more easily, by weighing the filled silo at the beginning and the end of the fermentation process (C). The figures obtained by B and C are substantially smaller than those by A. This difference represents a certain amount of CO_2 which is retained within the silage. The objective of this paper is to deduce a procedure for estimating the amount of this retained CO_2 so that the results obtained by method C (or B) can be corrected.

Materials and methods Berg (1971) carried out extensive and well documented investigations on the formation of gaseous fermentation products in silages from different kinds of crops at different DM contents. By means of special techniques he measured several gas fractions: the CO_2 that spontaneously left the silo (I), the CO_2 escaped from the silo under evacuation (II), the CO_2 released during drying the silage samples (III) and finally a small amount of other gases (IV) which spontaneously left the silo but were not absorbed in KOH. The total of fractions I to IV is the fermentation loss. The sum of I and IV is the weight loss of the silo during fermentation. The sum of II and III represents the CO_2 retained in the silage. Table 1 shows the mean results from 70 of such experimental balances. These data were used to estimate the retained CO_2.

Table 1 Fractions of gas produced in the silo*

Fraction	Amount as % DM ensiled	
	Mean	Range
CO_2 (I)	7.7	1.6 ...16.0
CO_2 (II)	1.3	0.8 ... 2.1
CO_2 (III)	1.2	0.4 ... 2.2
NO, H_2 etc. (IV)	0.4	0.2 ... 1.1
Total	10.6	4.4 ...19.8

Table 2 Retention of CO_2 in silage*

Crop	DM %	n	retained CO_2	
			g/kg H_2O	g/kg DM
Beet tops	11.8	4	4	26
Green rye	11.8	5	3	20
Cabbage	13.0	3	6	41
Red clover	14.1	5	6	36
Potatoes	15.0	5	3	15
Grass	15.6	5	6	31
Maize	16.7	4	5	23
Sugar beets	18.7	5	4	19
Lucerne	19.2	4	7	28
Potatoes	20.8	4	4	14
Grass	20.9	3	6	24
Sugar beets	22.5	3	5	16
Grass	34.9	3	15	29
Lucerne	39.8	5	15	23
Grass	50.5	12	27	27
Mean			7.7	24.8
Standard deviation			6.5	7.6
Coefficient of variance			84%	31%

*Source of data: Berg, 1971

Results Mean figures for the CO_2 retention (fraction **II** plus **III**) in the silage from the individual experiments are given in Table 2. The figures are listed in the order of DM content. The retained CO_2 was calculated per kg water or, alternatively, per kg DM ensiled. The CO_2 retention in the silo was not related to the quantity of water but to the quantity of DM. An approximately constant amount of CO_2 was retained per kg DM, obviously adsorbed by the surface structures of the solid phase in the silage. The variation in this amount may be explained by differences in the composition and microstructure of individual herbage materials. If only gramineae and legumes are included, the coefficient of variance decreased from 31 to 18%.

Conclusions The results of the present study show that the adsorbed CO_2 on average amounted to 2.5% of the ensiled DM. Consequently, in laboratory silos without air entrance and effluent outflow the fermentation losses (*FL*) can be calculated from the weight difference between the beginning and end of fermentation (*WD*) by the following equation:

$$FL[\%] = 100 \frac{WD[g]}{DMensiled[g]} + 2.5$$

Reference

Berg, K. (1971). Die Trockensubstanzbestimmung von Silagen und die Erfassung flüchtiger, den Futterwert beeinflussender Verbindungen sowie Modellversuche zur Ermittlung der Gärverluste. Dissertation, Deutsche Akademie der Landwirtschaftswissenschaften, Berlin, 97-145.

Development of a method for the fast and complete assessment of quality characteristics in undried grass silages by means of an NIR-diode array spectrometer

H. Gibaud[1], C. Paul[1], J.M. Greef[1] and B. Ruser[2]
[1]Institute of Crop and Grassland Science, Federal Agricultural Research Centre, Bundesallee 50, D-38116 Braunschweig, Germany, Email: helene.gibaud@fal.de, [2]Pioneer Hi-Bred Northern Europe GMBH, Apensener Str. 198, D-21614 Buxtehude, Germany

Keywords: fresh grass silage analysis, fermentative quality, nutritive quality, NIR-diode array spectrometer

Introduction Traditionally, the determination of grass silage is very time consuming and needs a lot of manpower and chemicals. The advantages of conventional laboratory NIRS instruments are well known but their disadvantage lies in their lacking suitability for on-farm use. A new type of spectrometer based on diode arrays may be used for this purpose. However, these new instruments still need to be calibrated for an accurate estimate of the fermentative and nutritive value of wet and unchopped grass silage.

Materials and methods Stored grass silage samples from the North of Germany covering four cuts of each of the years from 2000 to 2003 were available for calibration. For NIRS-analysis the frozen silages were thawed, equilibrated to room temperature and then carefully mixed. NIRS measurements were performed using a Corona 45 NIR high resolution diode array spectrometer (equipped with 256 InGaAs-diodes in the range 960-1690 nm) of Carl Zeiss Jena GmbH. Samples were measured using the turntable sample presentation device in Petri dishes of 200 mm diameter from below and the mean of 2x2 replicate spectra calculated. After transformation of the averaged spectra into WinISI format, a range of scatter correction procedures was performed. Multivariate data analysis of math treated spectra (1,6,6,1) and reference parameters were performed using modified partial least squares regression with 8 terms and a single pass for outlier removal (WinISI 1.50). The original calibration set consisted of 149 representative samples of all four harvest years. For truly independent validation 57 new samples from the last harvest of 2003 with a lower than average silage quality were used. The following reference parameters were employed and expressed as % of dry matter (DM): CP (Kjeldahl-N x 6.25), WSC (Anthrone method), CF (Weende method), ADF/NDF (modified van Soest), pH (electrometrically by a pH-meter), short chain fatty acids (HPLC from an acid silage extract). DM was assessed by oven drying at 105°C and corrected for losses of volatiles according to Weissbach & Kuhla (1995).

Table 1 Description and statistics of calibration and validation

		Calibration					Validation			
Parameter	Scatter	Mean	sd	SEC	R^2	Mean	sd	SEP(C)	Bias	R^2
DM	MSC	36.7	8.0	1.1	0.98	38.9	10.9	1.7	0.3	0.98
CP	None	17.0	2.7	1.5	0.69	15.7	2.8	1.4	1.1	0.78
WSC	None	3.0	2.8	1.4	0.75	4.1	3.5	1.6	-0.1	0.82
CF	MSC	27.1	3.3	1.4	0.82	27.3	2.7	1.6	-1.3	0.67
ADF	None	32.0	3.7	1.7	0.80	32.6	4.0	2.2	-2.0	0.72
NDF	None	51.3	5.8	2.5	0.81	53.1	5.0	3.0	-1.7	0.63
PH	SNV-D	4.4	0.4	0.2	0.76	4.7	0.5	0.3	0.0	0.66
Lactic acid	None	7.1	4.0	1.6	0.84	4.5	2.7	1.5	-0.1	0.73
Butyric acid	MSC	0.3	0.6	0.4	0.55	1.1	1.1	0.7	0.4	0.65

Scatter correction: None (no scatter correction); SNV-D (standard normal variate after de-trending); MSC (multiple scatter correction) dry matter (% of fresh matter), CP: crude protein, WSC: water soluble carbohydrate, CF: crude fibre, ADF: acid detergent fibre, NDF: neutral detergent fibre

Results Unlike in classic NIRS measurements on dried and ground forages, our samples contain a large amount of water which may interfere with the NIR measurement of relatively weak N-H and C-H-bands. Nevertheless, our initial calibration for fermentative and nutritive parameters of grass silage performed satisfactorily (Table 1). It showed potential even in a validation set which deviated from the calibration set by nature of a high proportion of poorly fermented silages.

Conclusions Further improvements in the analytical potential of the above method are anticipated by enhancing the variation in the calibration set. This will allow more complex NIR regression models to be developed so that both the systematic and random error component can be reduced.

References
Weissbach, F. & S. Kuhla (1995). Substance losses in determining the dry matter content of silage and green fodder: arising errors and possibilities of correction. *Übers Tierern. DLG-Verlag*, 23, 189-214

Prediction of red clover content in mixed swards by near-infrared reflectance spectroscopy

B. Deprez[1], D. Stilmant[2], C. Clément[2], C. Decamps[1] and A. Peeters[1]

[1]Laboratory of Grassland Ecology, Catholic University of Louvain, Place Croix du Sud 5 bte 1, B-1348 Louvain-la-Neuve, Belgium, Email: deprez@ecop.ucl.ac.be, [2]Farming System Section, Walloon Centre of Agricultural Researches, Rue du Serpont 100, B-6800 Libramont, Belgium

Keywords: red clover, legume percentage, NIRS

Introduction Because of the legume fixation capacity, their high protein content, digestibility and intake characteristics, more and more attention is paid to grassland clover content. In field experiments, clover content must often be determined, for example to quantify nitrogen flux or the best practices to manage such species (Stilmant *et al.*, 2004). However hand sorting of clover and grass, even if accurate, is time-consuming and has a high labour cost. In comparison, accuracy of visual estimation of clover content, directly in the field, varies according to training and experience. Near-infrared reflectance spectroscopy (NIRS) has been proposed as a method for the rapid determination of sward botanical (Petersen *et al.*, 1987; Pitman *et al.*, 1991) and morphological composition (Leconte *et al.*, 1999; Stilmant *et al.*, 2005). This paper describes the performance of a NIRS calibration developed to characterise red clover (*Trifolium pratense*) content when associated to different grass species and this at different phenological stages.

Materials and methods Plant material used to set up NIRS calibration was collected in three swards, located on loamy soil. In each sward red clover (cv. Merviot) was associated to different grass species : perennial ryegrass (*Lolium perenne*) cv. Merlinda, hybrid ryegrass (*Lolium hybridum*) cv. Barsilo and cocksfoot (*Dactylis glomerata*) cv. Lupré. The samples were taken in May, July, August and October by cutting the sward 7 cm above ground level. Samples were hand sorted into clover and grass fractions. These fractions were dried, weighed (G %) and then ground and remixed before being submitted to NIRS analysis (NIRSystem monochromator 5000). Spectral data, in the range of 1100–2500 nm scanned at 2 nm steps, were correlated to red clover content. Calibrations were developed according to the Partial Least Square procedure with cross validation using the ISI (Infrasoft International) software. The spectra of 647 samples were used for calibration and cross validation, while 45 samples were kept for independent validation.

Results and conclusions Among the 647 samples, the red clover content ranged from 0 to 100%, with a mean of 59.5%. The R^2 was 0.99 in calibration (Standard Error of 2.74%) as in cross validation (Standard Error of 2.87%) and the ratio of the standard deviation of the initial sample set on the standard error in cross validation (SD/SECV) was 9.4 (Williams, 2004). The performance of this calibration is sufficiently precise and reliable to quantify red clover content in a wide range of situations. This was confirmed by the results obtained on the independent validation set (Table 1).

Table 1 Statistics of the independent validation, including number of samples (N) and standard error of prediction (SEP)

N	Mean	SEP (%)	Bias	Slope	R^2
45	67.18	2.88	0.73	0.99	0.98

References

Leconte, D., P. Dardenne, C. Clément, & Ph. Lecomte (1999). Near infrared determination of the morphological structure of ryegrass swards. In: Davies A.M.C. and Giangiacomo R. (eds) *Near Infrared Spectrometry : Proceedings of the 9th International Conference.* NIR Publications, 41-44.

Petersen, J.C., F.E. Barton, W.R. Windham & C.S. Hoveland (1987). Botanical composition definition of tall fescue-white clover mixtures by near-infrared reflectance spectroscopy. *Crop Science*, 27, 1077-1080.

Pitman, W.D., C.K. Piacitelli, G.E. Aikenj, & F.E. Barton (1991). Botanical composition of tropical grass-legume pastures estimated with near-infrared reflectance spectroscopy. *Agronomy Journal*, 83, 103-107.

Stilmant, D., V. Decruyenaere, C. Clément & N. Grogna (2005). The use of near infrared reflectance spectroscopy (NIRS) to follow legumes leaf/stem ratio during drying. *XXth International Grassland Congress* (accepted).

Stilmant, D., V. Decruyenaere, J. Herman & N. Grogna (2004). Hay and silage making losses in legume-rich swards in relation to conditioning. *In: Land Use Systems in Grassland Dominated Regions,* A Lüscher, B. Jeangros, W. kesler, O. Huguenin, N. Millar, D. Suter (Eds). *Grassland Science in Europe*, 9, 939-941.

Williams, P. (2004). Near-Infrared Technology – Getting the Best Out of Light. A Short Course in the Practical Implementation of Near-infrared Spectrometry for the User. PDK Grain, Manitoba, Canada.

Keyword index

16S-rDNA-based marker	217
2,3-butanediol	261
access time	142
acetic acid	231
additive	140, 142, 197, 198, 227, 228, 229
additives	174, 200
ad-libitum concentrates	166
aerobic deterioration	225, 271
aerobic stability	97, 167, 221, 222, 224, 226, 231, 232, 233, 238, 241, 256, 262
alfalfa	156, 181
alfalfa-grass mixture	197
ammonium formate	198
anaerobic digestion	216
applicator	218
bacteria	248
bacterial additives	171
bacterial inoculant	241
bag silo	245
bagged silage	250
bale silage	193
baleage	247
baled silage	248, 249, 251, 252
bales	209
barley	144, 174
beef cattle	51, 159, 160
beef production	153
bi-crop	179
bi-crop silage	144
big bale	238
birdsfoot trefoil	163, 170
blood meal	159
bog grasslands	240
brassica	155
butyric acid	206, 213
butyric acid bacteria	147, 271
by-products	167
caffeic acid	210
capacity	192
carcass gain	65
cattle	154, 166
cell wall structure	164
chemical additive	158
chop length	140, 156
chopping length	137, 173
clostridia	213
clostridia spores	199, 206
Clostridium tyrobutyricum	271
clover	149
cob	175
cocksfoot varieties	214
coffee grounds	187
compaction	256
composition	176
concentrate supplementation	51
condensed tannins	163
conjugated linoleic acid	145
conservation	154
correlation	208
cost of production	65
costs	83
crimped	153
cultivar	175, 176
daily gain	264
dairy cows	35, 137, 140, 142, 143, 145, 146, 149, 183, 184, 259
dairy ewes	156
dairy performance	173
degradation	237
density	245
digestibility	171
diode array	109
DM intake	257
dough stage	138
D-value	139
education tool	148
effluent	97
endophytes	230
engineering	83
ensilability	214, 250
ensiled barley	184
ensiling	155, 170
ensiling additives	230
enterobacterium	262
environment	121
enzymes	237
ethanol	261
evaluation	268
farm management	147
fatty acids	158
feed quality	185
feed value	182
feeding	138
feeding value	19
fermentation	177, 182, 197, 200, 203, 238, 239, 256
fermentation losses	263, 278
fermentation parameters	277
fermentation process	213, 275
fermentation quality	193, 208, 213, 214, 269, 275
fermentative quality	279
fermented whole crop wheat	165
field bean	182
field pea	205, 272
forage breeding	121
forage conservation systems	19
forage maize	165, 168
forage paddy rice	201
forage quality	168
forage variety	263
forages	109, 148, 273, 274
formic acid	171, 198, 222
free-choice	188
fresh grass silage analysis	279
FT-NIR	277
fungi	232, 252
fungus	249

gains of heifers 240
galega 267
gas transmission rate 251
gases 97
grain silage 178
grain sorghum 226
grass 149, 210, 260
grass hay 145
grass silage 19, 35, 51, 65, 137, 153, 184, 198, 199,
    206, 209, 211, 239, 241, 258, 268, 276
green chopped sugarcane 257
green material 208
green tea polyphenol 215
green tea waste 215
growth 155
harvest date 175, 176
harvesting 156
harvesting silage 192
hay 188, 240
haylage 188, 247
high moisture maize 157
horses 188, 209
in situ 267
in vitro 141, 267, 270
in vitro fermentation 204, 207
in vitro method 225
in vivo 267
indigestible fibre 267
indigestible NDF 276
inoculant   158, 201, 204, 207, 223, 233, 238, 239
intake 137, 140, 180, 231
ketosis 146
L. buchneri 202, 231
laboratory experiments 278
lactic acid 215
lactic acid bacteria   199, 201, 205, 206, 216, 226,
    241, 272
Lactobacillus buchneri 261
Lactobacillus plantarum 217
lamb production 155
legume 139, 155, 197, 205, 260, 272
legume percentage 280
legume silage 164, 276
listeria 262
liveweight gain 180
Lolium perenne 224
losses 202, 223, 245, 256
Lotus species 170
Lotus varieties 170
lucerne 204, 207, 212, 245, 267
lucerne silage 233
maize 160, 166, 176, 179, 186, 245
maize silage   65, 145, 173, 177, 232, 255, 257, 271
maize whole crop silage 216
maturity 139, 164, 181, 185
meadow fescue 230
meat 158
meat quality 153, 160
mechanical treatment 250
mechanisation 83

Medicago sativa 247
metabolisable energy 268
methane yield 216
microbial protein supply 144
microbial protein synthesis 35
microbial yields 172
microflora 208
microorganisms 248
milk 145
milk production 144, 163, 165
milk stage 138
milk yield 65, 257
mixed ration 167
moist grain 154
moisture 255, 258
moulds 193, 246, 252, 270
mulch 175, 176
mycotoxins 177, 209, 232, 270
N use efficiency 141
N utilisation 163
naked oats 137
near infrared reflectance spectroscopy 109
N-efficiency 172
Neotyphodium 230
NIR-diode array spectrometer 279
NIRS 276, 280
nitrogen 159
nitrogen management 168
nitrogen partitioning 138
nitrogen solubility 274
nutrient balance 35
nutrient management 148
nutrients 181
nutritional quality 226
nutritive quality 279
nutritive value 154, 187, 202, 222, 240, 241
o-diphenols 210, 212
packing rotor 250
Panicum maximum 258
partial digestibility 157
participatory research 168
particle size 274
partitioning 159
pathogens 121
pea 144
Penicillium roqueforti 252
permeation coefficient 251
pH 225, 262
plant peptidases 273
plasma metabolite 139
plastic film 193
plastic wrapping 246
polyethylene film 251
polyphenol oxidase 212
polyphenol oxidase (PPO) 210
power need 192
prediction equation 268
propionic acid bacteria 221
protein degradation 35
protein efficiency 138

protein supplementation 184
proteolysis 210, 212, 272, 273, 274
Q-PCR 217
quantitative PCR 217
rate of digestion 164
raw milk 147
red clover 200, 203, 212, 267, 280
residual water soluble carbohydrates (WSC) 225
risk assessment 147
round baler 180
roundbales 246
rumen 141, 273, 274
rumen metabolism 172
ruminant 121
Rusitec 239
ryegrass 237
safflower 169
Schizophyllum commune 249
Sesbania cannabina 179
sheep 268
silage 83, 140, 141, 142, 156, 166, 167, 169, 174,
    179, 186, 188, 200, 201, 203, 208, 215, 217,
    221, 222, 224, 227, 228, 229, 230, 237, 240,
    246, 247, 259, 260, 261, 269, 270, 278
silage additive 177, 185, 206
silage feeding systems 19, 143
silage fermentation 146, 187, 264
silage harvesting system 191
silage inoculant 218
silage juice 277
silage quality 197, 275
silo losses 97, 258
small-scale technique 260
sodium bicarbonate 142
sorghum 178, 186
spores 147, 271

stage of development 205, 226, 272
stage of maturity 138, 180
starch 175
sugar 149
sugarcane silage 257, 263
sustainability 121
tannin 178
temperature 255
timothy 200
total digestibility 157
toxicity 270
transport equipment 246
tropical and temperate grass 259
tropical grass 264
tropical grass silage 97
ultra-low volume 218
urea 153, 263
urea-treated whole crop wheat 165
voluntary intake 264, 269
water soluble carbohydrate 141, 149, 211, 224, 227,
    228, 229
wheat 153, 154, 174
whole crop 174
whole crop barley silage 180
whole crop cereal silage 185
whole crop pea silage 183
whole crop rice 261
whole crop silage 171
whole crop wheat 160, 166
whole crop wheat silage 65
wholeseed soybean 157
wilt 227, 229
wilting 237, 258
yeast 193, 223, 225, 232, 248, 252
yield 176, 186

Author index

Acosta Aragón, Y.	178	Dorszewski, P.	230
Adelantado, C.	232	Doyle, E.M.	248
Adesogan, A.T.	255	Driehuis, F.	147, 271
Adjolohoun, S.	260	Elizalde, H.F.	156
Agneesens, R.	260	Ennahar, S.	201
Agnew, R.E.	109, 268	Ericson, L.	182
Ahvenjärvi, S.	139, 164	Ferris, C.P.	143
Almeida, E.O.	223	Filho, S.G.T.	257
Amaral, R.C.	256	Filya, I.	204, 207, 221, 222
Antoniazzi, S.	205, 272	Forristal, P.D.	248, 249, 251, 252, 83
Argamentería, A.	146	Foster, W.	247
Arvidsson, H.	192	Frank, B.	180
Arvidsson, K.	182	Fraser, M.D.	155, 267
Avasi, Z.	233	Frost, J.P.	143, 191
Bach, A.	232	Fukuyama, K.	264
Bakewell, E.L.	211	Fuller, H.	249, 252
Barbet-Massin, V.	259	Fychan, R.	155, 170
Bar-Tal, A.	169	Fyhri, T.	246
Barth, S.	217	Gabel, M.	178
Bellus, Z.	186	Gai, V.F.	157
Bernardes, T.F.	202, 223, 256	Gallo, M.	203
Berthiaume, R.	159	Gamburg, M.	169
Bertilsson, J.	138, 149	Garcia, G.R.	262
Binnie, R.C.	143, 191	García, R.	277
Blūzmanis, J.	208	Gibaud, H.	279
Borreani, G.	193, 205, 226, 272	Grajewsk, J.	177, 209
Branco, A.F.	157	Greef, J.M.	224, 225, 279
Braun, R.	216	Grimaud, P.	259
Brener, S.	169	Gubała, A.	209, 239
Brink, G.E.	210	Harrison, J.H.	148
Brito, A.F.	172	Hart, K.J.	183
Broderick, G.A.	163, 172	Hatfield, R.D.	210, 212
Buldgen, A.	260	Hattori, H.	167, 261
Cai, Y.	187, 201	He, Z.	237
Calvo, M.A.	232	Heikkilä, T.	140, 142
Camargo, M.S.	257	Helembai, J.	186
Cavallarin, L.	205, 272	Hellämäki, M.	276
Cecato, U.	157	Herrmann, H.-H.	218
Chen, Y.	169	Hino, N.	187, 201
Cherney, D.J.R.	168	Holmes, B.J.	245
Cherney, J.H.	168	Holzer, M.	206
Chunsheng, B.	237	Howard, H.	227, 228, 229
Cieślak, A.	239	Huhtanen, P.	35, 267, 276
Clément, C.	280	Huntington, J.A.	183
Clipson, N.J.W.	248	Hymes Fecht, U.C.	163
Coan, R.M.	262	Idler, C.	217
Contato, E.D.	262	Iglesias, C.	232
Contreras-Govea, F.E.	204, 207	Jaakkola, S.	184, 198
Cox, W.J.	168	Janusckiewicz, E.R.	223
Crowley, J.C.	175, 176	Jatkauskas, J.	158, 238
Davies, D.R.	121, 141, 211	Jianguo, H.	237
de Jong, P.	147	Jobim, C.C.	157
de la Roza, B.	146	Jones, R.	155, 170
de la Roza-Delgado, B.	277	Junqueira, M.C.	257, 258
Decamps, C.	280	Kaiser, E.	213, 275
Delacollette, M.	260	Kaldmäe, H.	200
Deprez, B.	280	Kangasniemi, R.	184
Devash, L.	169	Karabulut, A.	221, 222

Karp, V.	144	Nousiainen, J.	267, 276
Kärt, O.	200	Nussio, C.M.B.	263
Kawamura, O.	264	Nussio, L.G.	97, 257, 258, 263
Keady, T.W.J.	65, 153, 160	Nyholm, L.	276
Keane, G.P.	175, 176	O'Brien, M.	249, 252
Kelemen, Zs.	186	O'Kiely, P.	19, 154, 166, 175, 176, 217, 224, 227,
Kilpatrick, D.J.	153, 160, 165		228, 229, 248, 249, 251, 252, 83
Kim, S.C.	255	O'Mara, F.P.	154, 166, 227, 228, 229
Kingston-Smith, A.H.	121	Ogawa, M.	187, 201
Kirchmayr, R.	216	Olsson, I.	171
Kita, K.	179, 215	Olt, A.	200, 267
Kleinmans, J.	231	Orosz, Sz.	186
Klocke, M.	217	Osmane, B.	208
Knicky, M.	138, 174, 180	Ott, E.M.	178
Kondo, M.	179, 215	Owen, T.R.	218
Kramer, W.	206	Pahlow, G.	199, 218, 224, 225, 227, 228, 229
Krizsan, S.J.	269	Park, R.S.	109
Kuoppala, K.	139, 164	Patterson, D.C.	143, 165
Laffin, C.	251	Paul, C.	267, 279
Lafrenière, C.	159	Pauly, T.	224, 227, 250, 270
Landau, S.Y.	169	Paziani, S.F.	258
Lankveld, J.M.G.	147	Pedroso, A.F.	258
Larraín, R.	273	Peeters, A.	280
Lättemäe, P.	197	Peláez, M.	146
Leemans, D.K.	141, 211	Perez Lopez, C.	216
Leinonen, A.-R.	144	Pichard, G.	273, 274
Lingvall, P.	171, 174, 180, 192	Pichler, H.	216
Little, E.M.	175, 176	Podkówka, L.	230
Lively, F.O.	153, 160	Polip, I.	213
Loures, D.R.S.	258	Porter, M.G.	109
Mahlkow, K.	173	Potkański, A.	177, 209, 239
Mari, L.J.	257, 258, 263	Pursiainen, P.	144
Marita, J.M.	210	Qingfang, X.	237
Marley, C.L.	155, 170	Queiroz, O.C.M.	257, 263
Marley, G.	218	Raczkowska-Werwińska, K.	177, 209, 239
Martens, S.D.	225	Rajčáková, Ľ.	203
Martínez, A.	277	Randby, Å.T.	137, 246, 269
Martinsson, K.	182	Reis, R.A.	202, 223, 256, 262
Mathies, E.	199, 206	Ribeiro, J.L.	257, 258, 263
Mayne, C.S.	19	Rinne, M.	139, 164, 267, 276
Mayrhuber, E.	206	Rodrigues, A.A.	263
McEniry, J.	248	Rodríguez, M.L.	146
McGee, M.	51	Roth, A.P.T.P.	202
McNally, G.M.	251	Roth, M.T.P.	223
Merry, R.J.	121, 141, 211, 224, 227	Ruser, B.	231, 279
Mertens, D.R.	207	Rustas, B.	180
Miklaszewska, B.	177, 209	Saarisalo, E.	184, 198
Mikołajczak, J.	145, 230	Sánchez, D.	277
Mlynár, R.	203	Santos, M.C.	257, 263
Moloney, A.P.	154	Santos, P.M.	263
Moss, B.W.	160	Schmidt, P.	257, 258, 263
Muck, R.E.	163, 204, 207, 212, 245	Schocken-Iturrino, R.P.	202, 262
Muhonen, S.	171	Sei, S.	264
Müller, C.E.	188	Selwet, M.	209
Mundt, K.	217	Shingfield, K.J.	35
Nadeau, E.	185	Sinclair, L.A.	183
Nennich, T.D.	148	Siqueira, G.R.	202, 223, 256
Neureiter, M.	216	Small, C.M.	251
Niimi, M.	264	Soldado, A.	277
Nishino, N.	167, 237, 261	Solyakov, A.	270

Songisepp, E.	200
Souza, D.P.	257, 263
Stacey, P.	154
Staszak, E.	145, 230
Stilmant, D.	280
Sucu, E.	221, 222
Sullivan, M.L.	212
Sundberg, M.	250
Szucs, J.P.	233
Szumacher-Strabel, M.	209, 239
Tabacco, E.	193, 205, 226, 272
Tamm, U.	197
Tapia, C.	273, 274
te Giffel, M.C.	147, 271
Thaysen, J.	173, 199
Theodorou, M.K.	121
Thomas, P.	259
Toivonen, V.	140, 142
Trivelin, M.O.	257
Tuori, M.	144, 267
Twaruzek, M.	177
Tyrolova, Y.	181
Undersander, D.	247
Vanhatalo, A.	139, 164
Verrier, D.	259
Vicente, F.	146
Vieira, L.G.	257
Vissers, M.M.M.	147
Vrotniakiene, V.	158, 238
Vyborna, A.	181
Wada, H.	167
Wallsten, J.	180
Walsh, K.	166
Weimer, P.J.	207
Weinberg, Z.G.	169
Weiß, K.	213, 275
Weissbach, F.	278
Wilkinson, R.G.	183
Wood, T.	247
Wróbel, B.	240, 241
Wyss, U.	214
Xu, C.	187, 201
Yan, T.	268
Yanagisawa, J.	179
Yokota, H.	179, 215
Yoshida, N.	187, 201
Zastawny, J.	240, 241
Zerényi, E.	186
Zhu, Y.	237
Zopollatto, M.	257, 258, 263
Żurek, H.	240

Printed in the United States
by Baker & Taylor Publisher Services